Preface to the Second Edition

In the seven years which have passed since I finished the first edition of this book, much has happened in the field of sedimentary geology. The great leap forward in the study of sedimentary structures and environments seems to have come to an end. Environmental interpretation is now largely a matter of routine.

On the other hand, there has been a renaissance in sedimentary petrology, aided by the scanning electron microscope, and an improved understanding of geochemistry.

The second edition reflects these changes. The whole book has been reset, minor improvements have been made throughout, and references have been updated.

Chapter 8 (Environment and Facies) has suffered least modification since there is another excellent book which covers this topic (possibly even two).

It is Chapters 4 and 5 on the allochthonous and autochthonous sediments which have undergone most change. The sections on the diagenesis of sandstones and carbonates have been completely rewritten. These are now illustrated with thin section photographs and scanning electron micrographs. Photos have also been added to illustrate the text of other chapters where appropriate.

I should like to thank all those who offered comments and advice on the preparation of this new edition, especially Professor R. G. C. Bathurst. To them the credit for improvements, but not the blame for any errors or omissions.

R. C. Selley
June 1982

An Introduction to Sedimentology

second edition

R. C. Selley

Department of Geology, Royal School of Mines, Imperial College for Science and Technology, London

1982 Academic Press

A Subsidiary of Harcourt Brace Jovanovich, Publishers
London New York
Paris San Diego San Francisco
São Paulo Sydney Tokyo Toronto

ACADEMIC PRESS INC. (LONDON) LTD.
24/28 Oval Road,
London NW1

United States Edition published by
ACADEMIC PRESS INC.
111 Fifth Avenue
New York, New York 10003

Selley, Richard C.
Introduction to sedimentology.—2nd ed.
1. Sedimentation and deposition
I. Title
551.3′04 QE571

ISBN 0-12-636360-9 Hardback
ISBN 0-12-636362-5 Paperback

Phototypeset by Dobbie Typesetting Service, Plymouth
Printed in Great Britain by St Edmundsbury Press, Bury St Edmunds

Preface to the First Edition

Sedimentology, the study of sediments, has grown rapidly within the last quarter century. Initially the impetus for this development came from the oil industry. More recently sedimentology has been carried on the crest of the oceanography wave.

There are many excellent books which synthesize various aspects of the field of sedimentology; few, perhaps wisely, have attempted an overview of the subject. "An Introduction to Sedimentology" attempts such an overview, but within certain specific and limited objectives. I assume the reader to have attended an introductory course in geology, but hope that this book will also be of use to qualified geologists whose main interest is in other fields, particularly petroleum geology.

This book is written for geologists whose principal interest is ancient sediments. Modern depositional processes and products are discussed, not as an end in themselves, but only in so far as they aid our understanding of their ancient counterparts.

There are few full-time sedimentologists in the world; yet many professional geologists require a working knowledge of sedimentology as part of their stock in trade. I have written this text for students and practising geologists who wish to acquire such a background.

The book opens with an introductory chapter which places sedimentology within the context of the physical, chemical and biological sciences and discusses its relationship with the other branches of geology.

Chapter 2 discusses the physical properties of sediments. Attention is given not only to particles, but also to the porosity and permeability of sediments in bulk.

Chapter 3 reviews the processes of weathering which form sediments and the sedimentary cycle.

Chapters 4 and 5 describe the petrography and diagenesis of sediments. These are no substitute for a course of study in practical petrography aided by one of several excellent texts now available. The objective of these two chapters is to show the relationship between diagenesis and porosity development and to provide a minimal background to the subsequent chapters.

Chapter 6 gives a brief qualitative review of the processes of sediment transport and deposition. Chapter 7 describes the resultant sedimentary structures and shows how they may be used in facies analysis.

Chapter 8 is concerned with facies analysis and environmental interpretation. It shows how studies of modern sediments may be used to define a series of depositional models which have occurred repeatedly through geological time.

This leads on logically to Chapter 9, a discussion of sedimentary basins. Various types of basin are defined, described and illustrated. Basin topology and evolution are integrated with the concepts of plate tectonics.

The book ends with a selective review of some of the applications of sedimentology. Particular emphasis is placed on the search for oil and gas and for sedimentary metalliferous deposits.

In conclusion, then, I have tried to write a book which, while not spanning the whole field of sedimentology, covers those aspects which are of particular importance to practising geologists.

R. C. Selley
August 1975

Contents

5 Autochthonous Sediments

6 Transportation and Sedimentation

7 Sedimentary Structures

8 Environments and Facies

9 Sedimentary Basins

10 Applied Sedimentology

Acknowledgements

When this book was first conceived it was to have been co-authored by Dr D. J. Shearman of Imperial College. Unfortunately, due to other commitments, Dr Shearman was unable to write his full share of the book and he withdrew from full co-authorship. Nevertheless, he wrote the section on coal and parts of the sections on carbonate diagenesis and evaporites. It is a pleasure to acknowledge this contribution and his continuous enthusiasm for the project as a whole.

Similarly I am grateful to my colleagues in industry and in universities who critically reviewed chapters from their several viewpoints. Dr H. Reading in particular shouldered much of this burden on the academic side.

I am also grateful to the following for permission to use various figures; the American Association of Petroleum Geologists (Figs 62, 69, 72, 76, 134, 141, 144, 149, 150, 151, 158, 162, 168–173, 175, 176, 183, 184, 186–191 and 193); Elsevier Publishing Corporation (Figs 78, 112 and 179); the Institute of Mining and Metallurgy (Fig. 79); Houston Geological Society (Fig. 139); Chicago University Press (Figs 122, 153 and 157); the Council of the Geological Society of London (Figs 15, 148 and 180); the United States Geological Survey (Fig. 133); the Geological Survey of Western Australia (Fig. 154); the Society of Economic Paleontologists and Mineralogists (Figs 28, 67, 86, 87, 94 and 141–143); the Council of the Geologists' Association (Figs 32, 33, 34, 35, 43, 116 and 147); the Institute of Petroleum (Fig. 65) and Geologie Mijnbouw (Fig. 199).

1 Introduction

I. INTRODUCTION AND HISTORICAL REVIEW

A sediment is "what settles at the bottom of a liquid: dregs: a deposit" (Chambers Dictionary, 1972 edition). Sedimentology is the study of sediments. Few sedimentologists would accept the preceding definition because we like to include both eolian deposits and chemical precipitates as sedimentary rocks. The limits of the field of sedimentology are, therefore, pleasantly ill-defined.

The purpose of this chapter is to introduce the field of sedimentology and to place it in the scheme of science, both the basic sciences of chemistry, physics and biology and, more parochially, within the field of geology.

It is hard to trace the historical evolution of sedimentology. Arguably among the first practitioners must have been the Stone Age flint miners of Norfolk who, as seen in Grimes cave, mined the stratified chert bands to make flint artifacts (Shotton, 1968). Subsequently civilized man must have noticed that other useful economic rocks, such as coal, building stone and so on, occurred in planar surfaces in the earth's crust that cropped out in a predictable manner. It has been argued that the legend of the "Golden Fleece" suggests that sophisticated flotation methods were used for alluvial gold mining in the fifth century BC (Barnes, 1973).

From the Renaissance to the Industrial Revolution the foundations of modern sedimentary geology were laid by men such as Leonardo da Vinci, Hutton and Smith. By the end of the nineteenth century the doctrine of uniformitarianism was firmly established in geological thought. The writings of Sorby (1853, 1908) and Lyell (1865) showed how modern processes could be used to interpret ancient sedimentary textures and structures.

Throughout the first half of the twentieth century, however, the discipline of sedimentology, as we now understand it, lay moribund. The sedimentary rocks were either considered fit only for microscopic study or as homes for fossils. During this period heavy mineral analysis and point counting were extensively developed by sedimentary petrographers.

Simultaneously, stratigraphers gathered fossils; wherever possible erecting more and more refined zones until they were too thin to contain the key fossils.

Curiously enough modern sedimentology was not born from the union of petrography and stratigraphy. It seems to have evolved from a union between structural geology and oceanography. This strange evolution deserves an explanation.

Structural geologists have always searched for criteria for distinguishing whether strata in areas of tectonism were overturned or in normal sequence. This is essential if regional mapping is to delineate recumbent folds and nappes. Many sedimentary structures are ideal for this purpose, particularly desiccation cracks, ripples and graded bedding. This approach reached its apotheosis in Shrock's volume "Sequence in Layered Rocks", written in 1948.

On a broader scale, structural geologists were concerned with the vast prisms of sediments which occur in geosynclinal furrows. A valid stratigraphy is a prerequisite for a valid structural analysis. Thus it is interesting to see that it was not a stratigrapher, but Sir Edward Bailey, doyen of structural geologists, who wrote the paper "New light on sedimentation and tectonics" in 1930. This defined the fundamental distinction between the sedimentary textures and structures of shelves and those of deep basins. It was this paper which also contained the germ of the turbidity current hypothesis.

The concept of the turbidity flow rejuvenated the study of sediments in the 1950s and early 1960s. While petrographers counted zircon grains and stratigraphers collected more fossils, it was the structural geologists who asked "how do thick sequences of flysch facies get deposited in geosynclines?". It was modern oceanography which provided the turbidity current as a possible mechanism (see p. 185). It is true to say that this concept rejuvenated the study of sedimentary rocks. Though in their enthusiasm geologists identified turbidites in every kind of facies, from the Viking sand bars of Canada to the alluvial Nubian sandstones of the Sahara (Anon).

Another stimulus to sedimentology came from the oil industry. The search for stratigraphically trapped oil led to a boom in the study of modern sediments. One of the first fruits of this approach was the American Petroleum Institute's "Project 51"; a multidisciplinary study of the modern sediments of the north-west Gulf of Mexico (Shepard *et al.*, 1960).

This was followed by many other studies of modern sediments both by oil companies, universities and oceanographic institutes. At last, hard data became available so that ancient sedimentary rocks could be interpreted by comparison with their modern analogues. The concept of the sedimentary model was born as it became apparent that there are, and always have been, a finite number of sedimentary environments which deposit characteristic sedimentary facies (see p. 265).

By the end of the 1960s sedimentology was firmly established as a discrete discipline of the earth sciences.

Through the 1960s the main focus of research was directed towards an understanding of sedimentary processes. By studying the bedforms and depositional structures of recent sediments, either in laboratory flumes or in the wild, it became possible to accurately interpret the environment of ancient sedimentary rocks (Laporte, 1979; Selley, 1978; Reading, 1978).

Through the 1970s the emphasis of sedimentological research gradually changed from the macroscopic and physical to the microscopic and chemical. Improved analytical techniques and the application of cathodo-luminescence and scanning electron microscopy gathered new data which could be interpreted with a better understanding of geochemistry.

This renaissance of petrography has enhanced our understanding of the relationship between diagenesis, pore fluids and their effects on the evolution of porosity and permeability in sandstones and carbonates. Similarly we are now beginning to understand the relationships between clay mineral diagenesis and the maturation of organic matter in hydrocarbon source beds.

II. SEDIMENTOLOGY AND THE EARTH SCIENCES

Table I shows the relationship between sedimentology and the basic sciences of biology, physics and chemistry.

The application of one or more of these fundamental sciences to the study of sediments gives rise to various lines of research in the earth sciences. These will now be reviewed as a means of setting sedimentology within its context of geology.

Biology, the study of animals and plants, can be applied to fossils in ancient sediments. Palaeontology may be studied as a pure subject which concerns the evolution, morphology and taxonomy of fossils. In these pursuits the fossils are essentially removed from their sedimentological context.

The study of fossils within their sediments is a fruitful pursuit in two ways. Stratigraphy is based on the definition of biostratigraphic zones and the study of their relationship to lithostratigraphic units (Shaw, 1964; Mathews, 1974). Sound biostratigraphy is essential for regional structural and sedimentological analysis.

The second main field of fossil study is to deduce their behaviour when

Table I
The various branches of sedimentology and their relationship to the fundamental sciences

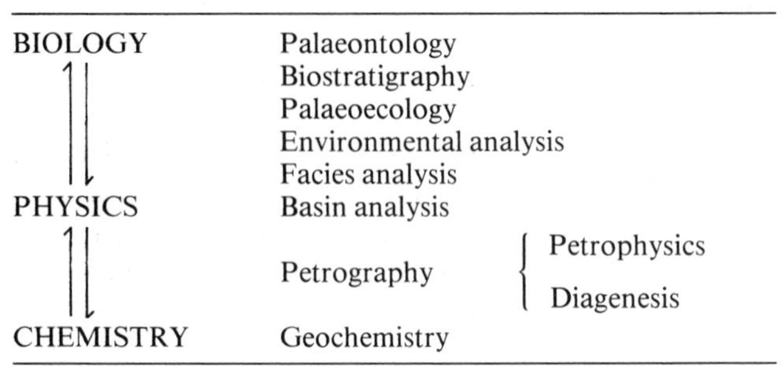

BIOLOGY	Palaeontology	
⇅	Biostratigraphy	
	Palaeoecology	
	Environmental analysis	
	Facies analysis	
PHYSICS	Basin analysis	
⇅	Petrography	{ Petrophysics / Diagenesis
CHEMISTRY	Geochemistry	

they were alive, their habitats and mutual relationships. This study is termed palaeoecology (Ager, 1963). Where it can be demonstrated that fossils are preserved *in situ* they are an important line of evidence in environmental analysis.

Environmental analysis is the determination of the depositional environment of a sediment (Selley, 1978).

Our review of sedimentology has now moved from the biological aspect to those facets which involve both biological, physical and chemical properties of sedimentary rocks. To determine the depositional environment of a rock it is obviously important to correctly identify and interpret the fossils which it contains. At a very simple level a root bed indicates a terrestrial environment, a coral reef a marine environment. Most applied sedimentology, however, is based on the study of rock chips from bore holes. In such subsurface projects it is micropalaeontology that holds the key to both stratigraphy and environment. The two aspects of palaeontology which are most important to sedimentology, therefore, are the study of fossils as rock builders (as in limestones) and micropalaeontology.

Aside from biology, environmental analysis is based also on the interpretation of the physical properties of a rock. These include grain size and texture as well as sedimentary structures. Hydraulics is the study of fluid movement. Loose boundary hydraulics is concerned with the relationship between fluids flowing over granular solids. These physical disciplines can be studied by theoretical mathematics, experimentally in laboratories, or in the field in modern sedimentary environments. Such lines of analysis can be applied to the physical parameters of an ancient sediment to determine the fluid processes which controlled its deposition (Allen, 1970).

Environmental analysis also necessitates applying chemistry to the study of sediments. The detrital minerals of terrigenous rocks indicate their source and pre-depositional history. Authigenic minerals can provide clues both of the depositional environment of a rock as well as its subsequent diagenetic history.

Environmental analysis thus involves the application of biology, physics and chemistry to sedimentary rocks.

Facies analysis is a branch of regional sedimentology which involves three exercises. The sediments of an area must be grouped into various natural types or facies, defined by their lithology, sedimentary structures and fossils. The environment of each facies is deduced and the facies are placed within a time-framework using biostratigraphy.

Like environmental analysis, facies analysis utilizes biology, chemistry and physics. On a regional scale, however, facies analysis involves the study of whole basins of sediment. Here geophysics becomes important, not just to study the sedimentary cover, but to understand the physical properties and processes of the crust in which sedimentary basins form.

Moving along the spectrum of Table I we come to the chemical aspects of sediments. It has already been shown how both environmental and facies analysis utilize knowledge of the chemistry of sediments. Petrology, or petrography, are terms which are now more or less synonymously applied to the microscopic study of rocks (Carozzi, 1960; Folk, 1968). These studies include petrophysics, which is concerned with such physical properties as porosity and permeability. More generally, however, they are taken to mean the study of the mineralogy of rocks.

Sedimentary petrology is useful for a number of reasons. As already pointed out, it can be used to discover the provenance of terrigenous rocks and the environment of many carbonates. Petrography also throws light on diagenesis: the post-depositional changes in a sediment. Diagenetic studies elucidate the chemical reactions which took place between a rock and the fluids which flowed through it. Diagenesis is of great interest because of the way in which it can destroy or increase the porosity and permeability of a rock. This is relevant in the study of aquifers and hydrocarbon reservoirs. Chemical studies are also useful in understanding the diagenetic processes which form the epigenetic mineral deposits, such as the lead–zinc sulphide and carnotite ores.

Lastly, at the end of the spectrum the pure application of chemistry to sedimentary rocks is termed sedimentary geochemistry (Degens, 1965). This is a vast field in itself. It is of particular use in the study of the chemical sediments, naturally, and of microcrystalline sediments which are hard to study by microscopic techniques. Thus the main contributions of sedimentary geochemistry lie in the study of clay minerals, phosphates and the evaporite rocks.

Organic geochemistry is primarily concerned with the generation and maturation of coal, crude oil and natural gas. Organic geochemistry combining biology and chemistry brings this discussion back to its point of origin.

The preceding analysis has attempted to show how sedimentology is integrated with the other geological disciplines. The succeeding chapters will continuously demonstrate how much sedimentology is based on the fundamental sciences of biology, physics and chemistry.

III. REFERENCES

Ager, D. V. (1963). "Principles of Palaoecology" McGraw-Hill, New York. 371pp.

Allen, J. R. L. (1970). "Physical Processes of Sedimentation" Allen and Unwin, London. 248pp.

Bailey, E. B. (1930). New light on sedimentation and tectonics. *Geol. Mag.* **67**, 77–92.

Barnes, J. W. (1973). Jason and the Gold Rush. *Proc. Geol. Ass.* **84**, 482–485.

Carozzi, A. (1960). "Microscopic Sedimentary Petrography" John Wiley, New York. 485pp.

Degens, E. T. (1965). "Geochemistry of Sediments: a Brief Survey" Prentice-Hall, New Jersey. 342pp.

Folk, R. L. (1968). "Petrology of Sedimentary Environments" Chapman and Austin, Texas. 170pp.

Laporte, L. F. (1979). "Ancient Environments" 2nd edn, Prentice-Hall, New Jersey. 163pp.

Lyell, C. (1865). "Elements of Geology" John Murray, London. 794pp.

Mathews, R. K. (1974). "Dynamic Stratigraphy" Prentice-Hall, New Jersey. 370pp.

Reading, H. G. (1978). "Sedimentary Environments and Facies" Blackwell Scientific, Oxford. 557pp.

Selley, R. C. (1978). "Ancient Sedimentary Environments" 2nd edn, Chapman and Hall, London. 287pp.

Shaw, A. B. (1964). "Time in Stratigraphy" McGraw-Hill, New York. 365pp. Phleger, F. B. and van Andel, T. H. (Eds).

Shepard, F. P., Phleger, F. B. and van Andel, T. H. (Eds) (1960). "Recent sediments, northwest Gulf of Mexico". Am. Ass. Petrol. Geol., Tulsa. 394pp.

Shotton, F. W. (1968). Prehistoric man's use of stone in Britain. *Proc. Geol. Ass.* **79**, 477–491.

Shrock, R. R. (1948). "Sequence in Layered Rocks" McGraw-Hill, New York. 507pp.

Sorby, H. C. (1853). On the oscillation of the currents drifting the sandstone beds of the south-east of Northumberland, and their general direction in the coalfield in the neighbourhood of Edinburgh. *Rep. Proc. geol. polytech. Soc. W. Riding, Yorkshire* 1852, 232–240.

Sorby, H. C. (1908). On the application of quantitative methods to the study of the structure and history of rocks. *Q. Jl geol. Soc. Lond.* **64**, 171–233.

2 Particles and Pores

A sediment is, by definition, a collection of particles, loose or indurated. Any sedimentological study commences with a description of the physical properties of the deposit in question. This may be no more than the terse description "sandstones", if the study concerns regional tectonic problems. On the other hand, it may consist of a multipage report, if the study is concerned with the physical properties of a specific sedimentary unit in a small area. It must be remembered that the analysis of the physical properties of a sedimentary rock should be adapted to suit the objectives of the project as a whole.

The study of the physical properties of sediments is an extensive field of analysis in its own right, and is of wide concern not only to geologists. In its broadest sense of particle analysis, it is of importance to the managers of sewage farms, manufacturers of leadshot, plastic beads and to egg graders.

Enthusiasts for this field of study are directed to the writings of Herdan (1953), Chayes (1956), Tickell (1965), Mueller (1967), Griffiths (1967) and Carver (1971).

The following account is a summary of this topic which attempts only to give sufficient background knowledge needed for the study of sedimentary rocks in their broader setting. This is in two parts. The first describes individual particles and sediment aggregates; the second describes the properties of pores—the voids between sediment particles.

I. PHYSICAL PROPERTIES OF PARTICLES

A. Surface Texture of Particles

The surface texture of sediment particles has often been studied, and attempts have been made to relate texture to depositional process.

In the case of pebbles, macroscopic striations are generally accepted as evidence of glacial action. Pebbles in arid eolian environments sometimes show a shiny surface, termed "desert varnish". This is conventionally attributed to capillary fluid movement within the pebbles and evaporation of the silica residue on the pebble surface. The folklore of geology records that wind-blown sand grains have opaque frosted surfaces, while water-laid sands have clear translucent surfaces. Kuenen and Perdok (1962) attributed frosting not to the abrasive action of wind and water but to alternate solution and precipitation under subaerial conditions. Electron microscope studies by Margolis and Krinsley (1971) show that wind-induced grain to grain impacts generate minute fractures which split the grain surface into upturned plates.

Krinsley and his co-workers have also described a variety of different types of quartz grain surface textures, using electron microscopes (Krinsley and Funnell, 1965; Krinsley and Cavallero, 1970; Krinsley and Doornkamp, 1973). They identified a number of different abrasion patterns which are characteristic of glacial, eolian and aqueous processes. At subcrop, and at outcrop in the tropics, these abrasional features are frequently modified by solution and by secondary quartz cementation. The surface textures of ancient quartz grains may thus reveal little or nothing of their depositional history.

B. Particle Shape, Sphericity and Roundness

Numerous attempts have been made to define the shape of sediment particles and to study the controlling factors of grain shape.

Pebble shapes have conventionally been described according to a scheme devised by Zingg (1935). Measurements of the ratios between length, breadth and thickness are used to define four classes: spherical (equant), oblate (disc or tabular), blade and prolate (roller), shown in Fig. 1. A more sophisticated scheme has been developed by Sneed and Folk (1958).

The shape of pebbles is controlled both by their parent rock type and by their subsequent history. Pebbles from slate and schistose rocks will tend to commence life in tabular or bladed shapes, whereas isotropic rocks, such as quartzite, are more likely to generate equant, subspherical pebbles. Traced away from their source pebbles diminish in size and tend to assume equant or bladed shapes (e.g. Plumley, 1948; Schlee, 1957; Miall, 1970). Attempts have also been made to relate pebble shape to depositional environment (Cailleux and Tricart, 1959). Sames (1966) has proposed criteria for distinguishing fluvial and littoral pebble samples using a combination of shape and roundness. The samples were restricted to isotropic rocks such as chert and quartzite.

Sand-sized particles are not amenable to measurement of long, medium and short axes. Their shape is generally measured by reference to a coefficient of sphericity. This is a measurement of the degree to which a grain approaches the shape of a sphere. Various coefficients of sphericity have been proposed by Wadell (1932) and Sneed and Folk (1958).

An additional property of particles is their roundness. This is a measure

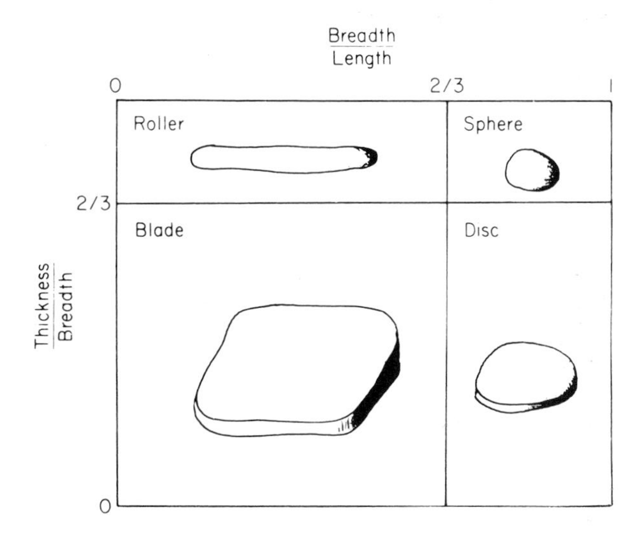

Fig. 1. A classification of pebble shapes. (After Zingg, 1935.)

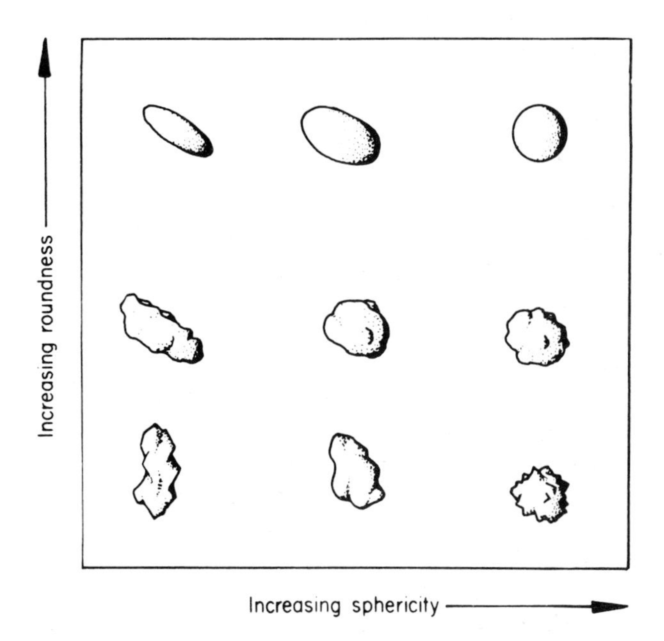

Fig. 2. To illustrate the independence of roundness and sphericity of sediment particles.

of the degree of curvature of corners. It is quite independent of shape. Figure 2 shows the independence of these two properties. Roundness scales have been defined by Russell and Taylor (1937) and by Powers (1953).

Considerable attention has been paid to the factors which control the roundness and sphericity of sediment particles. Many studies show the sphericity and roundness of sediments to increase away from their source area (e.g. Laming, 1966). Kuenen, in a series of papers, described the results of the experimental abrasion of both pebbles and sand by various eolian and aqueous processes (Kuenen, 1956a, b, 1959, 1960). These studies indicated that the degree of abrasion and shape change along rivers and beaches was due as much to shape sorting as to abrasion. The experiments themselves showed that eolian action was infinitely more efficient as a rounding mechanism than aqueous transportation over an equivalent distance. There is little evidence for chemical solution to be a significant rounding agent. This is shown by the angularity of very fine sand and silt. More recently, Margolis and Krinsley (1971) have demonstrated that the high rounding commonly seen in eolian sands is due to a combination of abrasion and of synchronous precipitation of silica on the grain boundary.

C. Particle Size

Size is perhaps the most striking property of a sediment particle. This fact is recognized by the broad classification of sediments into gravels, sands and muds.

It is easy to understand the concept of particle size. It is less easy to find accurate methods of measuring particles (Whalley, 1972).

1. Grade scales

Sedimentary particles come in all sizes. For communication it is convenient to be able to describe sediments as gravels, sands (of several grades), silt and clay.

Various grade scales have been proposed which arbitrarily divide sediments into a spectrum of size classes. The Wentworth grade scale is the one most commonly used by geologists (Wentworth, 1922). This is shown in Table II, together with the grade names and their lithified equivalents.

A common variation of the Wentworth system is the Phi (ϕ) scale proposed by Krumbein (1934). This retains the Wentworth grade names, but converts the grade boundaries into Phi values by a logarithmic transform:

$$\phi = -\log_2 d$$

where d is the diameter.

Table II
The Wentworth grade scale, showing correlation with ϕ scale and nomenclature for unconsolidated aggregates and for lithified sediments

ϕ values	Particle diameter (mm diam.)	Wentworth grades		Rock name
		Cobbles		
-6	64			Conglomerate
		Pebbles		
-2	4			
		Granules		Granulestone
-1	2			
		Very coarse		
0	1			
		Coarse		
1	0·5			
		Medium	sand	Sandstone
2	0·25			
		Fine		
3	0·125			
		Very fine		
4	0·0625			
		Silt		Siltstone
8	0·0039			
		Clay		Claystone

The relationship between the Wentworth and Phi grade scales is shown in Fig. 3.

2. *Methods of particle analysis*

The size of sediment particles can be measured from both unconsolidated and indurated sediment. The most common way is by visual estimation backed up, if necessary, by reference to a set of sieved samples of known sizes. With experience, most geologists can visually measure grain size within the accuracy of the Wentworth grade scale scheme, at least down to silt grade. Silt and clay can be differentiated by whether they are crunchy or plastic between one's teeth.

More sophisticated methods of analysis are available. Well-cemented sediments are measured from thin-section microscope study (Chayes, 1956). There is the quick and dirty method and the slow enthusiast's method. The first involves no more than measuring the diameter of the field of view of the microscope at a known magnification, counting the number of grains transected by the cross-wires, and dividing by twice the diameter. This process is repeated until tired, or statistical whims are satisfied. Then to calculate the average grain size of the thin section this formula is used:

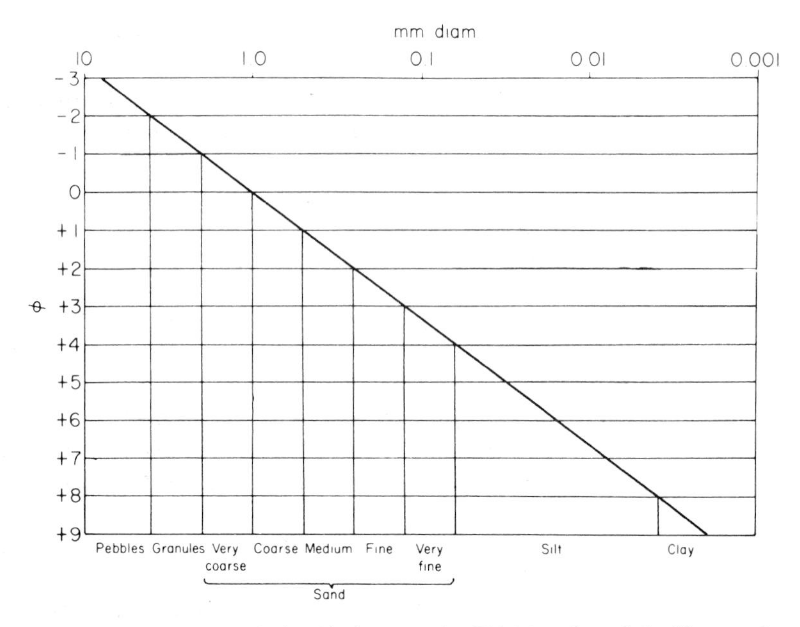

Fig. 3. Graph to show the relationship between the Phi (ϕ) scale and the Wentworth grade scale of particle measurement.

$$\text{average grain size} = \frac{\sum \frac{2d}{n}}{N}$$

where n is the number of grains cut by the cross-wires, d is the diameter of the field of view and N is the total number of fields of view counted.

This method is quick and fool-proof. It is adequate for most sedimentological studies which are aimed at broader aspects of facies analysis. It gives only the average grain size and does not describe the sorting of the sample. This method is quite inadequate for detailed granulometric studies. Then it is necessary to measure the length and/or some other size parameter of individual grains. This is not only very time consuming, but care must be taken over the orientation of the thin section and of the selection of samples (Chayes, 1956). It is improbable that the thin section cuts the longest axis of every particle, so there is an inherent bias to record axes which are less than the true long axis.

Friedman has empirically derived methods for converting grain size data from thin sections to sieve analyses (Friedman, 1958, 1962a). Ideally grain size studies of thin sections should be carried out using cathodoluminescence rather than under polarized light (see p. 101). This will clarify the extent of pressure solution and secondary cementation. These may have considerably modified both grain size and grain shape.

Claystones and siltstones are not amenable to size analysis from an optical microscope. Their particle size can be measured individually by electron microscope analysis.

There are many methods of measuring the particle size of unconsolidated sediment. The choice of method depends largely on the particle size.

Boulders, cobbles and gravel are best measured manually with a tape measure or ruler.

Sands are most generally measured by sieving. The basic principles of this technique are as follows. A sand sample of known weight is passed through a set of sieves of known mesh sizes. The sieves are arranged in downward decreasing mesh diameters. The sieves are mechanically vibrated for a fixed period of time. The weight of sediment retained on each sieve is measured and converted into a percentage of the total sediment sample (for additional information see ASTM, 1959). This method is quick and sufficiently accurate for most purposes. Essentially it measures the maximum girth of a sediment grain. Long thin grains are recorded in the same class as subspherical grains of similar girth. This fact is not too important in the size analysis of terrigenous sediments because these generally have a subovoid shape. Skeletal carbonate sands, however, show a diverse range of particle shapes.

This factor is overcome by another method of bulk sediment analysis termed elutriation, or the settling velocity method. This is based on Stokes law which quantifies the settling velocity of a sphere thus:

$$w = \left[\frac{(P_1 - P)g}{18\mu} \right] d^2$$

where w is the settling velocity, $(P_1 - P)$ is the density difference between the particle and the fluid, g is the acceleration due to gravity, μ is the viscosity and d is the particle diameter.

A derivation for Stokes law will be found in Krumbein and Pettijohn (1938, p. 95). Its significance in sediment transport studies will be returned to later (p. 178).

The method involves disaggregating the sediment and dispersing the clay fraction with an antiflocculent and tipping it into a glass tube full of liquid. The sediment will accumulate on the bottom in order of decreasing grain size, gravel first and clay last. There are several methods of measuring the time of arrival and volume of the different grades.

The elutriation method is widely used as it is fast and accurate, and can be used for particles from granule to clay grade. It necessitates making several assumptions about the effect of particle shape and surface friction on settling velocity.

Table III summarizes the various methods of particle size analysis.

Table III
Methods of particle size analysis

Induration	Sediment grade	Method	
Unconsolidated	Boulders, Cobbles, Pebbles	Manual measurement of individual clasts	
	Granules, Sand, Silt, Clay	Sieve analysis or elutriation of bulk samples	
Lithified	Boulders, Cobbles, Pebbles	Manual measurement of individual clasts	
	Granules, Sand, Silt, Clay	Thin section measurement X-ray analysis	Quick: grain counts on cross-wires Slow: individual grain micrometry

3. *Presentation of particle size analyses*

The method of displaying and analysing granulometric analyses depends on the purpose of the study. Both graphic and statistical methods of data presentation have been developed.

Starting from the tabulation of the percentage of the sample in each class (Table IV), the data may be shown graphically in bar charts (Fig. 4). A more usual method of graphic display is the cumulative curve. This is a graph which plots cumulative percentage against the grain size (Fig. 5). Cumulative curves are extremely useful because many sample curves can be plotted on the same graph and differences in sorting are at once apparent. The closer a curve approaches the vertical the better sorted it is, as a major percentage of sediment occurs in one class. Significant percentages of coarse and fine end-members show up as horizontal limbs at the ends of the curve. The grain size for any cumulative percent is termed a percentile (i.e. one talks of the 20 percentile, and so on). It is a common practice to plot grain size curves on probability paper. The great advantage of this is that samples with normal Gaussian grain size distributions plot out on a straight line.

Table IV
Grain size data of modern sediments tabulated by weight % and cumulative %

Grain size	Sample A		Sample B		Sample C		Sample D		Sample E	
	Weight (%)	Cumulative (%)	Weight (%)	Cumulative (%)	Weight (%)	Cumulative (%)	Weight (%)	Cumulative (%)	Weight (%)	Cumulative (%)
Granules	0·00	0·00	0·79	0·79	0·00	0·00	0·00	0·00	3·87	3·87
Very coarse	0·00	0·00	1·10	1·84	0·00	0·00	0·05	0·05	22·75	26·62
Coarse	0·02	0·02	1·66	3·50	0·00	0·00	0·44	0·49	50·68	77·30
Medium	9·00	9·02	3·90	7·40	0·00	0·00	51·89	52·38	20·24	97·54
Fine	90·26	99·28	10·65	18·05	0·00	0·00	47·57	99·95	2·20	99·74
Very fine	0·72	100·00	13·68	31·73	1·09	1·09	0·07	100·02	0·24	99·98
Silt	0·00	100·00	35·47	67·26	45·87	46·96	0·01	100·03	0·02	100·00
Clay	0·00	100·00	32·73	100·00	53·04	100·00	0·00		0·00	
Total	100·00		99·98		100·00		100·03		100·00	

A: Eolian coastal dune, Lincs., England. B: Pleistocene glacial till, Lincs., England.
C: Abyssal plain mud, Bay of Biscay. D: Beach sand, Cornwall, England. E: River sand, Dartmoor, England.
Data supplied by courtesy of G. Evans.

The sorting of a sediment can also be expressed by various statistical gymnastics. Simplest of these is the measurement of central tendency, of which there are three commonly used parameters: the median, the mode and the mean. The median grain size is that which separates 50% of the sample from the other, i.e. the median is the 50 percentile. The mode is the largest class interval. The mean is variously defined, but a common formula is the average of the 25 and 75 percentiles.

The second important aspect of a granulometric analysis is its sorting or the measure of degree of scatter, i.e. the tendency for the grains to all be of one class of grain size. This is measured by a sorting coefficient. Several formulae have been proposed. The classic Trask sorting coefficient is calculated by dividing the 75 percentile by the 25 percentile.

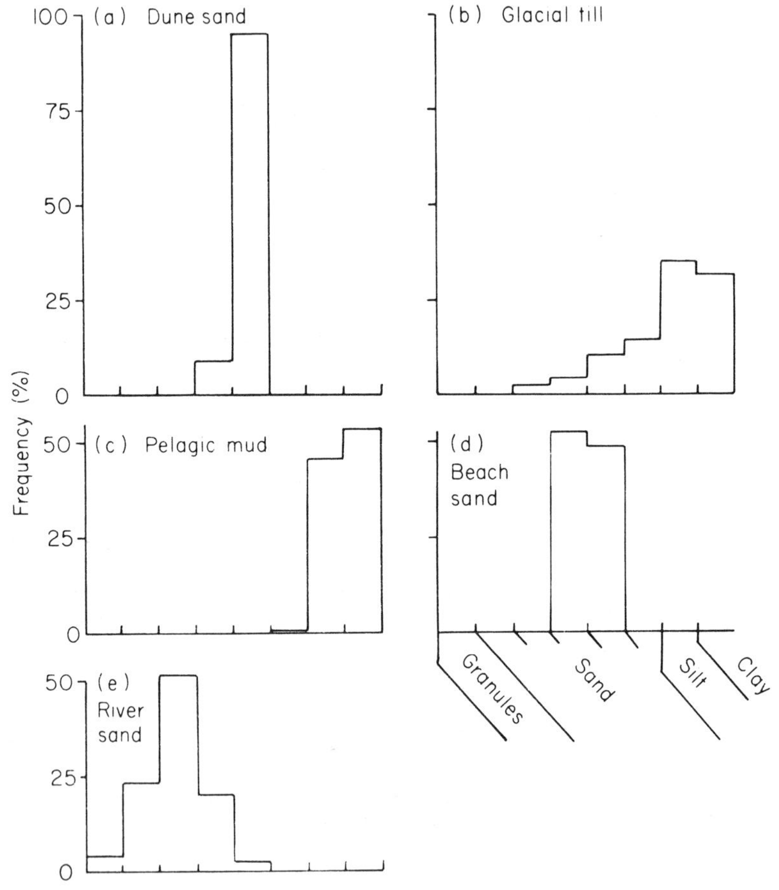

Fig. 4. Bar charts or histograms showing the frequency per cent of the various sediment classes recorded from the samples in Table IV.

A third property of a grain size frequency curve is termed "kurtosis", or the "degree of peakedness". The original formula proposed for kurtosis is as follows (Trask, 1930):

$$k = \frac{P_{75} - P_{25}}{2(P_{90} - P_{10})}$$

where P refers to the percentiles.

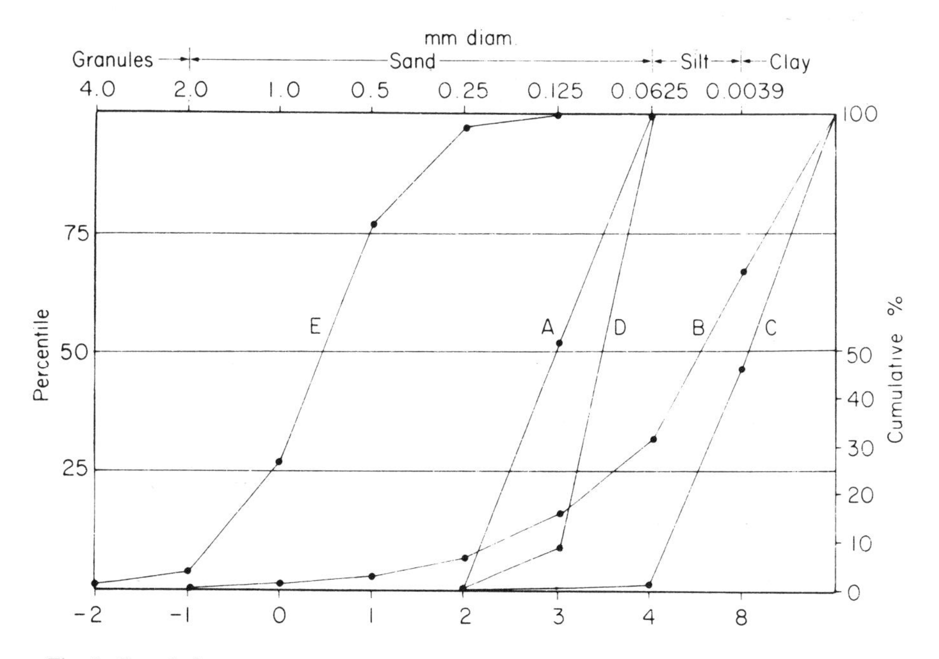

Fig. 5. Cumulative percent plotted against grain size for the samples recorded in Table IV. This method of data presentation enables quick visual comparisons to be made of both grain size and sorting. The steeper the curve the better sorted is the sample. Most statistical granulometric analyses require more class intervals to be measured than the Wentworth grade scale classes shown here.

Curves which are more peaked than the normal distribution curve are termed "leptokurtic"; those which are saggier than the normal are said to be "platykurtic" (Fig. 6).

The fourth property of a granulometric curve is its skewness, or degree of lop-sidedness. The Trask coefficient of skewness is (ibid.):

$$SK = \frac{P_{25} \times P_{75}}{P_{50}^{2}}.$$

Samples weighted towards the coarse end-member are said to be positively skewed, samples weighted towards the fine end are said to be negatively skewed (Fig. 7).

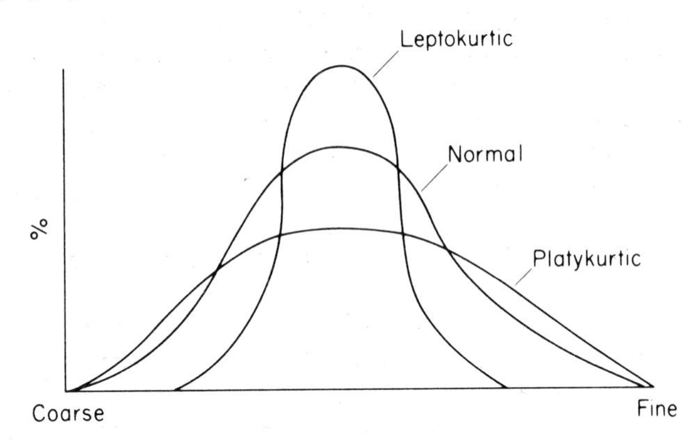

Fig. 6. Illustrative of the parameter of "peakedness" (kurtosis) in grain size distributions.

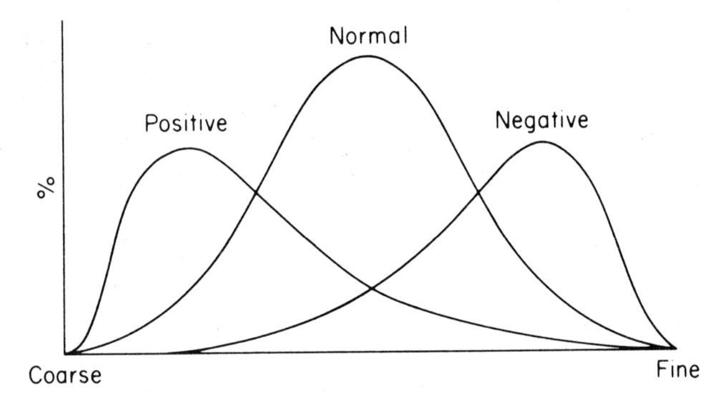

Fig. 7. Illustrative of the parameter of skewness (lopsidedness) in grain size distributions.

These are the four statistical coefficients which are commonly calculated for a granulometric analysis. In summary these consist of a measure of central tendency (including median, mode and mean); a measure of the degree of scatter or sorting; kurtosis, the degree of peakedness; and skewness, the lopsidedness of the curve. These concepts and formulae were originally defined by Trask (1930). More sophisticated formulae to describe these parameters have been proposed by Inman (1952) and by Folk and Ward (1957), and these are generally used by modern grain size buffs.

4. *Interpretation of particle size analyses*

The methods of granulometry and the techniques of displaying and statistically manipulating these data have now been described. Let us consider the value of this approach and its interpretation and application.

First of all there are many cases where it is quite inadequate to describe a sediment as "a medium-grained well-sorted sand". Within the sand and gravel industry rigid trade descriptions of marketable sediments are required. This includes hard core, road aggregates, building sands, sewage filter-bed sands, blasting sand, and so on. Sand and gravel for these and many other uses require specific grain size distributions. These must be accurately described using statistical coefficients such as those described above.

Within the field of geology, accurate granulometric analyses are required for petrophysical studies which relate sand texture to porosity and permeability (p. 34). The selection of gravel pack completions for water wells also requires a detailed knowledge of the granulometry of the aquifer.

Beyond these purely descriptive aspects of granulometry there is one interpretive aspect which has always led geologists on. This is the use of grain size analysis to detect the depositional environment of ancient sediments. The philosophy behind this approach is that modern environments mould the granulometric parameters of sediment populations; consider for example the manifest difference between a glacial till and a desert dune sand. If these differences can be quantitatively distinguished, then it should be possible to compare a granulometric analysis of an ancient sediment sample with modern ones and, by finding the closest match, find its environment.

This avenue of research has not been notably successful, despite intensive efforts. Modern environments whose sediment granulometry have been extensively studied include rivers, beaches and dunes. Reviews of this work have been given by Folk and Ward (1957), Friedman (1961, 1967), Folk (1966) and Moiola and Weiser (1968). It is apparent that statistical coefficients can very often differentiate sediments from modern environments. For example, a number of studies show that beach and dune sands are negatively and positively skewed respectively (e.g. Mason and Folk, 1958; Friedman, 1961; Chappell, 1967).

Several complicating factors have now emerged, however. Many sediments are actually combinations of two or more different grain size populations of different origins (Doeglas, 1946; Spencer, 1963). These admixtures may reflect mixing of sediments of different environments. It is more likely, however, that the presence of several populations in one sample reflects the action of different physical processes. For example, within Barataria Bay, Louisiana, multivariate analysis defined sediments into populations influenced by wind, wave, current and gravitational processes (Klovan, 1966). Yet these were all deposited in the same lagoonal or bay environment. Similarly Visher (1965) has demonstrated the variability of sediment type within fluvial channels and showed how these differences are related to sedimentary structure, i.e. to depositional process. The C–M diagrams of Passega (1957, 1964) are another approach to this same problem. Plots of C, the one percentile which measures the coarsest fraction, against M, the median, are most illuminating. They reveal different fields for pelagic suspensions, turbidites, bed load suspensions and so on (Fig. 8).

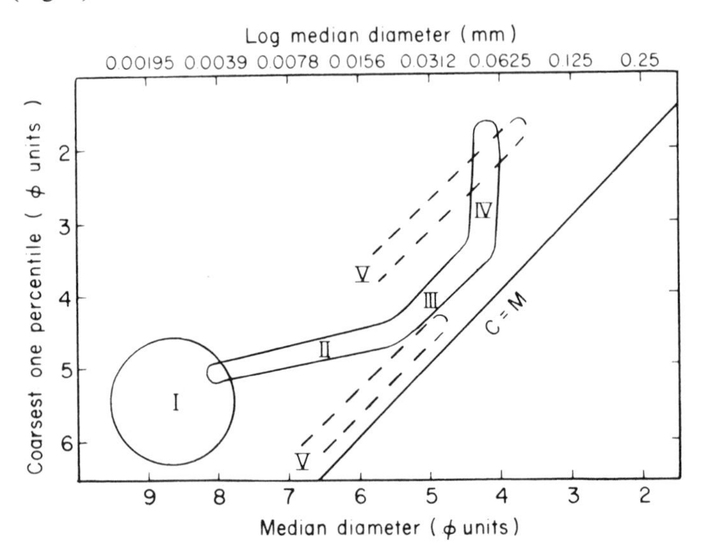

Fig. 8. C–M patterns, proposed by Passega (1957, 1964). Plot C, the 1 percentile, against M, the median 50 percentile. This segregates the products of several different sedimentary processes. I: pelagic suspension; II: uniform suspension; III: graded suspension; IV: bed load; V: turbidity currents.

A further problem of defining the granulometric characteristics of modern environments is that of inheritance. It has often been pointed out that if a fine-grained sand of uniform grain size is transported into a basin then that is the only granulometric type to be deposited regardless of environment or process. More specific examples of the effect of inheritance on sorting character have been described in studies of modern alluvium from Iran and the Mediterranean (Vita-Finzi, 1971).

Let us now apply these granulometric techniques to the environmental analysis of ancient sedimentary rocks. Several problems are at once apparent. Analyses of modern sediments show that the fine clay fraction is very sensitive to depositional process. In ancient sediments the clay matrix may have infiltrated after sedimentation, or may have formed diagenetically from the breakdown of labile sand grains (see p. 101). Clay can also be transported in aggregates of diverse size from sand up to clay boulders. Granulometric analyses of a fossil sediment may disaggregate such larger clay clasts to their constituent clay particles. Quartz grains may have been considerably modified by solution and cementation. Couple these factors with the problem of inheritance and the lack of correlation between depositional process and environment, and it can be seen that the detection of the depositional environment of ancient sediments from their granulometry is extremely difficult. Where other criteria are available, such as vertical grain size profiles, sedimentary structures and palaeontology, it is mercifully unnecessary.

II. POROSITY AND PERMEABILITY

A. Introduction

While it is true that geologists study rocks, much applied geology is concerned with the study of holes within rocks. The study of these holes or "pores" is termed "petrophysics" (Archie, 1950). This is of vital importance both in the search for oil, gas and ground water, and in locating regional permeability barriers which control the entrapment and precipitation of low temperature ore minerals (p. 393). It is also necessary for subsurface liquid waste disposal and gas storage schemes.

A sedimentary rock is composed of grains, matrix, cement and pores (Fig. 9). The grains are the detrital particles which generally form the

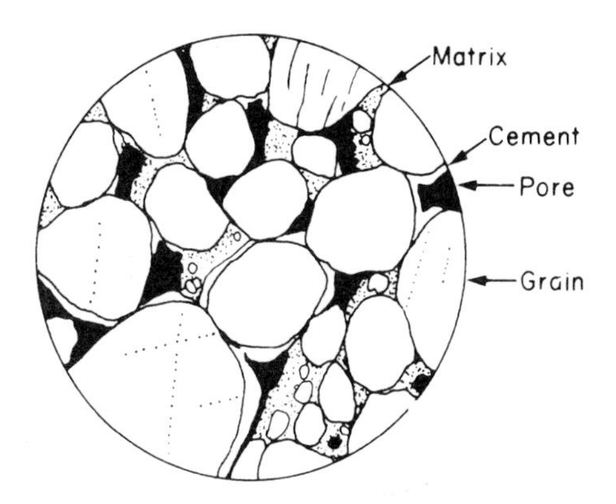

Fig. 9. A sedimentary rock is composed of grains and matrix of syndepositional origin; cement, of post-depositional origin, and pores. Pores are the voids not occupied by any of the three solid components.

framework of a sediment. Matrix is the finer detritus which occurs within the framework. There is no arbitrary size distinction between grains and matrix. Conglomerates generally have a matrix of sand, and sandstones may have a matrix of silt and clay. Cement is post-depositional mineral growth which occurs within the voids of a sediment. Pores are the hollow spaces not occupied by grains, matrix or cement. Pores may contain gases, such as nitrogen and carbon dioxide, or hydrocarbons such as methane. Pores may be filled by liquids ranging from potable water to brine and oil. Under suitable conditions of temperature and pressure, pores may be filled by combinations of liquid and gas.

The study of pore liquids and gases lies in the scope of hydrology and petroleum engineering. Petrophysics, the study of the physical properties of pores, lies on the boundary between these disciplines and sedimentary geology.

The geologist should understand the morphology and genesis of pores and, ideally, be able to predict their distribution within the earth's crust.

1. Definitions

The porosity of a rock is the ratio of its total pore space to its total volume, i.e. for a given sample: porosity = total volume − bulk volume. Conventionally porosity is expressed as a percentage. Hence:

$$\text{porosity} = \frac{\text{volume of total pore space}}{\text{volume of rock sample}} \times 100.$$

The porosity of rocks range from effectively zero in unfractured cherts to, theoretically, 100% if the "sample" is taken in a cave. Typically porosities in sediments range between 5–25%, porosities of 25–35% are regarded as excellent if found in an aquifer or oil reservoir.

An important distinction must be made between the total porosity of a rock and its effective porosity. Effective porosity is the amount of mutually interconnected pore spaces present in a rock. It is, of course, the effective porosity which is generally economically important, and it is effective porosity which is determined by many, but not all, methods of porosity measurement.

It is the presence of effective porosity which gives a rock the property of permeability. Permeability is the ability of a liquid or gas to flow through a porous solid. Permeability is controlled by many variables. These include the effective porosity of the rock, the geometry of the pores, including their tortuosity, and the size of the throats between pores, the capillary force between the rock and the invading fluid, its viscosity and pressure gradient.

Permeability is conventionally determined from Darcy's law using the equation:

$$Q = \frac{K\Delta A}{\mu \, . \, L}$$

where Q is the rate of flow in cm^3/s, Δ is the pressure gradient, A is the cross-sectional area, μ is the fluid viscosity in centipoises, L is the length and K is the permeability.

This relationship was originally discovered by H. Darcy in 1856 following a study of the springs of Dijon. Permeability is usually expressed in darcy units, a term proposed and defined by Wycoff *et al.* in 1934. One darcy is the permeability which allows a fluid of one centipoise viscosity to flow at one centimetre per second, given a pressure gradient of one atmosphere per centimetre.

The permeability of most rocks is considerably less than one darcy. To avoid fractions or decimals, the millidarcy is generally used, being one-thousandth of a darcy. The permeability of rocks is highly variable, both depending on the direction of measurement and vertically up or down sections. Permeabilities ranging from 10 to 100 millidarcies are good and above that are considered exceptionally high.

Figure 10 illustrates the concepts of porosity, effective porosity and permeability in different rock types.

Figure 11 shows the vertical variations of porosity and permeability which are found in a typical sequence of rock.

2. Methods of measurement of porosity and permeability

There are a number of different methods which may be used to measure the porosity and permeability of rocks. Many of these require the direct analysis of a sample of the rock in question. In bore holes, however, it is often possible to make accurate measurements of the porosity of a rock by indirect geophysical techniques. No comparable method can calculate permeability, but this can be discovered from production tests of the aquifer or hydrocarbon reservoir in question.

(i) Direct methods of porosity measurement

Measurement of porosity by direct methods of porosity measurement requires samples of the rock in question to be available for analysis. These may be hand specimens collected from surface outcrops; they may be bore-hole cores, or small plugs cut from cores.

For all the various methods of directly determining porosity it is necessary to determine both the total volume of the rock sample, and either the volume of its porosity or of its bulk volume. Most methods rely on the measurement of the porosity by vacuum extraction of the fluids contained within the pores. Such methods, therefore, measure not total porosity but effective porosity. This is not terribly important since it is the porosity of the interconnected pores which is of significance in an aquifer or hydrocarbon reservoir.

(ii) Indirect methods of porosity measurements

It is often impossible to obtain large enough samples for porosity analysis from rocks which, underground, hold water, oil or gas. The porosity of such host rocks must be known in any attempt to assess their economic potential. A number of methods are now available for measuring the

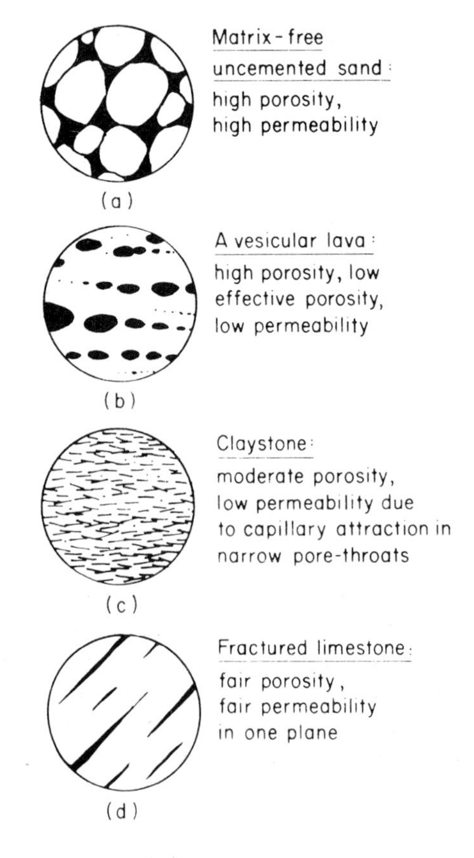

Fig. 10. Sketches of microscopic-thin sections to illustrate the relationship between total porosity, effective porosity and permeability. (Pores black, rock white.) The ideal aquifer or hydrocarbon reservoir (a) combines high effective porosity and high permeability.

porosity of rocks *in situ* when penetrated by a bore hole. These are based on measurements of various geophysical properties of the rock by a sonde—a complex piece of electronic equipment, which is lowered on a cable down the well bore. Different sondes are designed

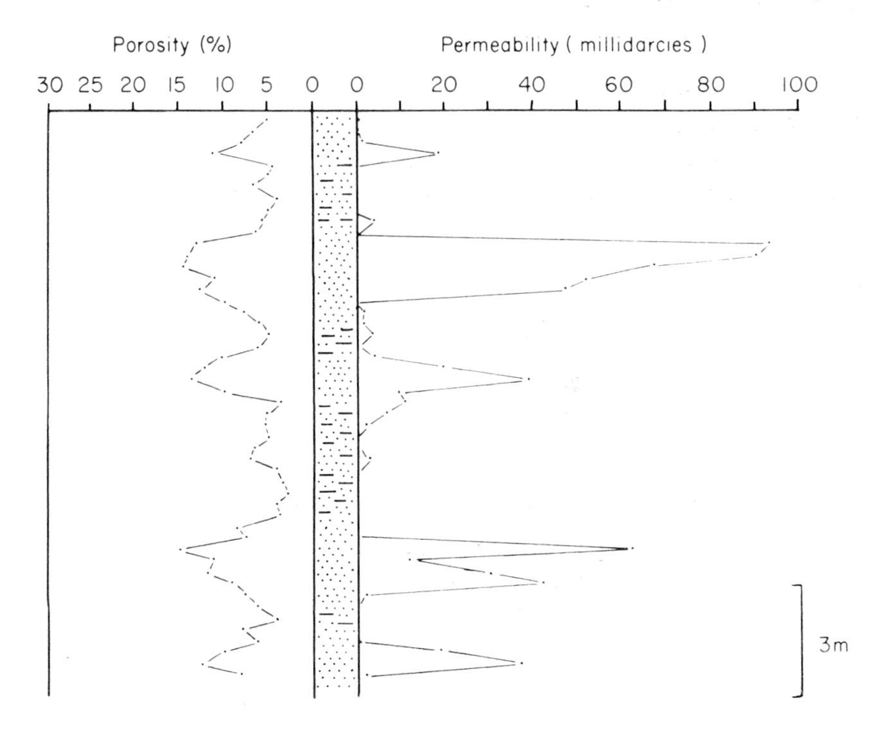

Fig. 11. Vertical section of a sandstone sequence to show the variations in porosity and permeability. The observed relationship between increasing porosity and permeability is not uncommon.

to measure different properties of the rock. These include its sonic velocity and bulk density. Sonic velocity is recorded continuously by use of an acoustic device in a sonde lowered down a well bore. The sonic velocity of the formation is recorded in micro-seconds per foot. Given the sonic velocity of the pore fluid and of the pure rock mineral (the sonic velocity of calcite is used for limestones, and of silica for sandstones, etc.) the porosity may be found thus:

$$\phi = \frac{t - t_{ma}}{t_f - t_{ma}}$$

where ϕ is the porosity %, t is the observed sonic velocity, t_{ma} is the sonic velocity of the matrix (i.e. non-pore rock), and t_f is the sonic velocity of the pore fluid.

Additional discussion of indirect methods of measuring porosity will be found in Lynch (1962). These geophysical well-logging techniques can give accurate measurements of porosity.

(iii) Direct method of permeability measurement

Permeability, like porosity, may be measured in the laboratory using a hand specimen or core sample of the rock to be analysed. The method of analysis involves pumping gas through a carefully dried and prepared rock sample. The apparatus records the length and cross-sectional area of the sample, the pressure drop across it, and the rate of flow for the test period. The permeability is then discovered by applying the Darcy equation (p. 23). The viscosity of gas at the temperature of the test can be found in printed tables of physical data.

(iv) Indirect methods of permeability measurement

Permeability, unlike porosity, cannot be measured by using geophysical sondes in bore holes. On the other hand, it can be calculated by recording the amount of gas or fluid which a known length of hole can produce over a given period of time. This is applicable both in hydrology where, unless the well is artesian, pump tests are carried out. These record the amount of water which can be extracted in a given period and the length of time required for the water table to return to normal if it was depressed during the test. Both the productivity under testing and the recharge time are measures of the permeability of the aquifer.

Oil and gas reservoirs typically occur under pressure. Permeability of the reservoir can be measured from drill-stem tests and from lengthier production tests. These measure the amount of fluid produced in a given period, the pressure drop during this time and the build-up in pressure over a second time interval when the reservoir is not producing. By turning Darcy's law inside out to apply to the shape of the bore hole, it is possible to measure the mean permeability of the reservoir formation over the interval which was tested.

B. Pore Morphology

1. Introduction and classification

Any petrophysical study of a reservoir rock necessitates a detailed description of the amount, type and genesis of its porosity. The quantitative measurement of porosity has been described in the previous section. The classification of the main types of porosity will now be discussed and followed by a description of the commoner varieties of pores. A large number of adjectives have been used to describe the different types of porosity present in sediments. Choquette and Pray (1970, pp. 244–250) have given a useful glossary of pore terminology.

The pores themselves may be studied by a variety of methods ranging from examination of rough or polished rock surfaces by hand-lens or stereoscopic microscope, through study of thin sections using a petrological microscope, to the use of the scanning electron microscope. Another effective technique of studying pore fabric is to impregnate the rock with a suitable plastic resin and then to dissolve the rock itself with an appropriate solvent. Examination of the residue gives some indication, not only of the size and shape of the pores themselves, but also of the throat passages which connect pores (e.g. Wardlaw, 1976). The minimum size of throats and the tortuosity of pore systems are closely related to the permeability of the rock.

These different observational methods show that there are a wide number of different types of pore systems.

Various attempts have been made to classify porosity types. These range from essentially descriptive schemes (e.g. Levorsen, 1967, p. 113), to those which combine descriptive and genetic criteria (e.g. Choquette and Pray, 1970), and those which relate the porosity type to the petrography of the host rock (e.g. Robinson, 1966).

The classification shown in Table V divides porosity into two main varieties which are commonly recognized (e.g. Murray, 1960). These are primary porosity fabrics, which were present immediately after the rock had been deposited, and secondary or post-depositional fabrics which formed after sedimentation by a variety of causes.

The main porosity types will now be described and illustrated.

Table V
A classification of porosity types

	Type	Origin
I. Primary or depositional	(a) Intergranular or Interparticle (b) Intraparticle	Sedimentation
II. Secondary or post-depositional	(c) Intercrystalline (d) Fenestral	Cementation
	(e) Moldic (f) Vuggy	Solution
	(g) Fracture	Tectonic movement, compaction or dehydration

2. Primary or depositional porosity

Primary or depositional porosity is that which, by definition, forms when a sediment is laid down. Two main types of primary porosity may be recognized.

(i) Intergranular porosity

Intergranular or interparticle porosity occurs in the spaces between the detrital grains which form the framework of a sediment (Fig. 12a). This is a very important porosity type. It is present initially in almost all sedimentary rocks. Intergranular porosity is generally progressively reduced by diagenesis in many carbonates, but is the dominant porosity type found in sandstones.

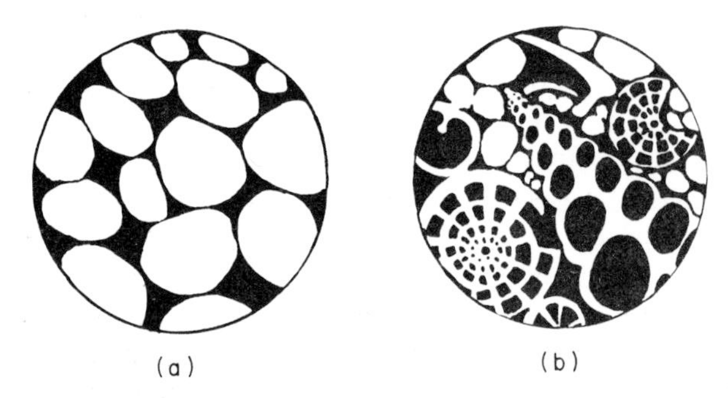

Fig. 12. Sketches of thin sections to illustrate primary, depositional porosity. (a) Intergranular porosity, commonly found in sandstones; (b) mixed intergranular and intra-particle porosity, typical of skeletal sands before burial and diagenesis.

The factors which influence the genesis of intergranular porosity and which modify it after deposition are discussed at length in a later section (p. 91).

(ii) Intraparticle porosity

In carbonate sands, particularly those of skeletal origin, primary porosity may be present within the detrital grains. For example, the cavities of molluscs, ammonites, corals, bryozoa and microfossils may all be classed as intraparticle primary porosity (Fig. 12b).

This kind of porosity is often diminished shortly after deposition by infiltrating micrite matrix. Furthermore, the chemical instability of the carbonate host grains often leads to their intraparticle pores being modified or obliterated by subsequent diagenesis.

3. Secondary or post-depositional porosity

Secondary porosity is that which, by definition, formed after a sediment was deposited. Secondary porosity is more diverse in morphology and more complex in genesis than primary porosity. The following main types of secondary porosity are recognizable.

(i) Intercrystalline porosity

Intercrystalline porosity occurs between the individual crystals of a crystalline rock (Fig. 13a). It is, therefore, the typical porosity type of the igneous and high-grade metamorphic rocks, and of some evaporites. Strictly speaking, such porosity is of primary origin. It is, however, most characteristic of carbonates which have undergone crystallization and is particularly important in recrystallized dolomites. Such rocks are sometimes very important oil reservoirs. The pores of crystalline rocks are essentially planar cavities which intersect obliquely with one another with no constrictions of the boundaries or throats between adjacent pores.

(ii) Fenestral porosity

The term "fenestral" porosity was first proposed by Tebbutt *et al.* (1965) for a "primary or penecontemporaneous gap in rock framework, larger than grain-supported interstices". This porosity type is typical of carbonates. It occurs in fragmental carbonate sands, where it grades into primary porosity, but is most characteristic of pellet muds and homogenous muds of lagoonal and intertidal origin. Penecontemporaneous dehydration, cementation and gas generation can cause depositional laminae to buckle and generate subhorizontal fenestral pores between the laminae (Fig. 13b). This type of fabric has been termed "loferite" by Fischer (1964) on the basis of a study of Alpine Triassic back reef carbonates. Analogous fabrics have been described from dolomite pellet muds in a similar setting from the Libyan Palaeocene (Conley, 1971).

The fenestrae in this last case are sometimes partially floored by lime mud, suggesting their penecontemporaneous origin.

A variety of fenestral fabric has long been known as "bird's-eye". This refers to isolated "eyes" up to a centimetre across which occur in some lime mudstones (Illing, 1954). These apertures have been attributed to organic burrows and to gas escape conduits. They are frequently infilled by crystalline calcite.

(iii) Moldic porosity

A third type of secondary porosity, generally formed later in the history of a rock than fenestrae, is moldic porosity.

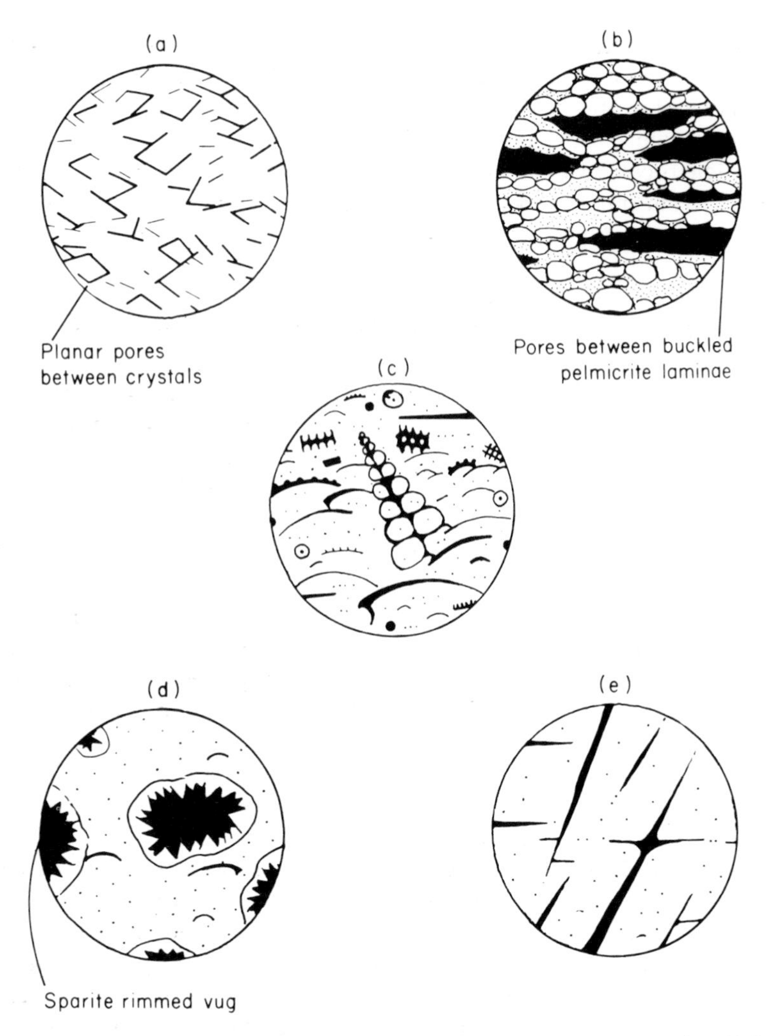

Fig. 13. Sketches of thin sections to illustrate the main kinds of secondary porosity. (a) Intercrystalline porosity, characteristic of dolomites; (b) fenestral porosity, characteristic of pelmicrites; (c) moldic porosity, formed by selective leaching, in this case of skeletal fragments (biomoldic); (d) vuggy porosity produced by irregular solution; (e) fracture porosity, present in many brittle rocks.

Molds are pores formed by the solution of primary depositional grains generally subsequent to some cementation. Molds are fabric selective. That is to say, solution is confined to individual particles and does not cross-cut cement, matrix and framework. Typically in any one rock it is all the grains of one particular type that are dissolved. Hence one may talk of oomoldic, pelmoldic or biomoldic porosity where there has been selective solution of ooliths, pellets or skeletal debris (Fig. 13c). The geometry and effective porosity and permeability of moldic porosity can thus be extremely varied. In an oomoldic rock, pores will be subspherical and of similar size. In biomoldic rocks, by contrast, pores may be very variable in size and shape, ranging from minute apertures to curved planar pores where shells have dissolved, and cylinders where echinoid spines have gone into solution.

(iv) Vuggy porosity

Vugs are a second type of pore formed by solution and, like molds, they are typically found in carbonates. Vugs differ from molds though because they cross-cut the primary depositional fabric of the rock (Fig. 13d). Vugs thus tend to be larger than molds. They are often lined by a selvedge of crystals. With increasing size vugs grade into what is loosely termed cavernous porosity. Choquette and Pray (1970, p. 244) proposed that the minimum dimension of a cavern is a pore which allows a man to enter or which, when drilled into, allows the drill string to drop by more than half a metre through the rotary table. Such cavernous porosity occurs in the Arab zone (Upper Jurassic) of the Abqaiq oil field in Saudi Arabia (McConnell, 1951). Some of the largest cavernous oil reservoirs on record occur in Texas, naturally, where caverns up to five metres high were reported in the Fusselman Limestone of the Dollarhide oil field (Stormont, 1949).

(v) Fracture porosity

The last main type of pore to be considered is that which occurs within fractures. Fractures occur in many kinds of rocks other than sediments. Fracturing, in the sense of a breaking of depositional lamination can occur penecontemporaneously with sedimentation. This often takes the form of microfaulting caused by slumping, sliding and compaction. Fractures in plastic sediments are instantaneously sealed. In brittle rocks, however, fractures may remain open after formation, thus giving rise to fracture porosity (Fig. 13e). This porosity type characterizes rocks which are strongly lithified and is, therefore, generally formed later in time than the other varieties of porosity.

It is important to note that fracture porosity is not only found in well-cemented sandstones and carbonates, but may also be present in shales, and igneous and metamorphic rocks.

Fracture porosity is much more difficult to observe and analyse than most other pore systems. Though fractures range from microscopic to cavernous in size, they are difficult to study in cores.

Fracture porosity can occur in a variety of ways and situations. Tectonic movement can form fracture porosity in two ways. Tension over the crests of compressional anticlines and compactional drapes can generate fracture porosity. Harris *et al.* (1960) mapped fracture orientation and intensity on anticlinal crests at outcrop. Their results showed the close relationship between gross structure and fracture pattern. On the symmetric Sheep Mountain anticline, fracture intensity is at its maximum on the crest. On the asymmetric Goose Egg Dome, however, fractures were best developed on the steepest limb of the structure.

Fracture porosity is also intimately associated with faulting and some oil fields show very close structural relations with individual fault systems. The Scipio fields of south-west Michigan are a case in point. These occur on a straight line of about 15 km. Individual fields are about 0·5 km wide. The oil is trapped in a fractured dolomitized belt within the Trenton Limestone (Ordovician). This fracture system was presumably caused by movement along a deep fault in the basement (Levorsen, 1967, p. 123).

Fracture porosity can also form from atectonic processes. It is often found immediately beneath unconformities. Here fractures, once formed by weathering, may have been enlarged by solution (especially in limestones) and preserved without subsequent loss of their porosity (p. 104).

(vi) Summary

The preceding account shows something of the diverse types and origins of pores. Essentially there are two main genetic groups of pore types.

Primary porosity is formed when a sediment is deposited. It includes intergranular or interparticle porosity, which is characteristic of sands, and intraparticle porosity found in skeletal lime sands.

Secondary porosity forms after sedimentation by diagenetic processes. Recrystallization, notably dolomitization, can generate intercrystalline porosity. Solution can generate moldic, vuggy and cavernous pores. Because such pores are often isolated from one another, permeability may be low. Fractures form in both unconsolidated and brittle sediments. In the first instance the fractures remain closed, but in brittle rocks fracture porosity may be preserved, enlarged by solution, or diminished by cementation. Fracture porosity occurs not just in indurated sediments, but also in igneous and metamorphic rocks.

It is important to note that many sedimentary rocks contain more than one type of pore. The combination of open fractures with another pore type is of particular significance. Fine-grained rocks, both shales, microcrystalline carbonates and fine sands, have considerable porosity. They often have very low permeabilities. The presence of fractures, however, can enable such rocks to yield up their contained fluids. The success of many oil and water wells in such formations often depends on whether they happen to penetrate an open fracture.

Recognition of the significance of fractures in producing fluids from high-porosity low-permeability formations has led to the development of artificial fracturing by explosive charges which simultaneously wedge the fractures open with sand, glass beads, etc. Similarly the productivity of fractured carbonate reservoirs may be increased by the injection of acid to dissolve and enlarge the fractures.

Figure 14 shows the relationship between pore type and petrophysics.

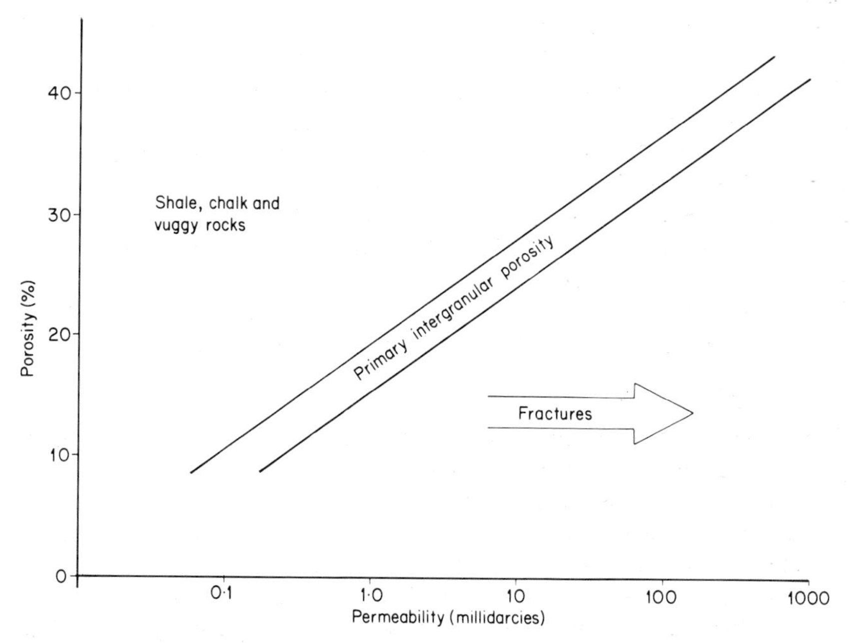

Fig. 14. Graph to show the relationship between petrophysics and pore type. For intergranular porosity seen in most sandstones, porosity generally plots as a straight line against permeability on a logarithmic scale. Shales, chalks and vuggy rocks tend to be porous but impermeable. Fractures increase permeability with generally little increase in porosity.

The effects of diagenesis on the porosity of sandstones and carbonates are discussed in Chapters 4 and 5 respectively. The rest of this chapter is concerned with the factors which control primary porosity at the time of deposition and with the initial loss of primary porosity by compaction.

C. The Origin of Primary Porosity

Primary intergranular porosity is a function of the depositional fabric at the time of sedimentation, modified by post-depositional compaction and subsequent diagenesis.

1. The relationships between porosity, permeability and texture

Beard and Weyl (1973) have shown that the porosity of a sediment when it has just been laid down is a result of five variables: grain size, sorting, grain shape (sphericity), grain roundness (angularity) and packing. A considerable amount of work has been done on the way that these five factors effect primary porosity. This work includes theoretical mathematical studies (Engelhardt and Pitter, 1951), experimental analyses of artificially made spheres, unconsolidated modern sediment, and even ancient rocks. The results of this work will now be summarized for the five parameters listed above.

(i) The effect of grain size on porosity and permeability

Theoretically porosity is independent of grain size. A mass of spheres of uniform sorting and packing will have the same porosity, regardless of the size of the spheres. The volume of pore space varies in direct proportion to the volume of the spheres (Fraser, 1935). Rogers and Head (1961), working with synthetic sands, showed that porosity is independent of grain size for well-sorted sands. This, the "ideal" situation, is seldom found in nature. Pryor (1973) analysed nearly a thousand modern sands and showed that porosity decreased with increasing grain size. River sands were the reverse, however, possibly due to packing differences. This trend is probably due to a number of factors which are only indirectly linked with grain size. Finer sands tend to be more angular and to be able to support looser packing fabrics, hence they may have a high porosity than coarser sands.

Whatever the cause of the relationship, it has been empirically shown that porosity generally increases with decreasing grain size for unconsolidated sands of uniform grain size. This relationship is not always true for lithified sandstones, whose porosity often increases with grain size. This may be because the finer sands have suffered more from compaction and cementation.

Permeability, by contrast, increases with increasing grain size (Fraser, 1935; Krumbein and Monk, 1942; Pryor, 1973). This is because in finer sediments the throat passages between pores are smaller and the higher capillary attraction of the walls inhibits fluid flow. This relationship is found in both unconsolidated and lithified sand.

(ii) The effect of sorting on porosity

A number of studies of sediments have shown that porosity increases with increasing sorting (Fraser, 1935; Rogers and Head, 1961; Pryor, 1973; Beard and Weyl, 1973). Krumbein and Monk (1942) and Beard and Weyl (1973) demonstrated that increasing sorting correlates with increasing permeabilities.

A reason for these relationships is not hard to find. A well-sorted sand has a high proportion of detrital grains to matrix. A poorly-sorted sand, on the other hand, has a low proportion of detrital grains to matrix. The finer grains of the matrix block up both the pores and throat passages within the framework, thus inhibiting porosity and permeability respectively.

Pryor's study (1973) of modern sands from different environments confirmed this relationship for river sands, but showed that beach and dune sands were anomalous in that their permeability increased with decreasing sorting.

Figure 15 summarizes the relationships between porosity, permeability, grainsize and sorting for unconsolidated sands.

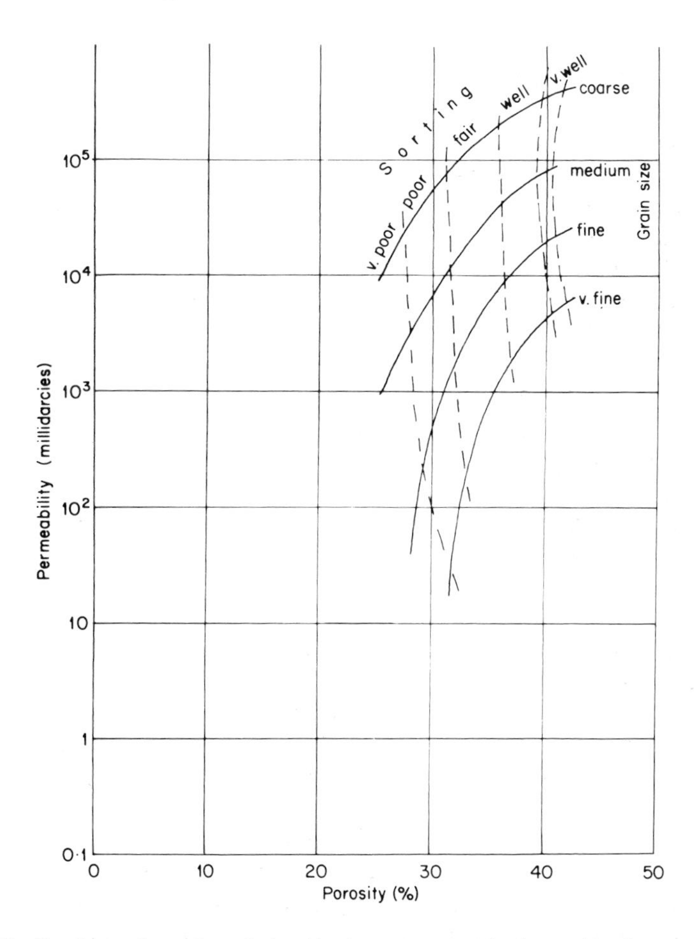

Fig. 15. Graph to show the relationship between petrophysics and sediment texture in unconsolidated clay-free sand. For explanation see text (from Nagtegaal, 1978; Beard and Weyl, 1973).

(iii) The effect of grain shape and roundness on porosity

Undoubtedly the shape and roundness of grains affects intergranular porosity. Little work has been done on this problem because of the time needed to measure these parameters in sufficiently large enough samples to be meaningful.

Fraser (1935) concluded that sediments composed of spherical grains have lower porosities than those with grains of lower sphericity. He attributed this to the fact that the former tend to fall into a tighter packing than sands of lower sphericity (see also Beard and Weyl, 1973).

(iv) The relationship between fabric and porosity

The way in which the particles of a sediment are arranged is termed the "fabric". There are two elements to fabric: grain orientation and grain packing. Since these are closely related to primary porosity, they will now be discussed.

Grain orientation. The orientation of sediment particles is generally discussed with reference to the axis of sediment transport (the flow direction) and to the horizontal plane.

The orientation of pebble fabrics is tolerably well known because their large size makes them relatively easy to measure. Sand-grain orientation has, until recently, been less well understood, possibly because it is more complex but probably because it is harder to measure and quantify.

One of the commonest features of gravel fabrics is "imbrication" in which the pebbles lie with their long axis parallel to the flow direction and dipping gently up-current. Isolated pebbles on channel floors are often imbricated too (Fig. 16). This is useful palaeocurrent indicator (see p. 242). Dispersed pebbles in diamictites (pebbly mudstones) also commonly show an alignment parallel to the direction of movement. This is true both of mud flow deposits and of glacial tills (see Lindsay, 1966 and Andrews and Smith, 1970 respectively).

Studies of the orientation of pebbles dispersed in fluvial sediments have yielded conflicting results. Many studies (reviewed in Schlee, 1957) show that long axes tend to parallel current flow. Some data from cross-bedded sands, however, show this orientation on the foreset, but record that on the topset and bottomset pebbles are elongated perpendicular to the current direction (Sengupta, 1966; Bandyopadhyay, 1971; Gnaccolini and Orombelli, 1971). For an additional review of this problem see Johansson (1965).

Studies of sand grain orientation were held back when it was only possible to measure each grain individually. Recently indirect methods of orientation measurement have evolved (Sippel, 1971; Shelton and Mack, 1970). These enable aggregate samples to be measured with speed.

It has generally been observed that grain orientation parallels flow direction in flat-bedded alluvial sands, turbidites and marine bars (Shelton

Fig. 16. Channel showing imbrication fabric of shale conglomerate on the channel floor. The clasts dip up-current in opposition to the down-current dip direction of the foresets. Cambro-Ordovician, Jordan.

and Mack, 1970; Martini, 1971; Von Rad, 1971). In channel sands, therefore, grains are orientated parallel to the axis of the sand body. These are coincident with the preferred permeability as shown by isopotential mapping (e.g. Busch, 1971, p. 1143). In linear bar sands, on the other hand, sand grains are aligned perpendicular to the bar axis by back-swash action. In this case the preferred direction of optimum permeability may be perpendicular to sand body trend (Fig. 17). Studies on the permeability

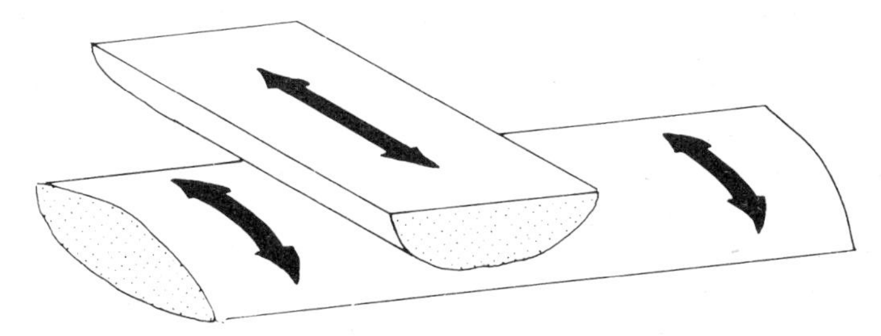

Fig. 17. Grain orientation and maximum permeability directions (arrowed) in channel and bar sand bodies, upper and lower respectively. In both cases these axes coincide with the depositional palaeoslope. In channels optimum permeability parallels the sand body trend. In bars it is perpendicular to sand body trend. (Based on data in Pryor, 1973, Figs 3 and 9.)

directions of modern river sands and marine beaches confirm this distinction (Pryor, 1973).

Grain packing. Graton and Fraser (1935) demonstrated how the porosity of a sediment varies according to the way in which its constituent grains are packed. They showed that theoretically there are six possible packing geometries for spheres of uniform size. These range from the loosest "cubic" style, with a theoretical porosity of 48%, to the closest rhombohedral packing with a theoretical porosity of 26%. These ideal situations never occur in nature. More realistic packing models are obtained when the analysis is based not on spheres, but on prolate spheroids which closer approximate real sand grains (Allen, 1970).

Though packing no doubt plays a major part in controlling the primary porosity of a sediment, this parameter has been one of the hardest to analyse in consolidated rocks. The reasons for this are threefold: difficulty of measurement, lack of knowledge of the control of environment and depositional process on packing, and the effect of post-depositional compaction.

A number of workers have suggested methods of measuring and quantifying packing. Emery and Griffiths (1954) proposed a packing index which is the product of the number of grain contacts observed in a thin-section traverse and the average grain diameter divided by the length of the traverse. Kahn (1956) renamed this packing index as packing proximity and revamped the original formula. Mellon (1964) proposed a horizontal packing intercept, which is the average horizontal distance between framework grains.

The problem with all of these different indices of packing is that it is extremely laborious to measure large enough samples, and sufficient specimens, to produce enough data to manipulate. Furthermore, these measures of packing are determined from lithified thin sections. It is possible to impregnate and thin-section unconsolidated sand, but it is unlikely that this can be done without disturbing the original packing.

In addition, Morrow (1971) has pointed out how the packing of a sediment fabric will change from type to type within and between adjacent laminae.

Because of these problems little is known of the relationship between packing and primary depositional porosity. Simplistically one might expect pelagic muds, turbidites and grain flows to be deposited with a looser packing than traction deposits, and, perhaps, cross-bedded sand to be more loosely packed than flat-bedded sand. There are few data to support this line of speculation. Pryor (1973) has shown that modern river sands are more loosely packed than beach and eolian dunes. The fact is that post-depositional compaction so reorients sand grains that packing can have little influence on the porosity of lithified sediment.

III. REFERENCES

Allen, J. R. L. (1970). The systematic packing of prolate spheroids with reference to concentration and dilatency. *Geologie Mijnb.* **49**, 211–220.

Allen, T. (1968). "Particle Size Measurement" Chapman and Hall, London. 248pp.

Andrews, J. T. and Smith, D. I. (1970). Statistical analysis of till fabric: methodology, local and regional variability. *Q. Jl geol. Soc. Lond.* **125**, 503–542.

Anon. (1972). Submarine lithification of ancient limestones. *Nature, Lond.* **237** (5354), 309.

Archie, G. E. (1950). Introduction to petrophysics of reservoir rocks. *Bull. Am. Ass. Petrol. Geol.* **34**, 943–961.

Arps, J. J. (1964). Engineering concepts useful in oil finding. *Bull. Am. Ass. Petrol. Geol.* **48**, 157–165.

ASTM (1959). Symposium on particle size measurement. *Spec. tech. Publs Am. Soc. Test. Mater.* **234**, 303pp.

Atwater, G. I. and Miller, E. E. (1965). The effect of decrease in porosity with depth on future development of oil and gas reserves in South Louisiana. *Bull. Am. Ass. Petrol. Geol.* **49**, 334 (Abs.).

Bandyopadhyay, S. (1971). Pebble orientation in relation to cross-stratification: a statistical study. *J. sedim. Petrol.* **41**, 585–587.

Bathurst, R. G. C. (1972). "Carbonate Sediments and Their Diagenesis" Elsevier, Amsterdam. 700pp.

Beales, F. W. and Keith, J. W. (1972). Limestone genesis and diagenesis. Rep. 24th int. geol. Cong. Montreal, 1972. Section 6, Stratigraphy and Sedimentology, 121–123.

Beard, D. C. and Weyl, P. K. (1973). The influence of texture on porosity and permeability of unconsolidated sand. *Bull. Am. Ass. Petrol. Geol.* **57**, 349–369.

Busch, D. A. (1971). Genetic units in delta prospecting. *Bull. Am. Ass. Petrol. Geol.* **55**, 1137–1154.

Cailleux, A. and Tricart, J. (1959). "Initiation à l'Étude des Sables et des Galets" Centre de Documentation, University of Paris.

Carver, R. E. (1971). "Procedures in Sedimentary Petrology" John Wiley, New York. 576pp.

Chafetz, H. S. (1972). Surface diagenesis of limestone. *J. sedim. Petrol.* **42**, 325–329.

Chappell, J. (1967). Recognizing fossil strand lines from grain-size analysis. *J. sedim. Petrol.* **37**, 157–165.

Chayes, F. (1956). "Petrographic Modal Analysis" John Wiley, New York. 113pp.

Chilingar, G. V. (1964). Relationship between porosity, permeability and grain-size and distribution of sands and sandstones. *In* "Deltaic and Shallow Marine Deposits" (L. M. J. U. Van Straaten, Ed.), 71–75. Elsevier, Amsterdam.

Chilingar, G. V., Mannon, R. W. and Rieke, H. (1972). "Oil and Gas Production from Carbonate Rocks" Elsevier, Amsterdam. 408pp.

Choquette, P. W. and Pray, L. C. (1970). Geologic nomenclature and classification of porosity in sedimentary carbonates. *Bull. Am. Ass. Petrol. Geol.* **54**, 207–250.

Conley, C. D. (1971). Stratigraphy and lithofacies of Lower Paleocene rocks, Sirte Basin, Libya. *In* "Symposium on the Geology of Libya" (C. Gray, Ed.), 127–140. University of Libya, Tripoli.

Cummins, W. A. (1962). The greywacke problem. *Lpool Manchr geol. J.* **3**, 51–69.

Dapples, E. C. (1972). Some concepts of cementation and lithification of sandstones. *Bull. Am. Ass. Petrol. Geol.* **56**, 3-25.

Darcy, H. (1856). "Les Fontaines Publiques de la Ville de Dijon" Dalmont, Paris. 674pp.

Doeglas, D. J. (1946). Interpretation of the results of mechanical analyses. *J. sedim. Petrol.* **16**, 19–40.

Doornkamp, J. C. and Krinsley, D. (1971). Electron microscopy applied to quartz grains from a tropical environment. *Sedimentology* **17**, 89–101.

Emery, J. R. and Griffiths, J. C. (1954). Reconnaissance investigation into relationships between behaviour and petrographic properties of some Mississippian sediments. *Bull. Miner. Inds Exp. Stn Penn. St. Univ.* **62**, 67–80.

Engelhardt, W. V. and Pitter, H. (1951). Uber die Zusammenhange zwischen Porosität, Permeabilität und Korngrösse bei Sanden und Sandsteinen. *Heidelb. Beitr. Miner. Petrogr.* **2**, 477–491.

Fischer, A. G. (1964). The Löfer cyclothems of the Alpine Triassic. —Symposium on Cyclic Sedimentation. *Bull. Kans. Univ. geol. Surv.* **169**, 107–150.

Folk, R. L. (1966). A review of grainsize parameters. *Sedimentology* **6**, 73–94.

Folk, R. L. and Ward, W. C. (1956). A study in the significance of grain size parameters. *J. sedim. Petrol.* **27**, 3–26.

Fraser, H. J. (1935). Experimental study of the porosity and permeability of clastic sediments. *J. Geol.* **43**, 910–1010.

Friedman, G. M. (1958). Determination of sieve-size distribution from thin section data for sedimentary petrological studies. *J. Geol.* **66**, 394–416.

Friedman, G. M. (1961). Distinction between dune, beach and river sands from their textural characteristics. *J. sedim. Petrol.* **31**, 514–529.

Friedman, G. M. (1962a). Comparison of moment measures for sieving and thin section data in sedimentary petrologic studies. *J. sedim. Petrol.* **32**, 15–25.

Friedman, G. M. (1962b). On sorting, sorting coefficients and log normality of grainsize distribution of sandstones. *J. Geol.* **70**, 737–753.

Friedman, G. M. (1965). Terminology of crystallization textures and fabrics in sedimentary rocks. *J. sedim. Petrol.* **39**, No. 3, p. 643.

Friedman, G. M. (1967). Dynamic processes and statistical parameters compared for size frequency distribution of beach and river sands. *J. sedim. Petrol.* **37**, 327–354.

Fruth, L. S., Orme, G. R. and Donath, F. A. (1966). Experimental compaction effects in carbonate sediments. *J. sedim. Petrol.* **36**, 747–754.

Fuchtbauer, H. (1967). Influence of different types of diagenesis on sandstone porosity. Proc. 7th Wld Petrol. Cong. Mexico, 353-369.

Garrison, R. E., Luternaur, J. L., Grill, E. V., MacDonald, R. D. and Murray, J. W. (1969). Early diagenetic cementation of Recent sands, Fraser River delta, British Columbia. *Sedimentology* **12**, 27–46.

Gnaccolini, M. and Orombelli, G. (1971). Orientazione dei ciottoli in un delta lacustre Pleistocenico della Brianza. *Riv. ital. Paleont. Stratigr.* **77**, 411–424.

Graton, L. C. and Fraser, H. J. (1935). Systematic packing of spheres, with particular reference to porosity and permeability. *J. Geol.* **43**, 785–909.

Griffiths, J. C. (1967). "Scientific Methods in Analysis of Sediments" McGraw-Hill, New York. 508pp.

Harris, J. F., Taulor, G. L. and Walper, J. L. (1960). Relation of deformational fractures in sedimentary rocks to regional and local structure. *Bull. Am. Ass. Petrol. Geol.* **44**, 1853–1873.

Herdan, G. (1953). "Small Particle Statistics" Elsevier, Amsterdam. 418pp.

Illing, L. V. (1954). Bahaman calcareous sands. *Bull. Am. Soc. Petrol. Geol.* **38**, 1–45.

Inman, D. L. (1952). Measures for describing the size distribution of sediments. *J. sedim. Petrol.* **22**, 125–145.

Johansson, C. E. (1965). Structural studies of sedimentary deposits. *Geol. För. Stockh. Förh.* **87**, 3–61.

Kahn, J. S. (1956). The analysis and distribution of the properties of packing in sand size sediments. *J. Geol.* **64**, 385–395.

Klein, G. de Vries (1963). Analysis and review of sandstone classifications in the North American geological literature. *Bull. geol. Soc. Am.* **74**, 555–576.

Klovan, J. E. (1966). The use of factor analysis in determining depositional environments from grainsize distributions. *J. sedim. Petrol.* **36**, 115–125.

Krinsley, D. and Cavallero, L. (1970). Scanning electron microscopic examination of periglacial eolian sands from Long Island, New York. *J. sedim. Petrol.* **4**, 1345–1350.

Krinsley, D. H. and Doornkamp, J. C. (1973). "Atlas of Quartz Sand Surface Textures" Cambridge University Press, Cambridge. 91pp.

Krinsley, D. H. L. and Funnell, B. M. (1965). Environmental history of sand grains from the Lower and Middle Pleistocene of Norfolk, England. *Q. Jl geol. Soc. Lond.* **121**, 435–462.

Krumbein, W. C. (1934). Size frequency distributions of sediments. *J. sedim. Petrol.* **4**, 65–77.

Krumbein, W. C. and Monk, G. D. (1942). Permeability as a function of the size parameters of unconsolidated sands. *Am. Inst. Min. Metall. Engrs* **1492**, 1–11.

Krumbein, W. C. and Pettijohn, F. J. (1938). "Manual of sedimentary petrography" Appleton-Century-Crofts, New York. 549pp.

Kuenen, P. H. (1956a). Experimental abrasion of pebbles 1: wet sand blasting. *Leid. geol. Meded.* **20**, 131–137.

Kuenen, P. H. (1956b). Experimental abrasion of pebbles 2: rolling by current. *J. Geol.* **64**, 336–368.

Kuenen, P. H. (1959). Experimental abrasion 3: fluviatile action. *Am. J. Sci.* **257**, 172–190.

Kuenen, P. H. (1960). Experimental abrasion 4: Eolian action. *J. Geol.* **68**, 427–449.

Kuenen, P. H. and Perdok, W. G. (1962). Experimental abrasion 5: frosting and defrosting of quartz grains. *J. Geol.* **70**, 648–659.

Laming, D. J. C. (1966). Imbrication, paleocurrents and other sedimentary features in the Lower New Red Sandstone, Devonshire, England. *J. sedim. Petrol.* **36**, 940–959.

Langres, G. L., Robertson, J. O. and Chilingar, G. V. (1972). "Secondary Recovery and Carbonate Reservoirs" Elsevier, Amsterdam. 250pp.

Lee, C. H. (1919). Geology and groundwaters of the western part of San Diego County, California. *Wat. Supply Irrig. Pap., Wash.* **446**, 121pp.

Lerbekmo, J. F. and Platt, R. L. (1962). Promotion of pressure solution of silica in sandstones. *J. sedim. Petrol.* **32**, No. 3, 514–519.

Levorsen, A. I. (1967). "The Geology of Petroleum" Freeman, Reading. 724pp.

Lindsay, J. F. (1966). Carboniferous subaqueous mass-movement in the Manning Macleay basin, Kempsey, New South Wales. *J. sedim. Petrol.* **36**, 719–732.

Lynch, E. J. (1962). "Formation Evaluation" Harper and Row, New York. 422pp.

McBride, E. F. (1963). A classification of common sandstones. *J. Sedim. Petrol.* **33**, 664–669.

McConnell, P. C. (1951). Drilling and production techniques that yield nearly 850,000 barrels per day in Saudi Arabia's fabulous Abqaiq field. *Oil Gas. J.* Dec. 20th, 1951, 197.

Manten, A. A. (1966). Note on the formation of stylolites. *Geologie Mijnb.* **45**, 269–274.

Margolis, S. V. and Krinsley, D. H. (1971). Submicroscopic frosting on eolian and subaqueous quartz sand grains. *Bull. geol. Soc. Am.* **82**, 3395–3406.

Martini, I. P. (1971). Grainsize orientation and paleocurrent systems in the Thorold and Grimsby sandstones (Silurian), Ontario and New York. *J. sedim. Petrol.* **41**, 225–234.

Martini, I. P. (1972). Studies of microfabrics: an analysis of packing in the Grimsby sandstone (Silurian), Ontario and New York State. Rep. 24th int. geol. Cong. Montreal, 1972. Section 6, Stratigraphy and Sedimentology, 415–423.

Mason, C. C. and Folk, R. L. (1958). Differentiation of beach, dune and eolian flat environments by size analysis; Mustang Island Texas. *J. sedim. Petrol.* **28**, 211–226.

Meade, R. H. (1966). Factors influencing the early stages of the compaction of clays and sands—Review. *J. sedim. Petrol.* **36**, 1085–1101.

Mellon, G. B. (1964). Discriminatory analysis of calcite and silicate—cemented phases of the Mountain Park sandstone. *J. Geol.* **72**, 786–809.

Miall, A. D. (1970). Devonian alluvial fans, Prince of Wales Island, Arctic Canada. *J. sedim. Petrol.* **40**, 556–511.

Middleton, G. V. (1972). Albite of Secondary origin in Charny sandstones, Quebec. *J. sedim. Petrol.* **42**, 341–349.

Moiola, R. J. and Weiser, D. (1968). Textural parameters: an evaluation. *J. sedim. Petrol.* **38**, 45–53.

Moore, C. H., Smitherman, J. E. and Allen, S. H. (1972). Pore systems evolution in a Cretaceous carbonate beach sequence. Rep. 24th int. geol. Cong. Montreal, 1972, Section 6, Stratigraphy and Sedimentology, 124–136.

Morrow, N. R. (1971). Small scale packing heterogeneities in porous sedimentary rocks. *Bull. Am. Ass. Petrol. geol.* **55**, 514–522.

Mueller, G. (1967). "Methods in Sedimentary Petrology" Hafner, London. 283pp.

Murray, R. C. (1960). Origin of porosity in carbonate rocks. *J. sedim. Petrol.* **30**, 59–84.

Nagtegaal, P. J. C. (1978). Sandstone-framework instability as a function of burial diagenesis. *J. geol. Soc. Lond.* **135**, 101–106.

Park, D. E. and Croneis, C. (1969). Origin of Caballos and Arkansas Novaculite Formations. *Bull. Am. Ass. Petrol. Geol.* **53**, 94–111.

Park, W. C. and Schot, E. H. (1968). Stylolites: their nature and origin. *J. sedim. Petrol.* **38**, 175–191.

Passega, R. (1957). Texture as characteristic of clastic deposition. *Bull. Am. Ass. Petrol. Geol.* **41**, 1952–1984.

Passega, R. (1964). Grainsize representation by C. M. Patterns as a geological tool. *J. sedim. Petrol.* **34**, No. 4, 830–847.

Perry, E. D. and Hower, J. (1970). Burial diagenesis in Gulf Coast pelitic sediments. *Clays Clay Miner.* **18**, 165–177.

Pettijohn, F. J. (1957). "Sedimentary Rocks" (2nd edition). Harper Bros, New York. 718pp.
Pettijohn, F. J., Potter, P. E. and Siever, R. (1972). "Sand and Sandstone" Springer-Verlag, Heidelberg. 618pp.
Plumley, W. J. (1948). Blackhill terrace gravels: a study in sediment transport. *J. Geol.* **56**, 526–577.
Potter, P. E. and Mast, R. F. (1963). Sedimentary structures, sand shape fabrics and permeability—I. *J. Geol.* **71**, 441–471.
Powers, M. C. (1953). A new roundness scale for sedimentary particles. *J. sedim. Petrol.* **23**, 117–119.
Pryor, W. A. (1973). Permeability-porosity patterns and variations in some Holocene sand bodies. *Bull. Am. Ass. Petrol. Geol.* **57**, 162–189.
Rieke, H. H. and Chilingar, G. V. (1973). "Compaction of argillaceous sediments" Elsevier, Amsterdam. 350pp.
Rittenhouse, G. (1971). Mechanical compaction of sands containing different percentages of ductile grains: a theoretical approach. *Bull. Am. Ass. Petrol. Geol.* **52**, 92–96.
Robinson, R. B. (1966). Classification of reservoir rocks by surface texture. *Bull. Am. Ass. Petrol. Geol.* **50**, 547–559.
Rogers, J. J. and Head, W. B. (1961). Relationship between porosity median size, and sorting coefficients of synthetic sands. *J. sedim. Petrol.* **31**, 467–470.
Russell, R. D. and Taylor, R. E. (1937). Roundness and shape of Mississippi River sands. *J. Geol.* **45**, 225–267.
Sames, C. W. (1966). Morphometric data of some recent pebble associations and their application to ancient deposits. *J. sedim. Petrol.* **36**, 126–142.
Schlee, J. (1957). Fluvial gravel fabric. *J. sedim. Petrol.* **27**, 162–176.
Sengupta, S. (1966). Studies on orientation and imbrication of pebbles with respect to cross-stratification. *J. sedim. Petrol.* **36**, 362–369.
Shelton, J. W. and Mack, D. E. (1970). Grain orientation in determination of paleocurrents and sandstone trends. *Bull. Am. Ass. Petrol. Geol.* **54**, 1108–1119.
Shinn, E. A. (1969). Submarine lithifaction of Holocene carbonate sediments in the Persian Gulf. *Sedimentology* **12**, 109–144.
Sippel, R. F. (1971). Quartz grain orientations—I (the photometric method). *J. sedim. Petrol.* **41**, 38–59.
Smith, R. E. (1969). Petrography-porosity relations in carbonate-quartz system, Gatesburg formation (Late Cambrian), Pennsylvania. *Bull. Am. Ass. Petrol. Geol.* **53**, 261–278.
Sneed, E. D. and Folk, R. L. (1958). Pebbles in the Lower Colorado River, Texas, a study in particle morphogenesis. *J. Geol.* **66**, 114–150.
Spencer, D. W. (1963). The interpretation of grain size distribution curves of clastic sediments. *J. sedim. Petrol.* **33**, No. 1, 180–190.
Stearns, D. W. and Friedman, G. M. (1972). Reservoirs in fractured rocks. *In* "Stratigraphic Oil and Gas Fields; Classification, Exploration Methods, and Case Histories" 82–106. Am. Ass. Petrol. Geol. Mem. No. 16.
Stormont, D. H. (1949). Huge caverns encountered in Dollarhide Field. *Oil Gas J.* April 7, 1949, 66–68.
Stout, J. L. (1964). Pore geometry as related to carbonate stratigraphic traps. *Bull. Am. Ass. Petrol. Geol.* **48**, 329–337.
Taylor, J. C. M. and Illing, L. V. (1969). Holocene intertidal calcium carbonate cementation, Quatar, Persian Gulf. *Sedimentology* **12**, 69–107.

Tebbutt, G. E., Conley, C. D. and Boyd, D. W. (1965). Lithogenesis of a distinctive carbonate fabric. *Wyo. Univ. Contrib. Geol.* **4**, No. 1.

Tickell, F. G. (1965). "The Techniques of Sedimentary Mineralogy" Elsevier, Amsterdam.

Trask, P. D. (1930). Mechanical analysis of sediment by centrifuge. *Econ. Geol.* **25**, 581–599.

Truex, J. N. (1972). Fractured shale and basement reservoir, Long Beach Unit, California. *Bull. Am. Ass. Petrol. Geol.* **56**, 1931–1938.

Van Houten, F. B. (1972). Iron and clay in tropical savanna alluvium, northern Columbia: a contribution to the origin of red beds. *Bull. geol. Soc. Am.* **83**, 2761–2772.

Visher, G. S. (1965). Fluvial processes as interpreted from Ancient and Recent Fluvial deposits. *In* "Primary Sedimentary Structures and their Hydrodynamic Interpretation" (G. V. Middleton, Ed.), *Spec. Publs Soc. econ. Palaeont. Miner., Tulsa.* **12**, 116–132.

Vita-Finzi, C. (1971). Heredity and environment in clastic sediments: silt/clay deplation. *Bull. geol. Soc. Am.* **82**, 187–190.

Von Rad, U. (1971). Comparison between "magnetic" and sedimentary fabric in graded and cross-laminated sand layers, southern California. *Geol. Rdsch.* **60**, 331–354.

Wadell, H. (1935). Volume, shape and roundness of quartz particles. *J. Geol.* **43**, 250–280.

Walker, K. R. (1964). Influence of depth, temperature and geologic age on porosity of quartzose sandstone. *Bull. Am. Ass. Petrol. Geol.* **48**, 1945–1946.

Wardlaw, N. C. (1976). Pore geometry of carbonate rocks as revealed by pore casts and capillary pressure. *Bull. Am. Ass. Petrol. Geol.* **60**, 254–257.

Wentworth, C. K. (1922). A scale of grade class terms for clastic sediments. *J. Geol.* **30**, 377–392.

Whalley, W. B. (1972). The description and measurement of sedimentary particles and the concept of form. *J. sedim. Petrol.* **42**, 961–965.

Wilson, R. C. L. (1966). Silica diagenesis in Upper Jurassic limestones of Southern England. *J. sedim. Petrol.* **36**, 1036–1049.

Wycoff, R. D., Botset, H. G., Muskat, M. and Reed, D. W. (1934). Measurement of permeability of porous media. *Bull. Am. Ass. Petrol. Geol.* **18**, 161–190.

Zingg, T. (1935). Beitrage zur Schotteranalyse. *Schweiz. miner. petrogr. Mitt.* **15**, 39–140.

3 Weathering and the Sedimentary Cycle

I. INTRODUCTION

The first chapter of this book reviewed the field of sedimentology, the second described the physical properties of sediments. Before proceeding to the analysis of sedimentary rocks, their petrography, transportation and deposition, it is appropriate to analyse the genesis of sediment particles.

A sedimentary rock is the product of provenance and process. This chapter is concerned primarily with the provenance of sediment; that is to say the pre-existing rocks from which it forms and the effect of weathering on sediment composition.

First, however, it is apposite to consider the place of sediment particles within the context of what may be called, for want of a better term, "the earth machine".

Consider a geologist idly hammering a piece of Continental Mesozoic "Nubian" Sandstone in a Saharan wadi. A piece of that sandstone falls from the cliff to the wadi floor. Within a second a sample of Mesozoic Nubian Sandstone has suddenly become part of a Recent alluvial gravel. The pebble deserves closer inspection. This sandstone clast is composed of a multitude of quartz particles. The precise provenance of these cannot be proved, but locally this formation overlies Upper Palaeozoic sandstones and infills ancient wadis cut within them. The Mesozoic sand grains are clearly, in part, derived from the Upper Palaeozoic sandstones. Similarly it can be shown that the Upper Palaeozoic sediment has been through several previous cycles of erosion and deposition. The sand grains first formed from the weathering of Pre-Cambrian granite.

This simple review introduces the concept of the sedimentary cycle and shows how sediments are frequently polycyclic in their history.

Let us now move from a specific event to the other extreme. The modern concept of plate tectonics is reviewed in Chapter 9. At this point it is necessary to briefly introduce it, however, to better place sand grains in perspective. The basic thesis of plate tectonics is that the earth's crust is formed continuously along linear zones of sea-floor spreading. These occupy the mid-oceanic rift valleys, areas of extensive seismic and volcanic activity. New crust, formed by vulcanicity, moves laterally away from the zone of sea-floor spreading, forming a rigid lithospheric plate. Simultaneously, at the distal edge of the plate, there is a linear zone of subduction marked by earthquakes, island arcs and, often, geosynclines. Here old crust is drawn down and digested in the mantle.

Sediment forms initially, therefore, from volcanic rocks on the mid-ocean ridges. It may then be recycled several times, as we have seen in the Sahara, but inexorably individual particles are gradually drawn to the edge of the plate to descent and be destroyed in the mantle.

We may liken the history of a sediment particle to a ball bouncing on a conveyor belt. Its history is one major cyclic event composed of many subsidiary ones.

The sedimentary cycle will now be examined in more detail.

II. THE SEDIMENTARY CYCLE

This section looks more closely at the sedimentary cycle in the smaller scale, leaving aside the major cycle of plate formation and destruction.

Classically the sedimentary cycle consists of the phases of weathering, erosion, transportation, deposition, lithifaction, uplift, and weathering again (Fig. 18). Only the upper part of this figure is essentially concerned with the sedimentary cycle. In the lower part mass movement occurs but not of sedimentary material.

Our prime concern at this stage is weathering, erosion, transportation and deposition. Weathering is the name given to the processes which break down rock at the earth's surface to form discrete particles (Ollier, 1969). Erosion is the name given to the processes which remove newly formed sediment from bedrock. This is followed generally by transportation and finally, when energy is exhausted, by deposition.

The processes and products of weathering are examined more closely in the next section. It is sufficient at this point to state that weathering is generally divided into biological, chemical and physical processes. Chemical weathering selectively oxidizes and dissolves the constituent minerals of a rock. Physical processes of weathering are those which bring about its actual mechanical disaggregation. Biological weathering is caused by the chemical and physical effects of organic processes on rock.

Erosion, the removal of new sediment, can be caused by four agents: gravity, glacial action, running water and wind.

The force of gravity causes the gradual creep of sediment particles and slabs of rock down hillsides, as well as the more dramatic avalanches.

Glacial erosion occurs where glaciers and ice sheets scour and abrade the face of the earth as they flow slowly downhill under the influence of gravity.

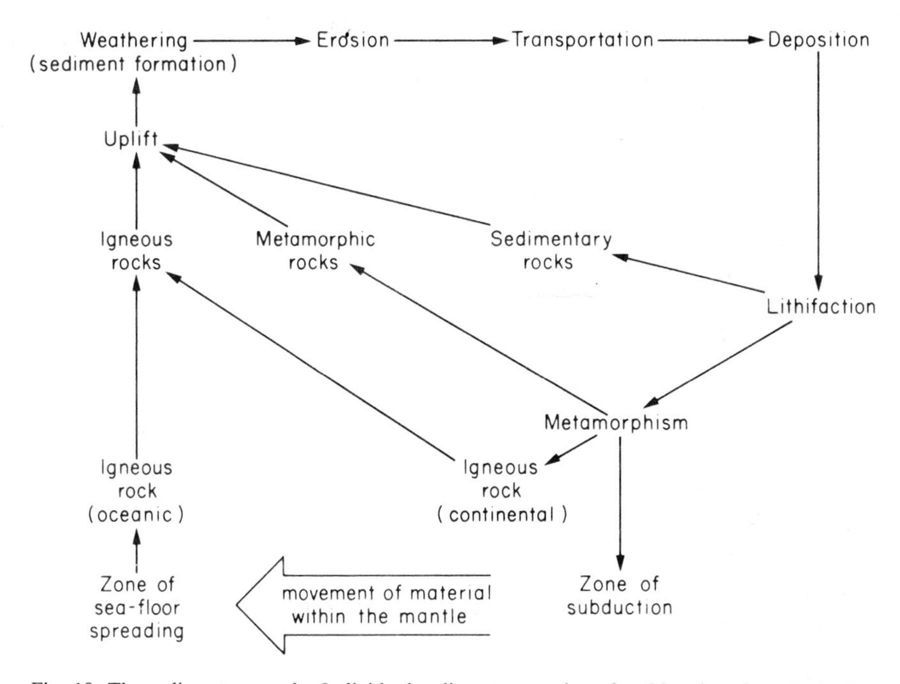

Fig. 18. The sedimentary cycle. Individual sedimentary grains of stable minerals, principally quartz, may be recycled several times before being destroyed by metamorphism.

Moving water is a powerful agent of erosion in a wide spectrum of geomorphological situations ranging from desert flash flood to riverbank scouring and sea-cliff undercutting.

The erosive action of wind, on its own, is probably infinitesimal. Wind, however, blowing over a dry desert, quickly picks up clouds of sand and sandblasts everything in its path for a height of a metre or so. Eolian sandblasting undercuts rock faces, carving them into weird shapes, and expedites the erosion of cliffs by gravity collapse and rainstorm.

It is important to note that it is an oversimplification to place erosion after weathering in the sedimentary cycle. Weathering processes need time for their effects to be noticeable on a rock surface. In some parts of the earth, notably areas of high relief, erosion may occur so fast that rock is not exposed to the air for a sufficient length of time to undergo any significant degree of weathering (Fig. 19). This point will be amplified in the next section.

Returning to the role of gravity, ice, water and wind, it is apparent that these are the agents both of erosion and of subsequent transportation of sediment.

The physical processes of these various transporting media are described on pp. 177–204. At this point, however, it is appropriate to point out the role played by these agents in the segregation of sediments.

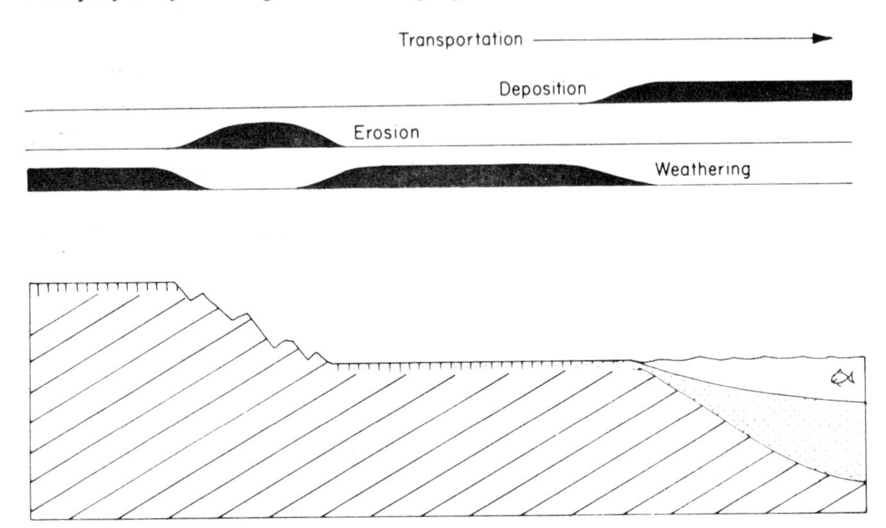

Fig. 19. Diagrammatic cross-section to show how weathering is favoured by low relief, because this is correlative with slow rates of erosion. It is not necessarily correlative with altitude.

The products of weathering are twofold: solutes and residua. The solutes are the soluble fraction of rocks which are carried in water. The residua are the insoluble products of weathering, which range in size from boulders down to colloidal clay particles. It is interesting to note the competency of the various transporting media to handle and segregate the products of weathering.

Gravity and ice, as seen in the work of avalanches and glaciers, are competent to transport all types of weathering product, solutes, and residual particles. They are, however, both inefficient agents of sediment segregation. Their depositional products, therefore, are generally poorly sorted boulder beds and gravels.

Water, by contrast, is a very efficient agent for carrying material in solution; it is less efficient, however, in transporting residual sediment particles. Current velocities are seldom powerful enough to carry boulders and gravels for great distances. For this same reason running water segregates sands from gravels and the colloidal clays from the detrital sands.

The sediments which are deposited from running water, therefore, include the sandstones, limestones and mudrocks.

Finally wind action is the most selective transporting agent of all. Wind velocities are seldom strong enough to transport sediment particles larger than about 0·35 mm diam. Eolian sediments are generally of two types, sands of medium–fine grade, which are transported close to the ground by saltation; and the silty loess deposits which are transported in the atmosphere by suspension.

This review of the agents of sediment transport shows that sediments are segregated into the main classes of conglomerates, sands, shales and limestones by natural processes. This is worth bearing in mind during the discussions of sedimentary rock nomenclature and classification which occur in the next two chapters.

Before proceeding to these, however, it is necessary to examine weathering processes in further detail.

III. WEATHERING

Weathering, as already defined, includes the processes which break down rock at the earth's surface to produce discrete sediment particles. Weathering may be classified into chemical, physical and biological processes. Chemical processes lead essentially to destruction of rock by solution. Physical processes cause mechanical fracture of the rock. Biological weathering is due to organic processes. These include both biochemical solution, brought about largely by the action of bacteria, and humic acids derived from rotting organic matter, as well as physical fracturing of rock such as may be caused by tree roots.

A. Biological Weathering and Soil Formation

Soil is the product of biological weathering. It is that part of the weathering profile which is the domain of biological processes. Soil consists of rock debris and humus, which is decaying organic matter largely of plant origin. Humus ranges in composition from clearly identifiable organic debris such as leaves and plant roots, to complex organic colloids and the humic acids.

It is doubtful that soils, as defined and distinguished from the weathering profile, existed before the colonization of the land by plants in the Devonian.

The study of soils, termed "pedology", is of interest to geologists in so far as it affects rock weathering and sediment formation. Pedology is, however, of particular importance to agriculture, forestry and to correct land utilization in general (Bunting, 1967).

Pedologists divide the vertical profile of a soil into three zones (Fig. 20). The upper part is termed the "A zone", or eluvial horizon. In this part of

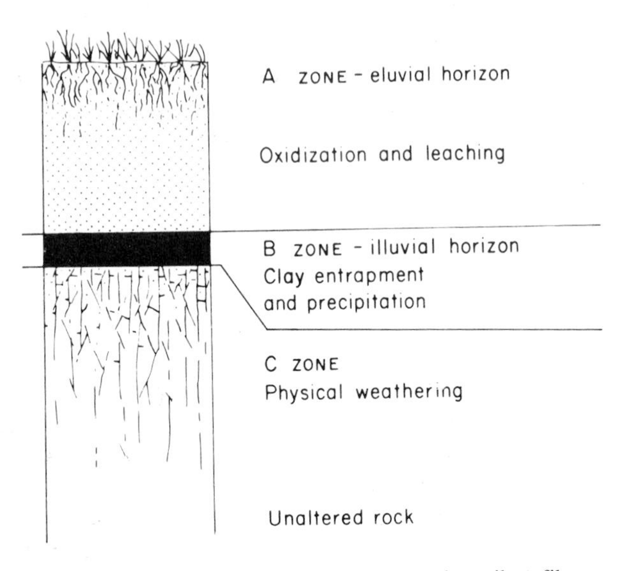

Fig. 20. Terminology and processes through a soil profile.

the profile, organic content is richest and chemical and biochemical weathering generally most active. Solutes are carried away by ground water. The fine clay fraction percolates downward through the coarser fabric supporting grains.

Below the "A zone" is the "B zone" or illuvial horizon. At this level downward percolating solutes are precipitated and entrap clay particles filtering down from the "A zone".

Below the illuvial horizon is the "C zone". This is essentially the zone where physical weathering dominates over chemical and biological processes. It passes gradually downward into unweathered bedrock.

The thickness of a soil profile is extremely variable and all three zones are not always present. Thus soil thickness depends on the rate of erosion, climatic regime and bedrock composition. As already seen, in areas of high relief, erosion can occur so fast that weathering and soil formation cannot develop. By contrast, in humid tropical climates granite can be weathered for nearly 100 m. This forms what is known as "granite wash" which passes, with the subtlest transition, from arkosic sand down to fresh granite.

Ancient granite washes occasionally make good hydrocarbon reservoirs because they may be highly porous in their upper part. The Augila oil field of Libya is a good example (Williams, 1968). Epidiagenesis, the formation of porosity by weathering at unconformities, is discussed in greater detail when sandstone diagenesis and porosity are described (p. 104).

Returning to biological weathering and soils, it is known that soil type is closely related to climate. If erosion is sufficiently slow for a soil profile to evolve to maturity there is a characteristic soil type for each major climatic zone, irrespective of rock type. Modern soils and their ancient counterparts will now be reviewed.

1. Modern soils

In polar climates true soil profiles do not develop due to the absence of organisms. A weathering mantle may be present, but this is frozen for much, if not all of the year. This is called permafrost.

In temperate climates leaching plays a dominant role. The "A zone" is intensely weathered, though it may support an upper peaty zone of plant material. The high pH of such soil inhibits or delays bacterial decay. The "B zone" may be deep but is typically well developed as a limonitic or calcareous hard-pan which inhibits drainage.

Soils of this general type include the "podsols" of cool temperate climates and the humus rich "tchernozems", or "black earths" of warm temperate zones.

In arid climates, by contrast, the downward percolation of chemicals by leaching is offset by the upward movement of moisture by capillary attraction. Precipitation of solutes occurs, therefore, at or close to the land surface. Organic content is very low. By this means are formed the "duricrusts" which are found at or close to the surface in many modern deserts (Woolnough, 1927). These hard crusts often show pisolitic and concretionary structures as the minerals are precipitated in colloform habits. Duricrusts are of several chemical types. The commonest are the "kankars" and "caliches" composed of calcite and dolomite; these are sometimes termed "calcretes" (Chapman, 1974). Less common are the siliceous duricrusts formed, generally, by chalcedony (silcretes). Ferrugenous duricrusts are termed ferricrete. In low-lying waterlogged areas in arid climates the duricrust is formed of evaporite minerals. These occur in salt marshes (sabkhas, see pp. 164 and 299) both inland and at the seaside.

Humid tropical climates form soils that are very rich in iron and kaolin. The composition and origin of laterite, as these ferrugenous soils are called, is described shortly (p. 57).

This review of modern soils, though brief, should be sufficient to show

how significant climate is in weathering and how it therefore helps to determine which type of sediment is produced in a particular area.

2. Fossil soils

It is not uncommon to find ancient soil horizons in the geological record. "Palaeosols", as they are termed, occur beneath unconformities in all types of rock, and within sedimentary sequences where they provide invaluable evidence of subaerial exposure of the depositional environment (Fig. 21).

The rocks immediately beneath an unconformity are often intensely weathered but erosion has generally removed the upper part of the soil profile. An interesting exception to this rule has been described by Williams (1969), who discussed the palaeoclimate which weathered the Lewisian gneisses prior to the deposition of Torridonian (Pre-Cambrian) continental sediments in Scotland.

Soil horizons within sedimentary sequences are especially characteristic of fluvial and delatic deposits. Many examples occur beneath coals and lignites. These are pierced by plant rootlets, and often are white in colour due to intensive leaching. Such palaeosols are often called "seat earths". Examples of such palaeosols are common in the Upper Carboniferous coal

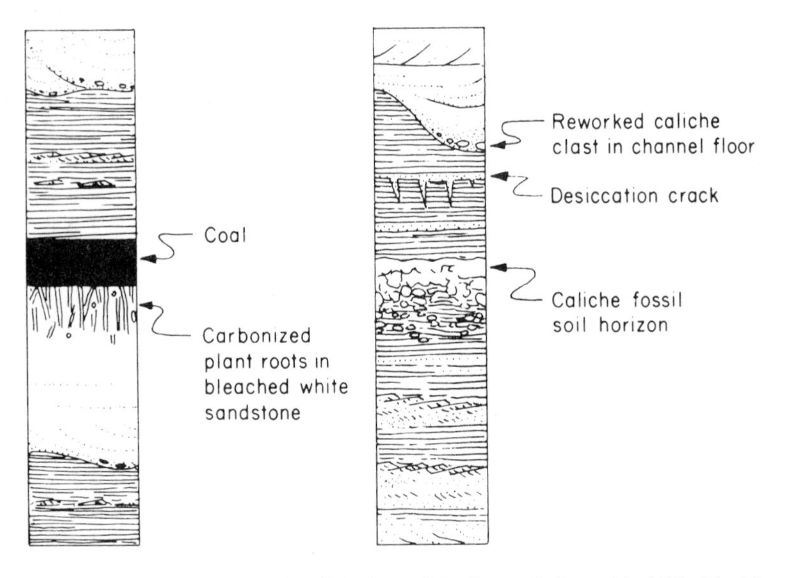

Fig. 21. Cross-sections through fossil soil (palaeosol) horizons. Left: coal bed (black) with seat earth beneath. This is a white, bleached, kaolinitic sandstone pierced by plant roots. Based on examples in the coal measures (Carboniferous) of England. Right: caliche soil horizon of nodular carbonate in red fluvial siltstone. Reworked caliche pebbles in the overlying channel sand testify to penecontemporaneous formation. Based on examples in the Old Red Sandstone (Devonian) of the Welsh borderlands.

measures of northern England. Two types of soil have been recognized. "Ganisters" are silica-rich horizons which are quarried for the manufacture of refractory bricks. "Tonsteins" are kaolin-rich seat earths. It has been argued both that these are normal desilicified soil horizons, and also that they may be intensely weathered volcanic ash bands (see p. 71).

Examples of soils in fluvial environments include the caliches of the Old Red Sandstone facies of the Devonian in the North Atlantic borderlands. These are sometimes termed "cornstones". They occur as nodules and concretionary bands within red floodplain siltstones. The occurrence of reworked "cornstone" pebbles in interbedded channel sands testifies to their penecontemporaneous origin. Friend and Moody-Stuart (1970) have described the caliche soils from the Old Red Sandstone and have discussed their probable genesis.

B. Physical Weathering

Three main types of physical weathering are generally recognized: freeze–thaw, isolation, hydration and dehydration and stress release.

Freeze–thaw weathering occurs where water percolates along fissures and between the grains and crystals of rock. When water freezes, the force of ice crystallization is sufficient to fracture the rock. The two halves of a fracture do not actually separate until the ice thaws and ceases to bind the rock together. Freeze–thaw weathering is most active, therefore, in polar climates and is most effective during the spring thaw.

Insolation weathering occurs by contrast in areas with large diurnal temperature ranges. This is typical of hot arid climates. In the Sahara, for example, the diurnal temperature range in winter may be 25°C. Rocks expand and contract in response to temperature. The diverse minerals of rocks change size at different rates according to their valuable physical properties. This differential expansion and contraction sets up stresses within rock. When this process occurs very quickly the stresses are sufficient to cause the rock to fracture. This is why insolation weathering is most effective in arid desert climates. In the author's personal experience this process was most dramatically experienced when trying to sleep on the slopes of the volcano Waw en Namus in the Libyan Sahara. Here black basalt sands are cemented by evaporite minerals. Sleep was impossible for several hours after sunset as the rock snapped, crackled and popped.

In climatic zones which experience alternate wet and dry seasons, a third process of physical weathering occurs. Clays and lightly indurated shales alternatively expand with water and develop shrinkage cracks as they dehydrate. This breaks down the physical strength of the formation; the shrinkage cracks increase permeability, thus aiding chemical weathering from rain water, whilst waterlogged clays may lead to landslides.

The fourth main physical process of weathering is caused by stress release. Rocks have elastic properties and are compressed at depth by the overburden above them. As rock is gradually weathered and eroded the overburden pressure decreases. Rock thus expands and sometimes fractures in so doing. Such fracturing is frequently aided by lateral downslope creep. Once stress-release fractures are opened they are susceptible to enlargement by solution from rainwater and other processes.

Stress release, insolation, hydration–dehydration and freeze–thaw are the four main physical processes of weathering. Stress release is ubiquitous in brittle rocks, insolation is characteristic of hot deserts, hydration–dehydration is typical in savannah and temperate climates, and freeze-thaw of polar climates.

C. Chemical Weathering

The processes of chemical weathering rely almost entirely on the agency of water. Few common rock-forming minerals react with pure water, evaporites excepted. Ground water, however, is commonly acidic. This is due to the presence of dissolved carbon dioxide from the atmosphere forming dilute carbonic acid. The pH is also lowered by the presence of humic acids produced by biological processes in soil. The principal chemical reactions to take place in rock weathering are oxidation and hydration. Numerous studies have been made of the rate of chemical weathering of different rock-forming minerals (e.g. Ruxton, 1968; Parker, 1970). This work suggests that the rate of relative mobility of the main rock-forming elements decreases from calcium and sodium to magnesium, potash, silicon, iron and aluminium. Rocks undergoing chemical weathering, therefore, tend to be depleted in the first of these elements and thus to show a relative increase in the proportions of iron oxide, alumina and silica.

The order in which minerals break down by weathering is essentially the reverse of Bowen's reaction series for the crystallization of igneous minerals (Table VI).

Chemical weathering separates rock into three main constituents: the solutes, the newly formed minerals and the residuum. The solute includes the elements such as the alkali metals, principally sodium and potassium, and the rare earths, magnesium, calcium and strontium. These tend to be flushed out of the weathering profile and ultimately find their way into the sea to be precipitated as limestones, dolomites and evaporites.

The residuum is that part of the rock which, when weathered, is not easily dissolved by ground water. As Table VI shows, the residuum may be expected to be largely composed of quartz (silica) and, depending on

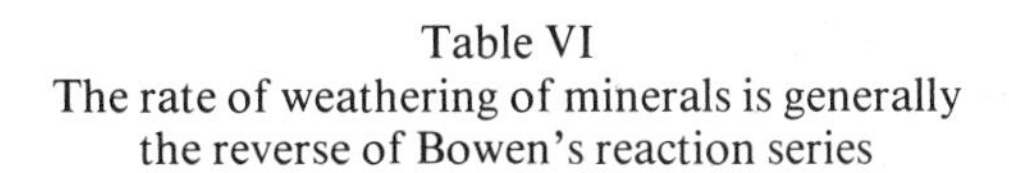
Table VI
The rate of weathering of minerals is generally the reverse of Bowen's reaction series

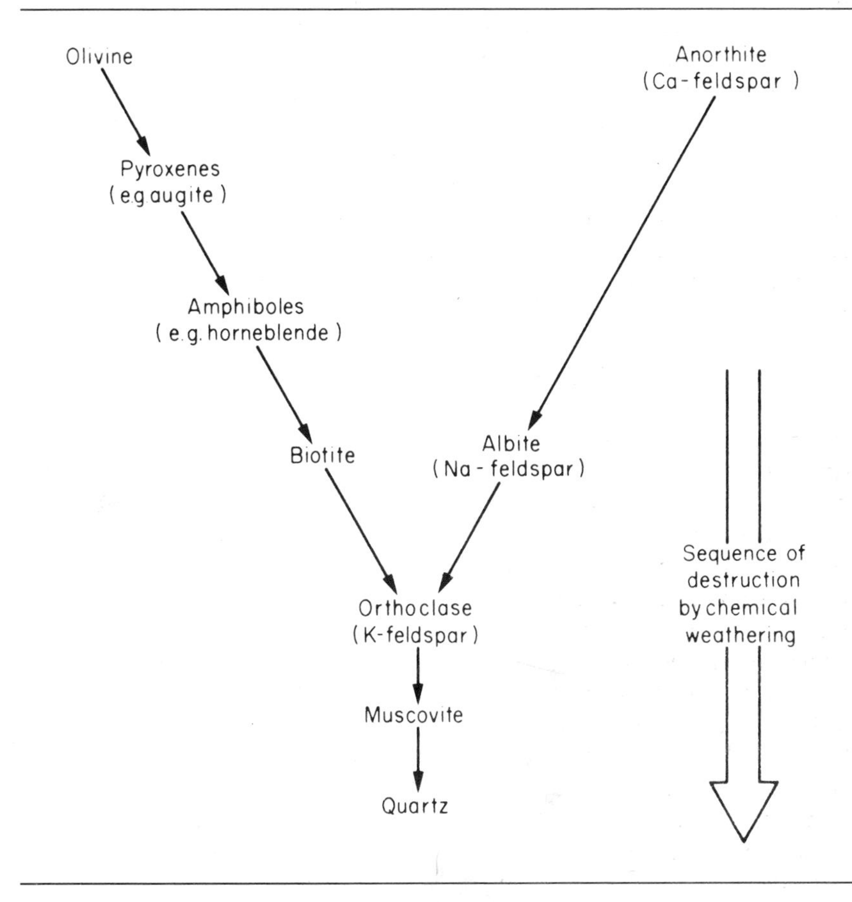

the degree of weathering, of varying amounts of feldspar and mica.

A further important result of chemical weathering is the formation of new minerals, principally the clays. The mineralogy and deposition of the clays is described in the next chapter. It is, however, necessary at this point to discuss their formation by weathering.

The clay minerals are a complex group of hydrated aluminosilicates. Their chemically fairly stable molecular structures scavenge available cations and attach them to their lattices in various geometric arrangements. The clay minerals are thus classified according to the way in which hydrated aluminosilicate combines with calcium, potassium, magnesium and iron.

During the early stages of weathering the mafic minerals (olivines, pyroxenes and amphiboles) break down to form the chlorite clays, rich in iron and magnesium. Simultaneously, weathering of the feldspars produces the smectites, illites and kaolin clays.

As weathering progresses the clays are partially flushed out as colloidal-clay particles, but can also remain to form a residual clay deposit. If weathering continues further still all the magnesium and calcium-bearing minerals are finally leached out.

The ultimate residuum of a maturely weathered rock consists, therefore, of quartz (if it was abundant in the parent rock), and newly-formed kaolin (the purest clay mineral, composed only of hydrated aluminosilicate), bauxite (hydrated alumina) and limonite (hydrated iron oxide).

For such a residuum to form by intense chemical weathering a warm humid climate is necessary, coupled with slow rates of erosion. These deposits, though thin, can form laterally extensive blankets which are often of considerable economic significance. Three main types of residual deposit are generally recognized and defined according to their mineralogy (Fig. 22). These will now be briefly described.

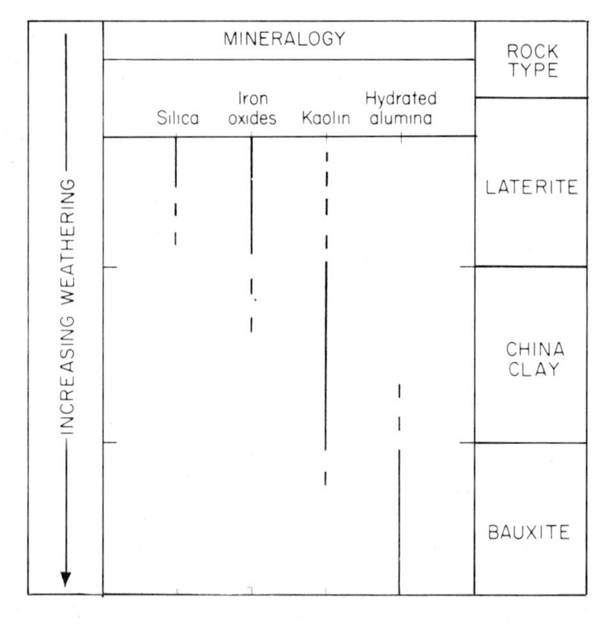

Fig. 22. Diagram to show the composition of the residual deposits formed by intensive chemical weathering.

1. Laterite

The term "laterite", from the latin *later*, a brick, was originally proposed by Buchanan (1807) to describe the red soils of parts of India, notably those developed on the plateau basalts of the Deccan. Subsequently the precise definition of the term has become somewhat diffuse (Sivarajasingham *et al.*, 1962).

As generally understood, laterite is the product of intensely weathered material. It is rich in hydrated iron and aluminium oxides, and low in humus, silica, lime, silicate clays and most other minerals. Analyses of Ugandan laterites given by McFarlane (1971) show between 40 and 50% iron oxide and between 20 and 25% of both alumina and silica. In physical appearance laterite is a red–brown earthy material which, though often friable, can harden rapidly on exposure to the atmosphere, a property useful in brick-making.

Laterites often show pisolitic and vermiform structures. The pisolites are rounded concentrically-zoned concretions up to a centimetre or more in diameter. Vermicular laterite consists of numerous subvertical pipes of hard laterite in a friable matrix.

Lateritic soils are widely distributed throughout the humid tropics occurring both in India and Africa, as already mentioned, and also in South America.

It has been argued already that residual deposits such as laterite require prolonged intensive chemical weathering for them to reach maturity. This will occur fastest in areas of low relief, where rates of erosion are low. High initial iron content in the parent rock is also a further significant factor. Thus laterites are often best developed on plateau basalts and basic intrusive rocks. Thin red laterite bands are commonly found separating fossil basalt lavas.

2. China clay

China clay is the name given to the rock aggregate of kaolin. This is the hydrated aluminosilicate clay mineral ($Al_2O_3.2SiO_2.2H_2O$). China clay occurs in three geological situations, only one of which is, strictly speaking, a residual weathering product. It is formed by the hydrothermal alteration of feldspars in granite, the Dartmoor granites of south-west England are an example (Bristow, 1969). China clay is formed by intensive weathering of diverse rock types, but particularly those which are enriched in aluminosilicates, such as shales, and acid igneous and metamorphic rocks.

Thirdly, china clay can be transported short distances from hydrothermal and residual deposits to be resedimented in lacustrine environments. Here the kaolin beds occur associated with sands and coals or lignites. The "ball clays'' of Bovey Tracey were deposited in an Oligocene lake which received kaolin from rivers draining the Dartmoor granites.

The residual kaolins, with which we are primarily concerned here, are very variable in composition. With increasing iron content they grade into the laterites and with desilicification they grade into the bauxites. Where kaolin forms as a residue on granitoid rocks large quantities of quartz may be present which must be removed before the kaolin may be put to use.

China clay is economically important in the ceramic and paper-making industry.

3. Bauxite

Bauxite is a residual weathering product composed of varying amounts of the aluminium hydroxides boehmite, chliachite, diaspore and gibbsite (Valeton, 1972). It takes its name from Les Baux near Arles, in France. The bauxite minerals are formed by the hydrolysis of clay minerals, principally kaolin:

$$\underset{\text{water}}{H_2O} + \underset{\text{kaolin}}{Al_2O_3.2SiO_2.2H_2O} \longrightarrow \underset{\text{aluminium hydroxide}}{Al_2O_3\,nH_2O} + \underset{\text{silicic acid}}{2SiO_2.2H_2O}$$

Bauxite thus needs the presence of pure leached clay minerals as its precursor. This occurs in two geological settings. Bauxite is often found overlying limestone formations, as, for example, in the type area of southern France, and in Jamaica. In these situations the calcium carbonate of the limestones has been completely leached away to leave an insoluble residue of clays. These have been desilicified to form the bauxite suite of minerals.

In the second situation, as seen in Surinam in South America (Valeton, 1973), bauxite occurs as a weathering product of kaolinitic sediments which forms a thin veneer of Pre-Cambrian metamorphic basement.

In both these geological settings it is apparent that the formation of bauxite is the terminal phase of a sequence of weathering which included kaolinization as an intermediate step.

Bauxite is of great economic importance as it is the only source of aluminium.

D. Economic Significance and Conclusion

From the preceding description of weathering it should be apparent that it is necessary to understand the processes of weathering to have some insight into sediment formation. To conclude this chapter it is appropriate to summarize some of the major points and to draw attention to the economic significance of weathering.

It is important to remember that weathering separates rock into an insoluble residue and soluble chemicals. The residue, composed largely of quartz grains and clay particles, ultimately contributes to the formation of the terrigenous sands and shales. The soluble products of weathering are ultimately reprecipitated as the chemical rocks, such as limestones, dolomites and evaporites. Most terrigenous sediments are the product of weathered granitoid rocks and pre-existing sediment. Basic igneous and volcanic rocks and limestones are very susceptible to chemical weathering and contribute little material to the insoluble residue. Figure 23 shows how, in a terrain of diverse rock types, only a few lithologies contribute to the adjacent sediment apron.

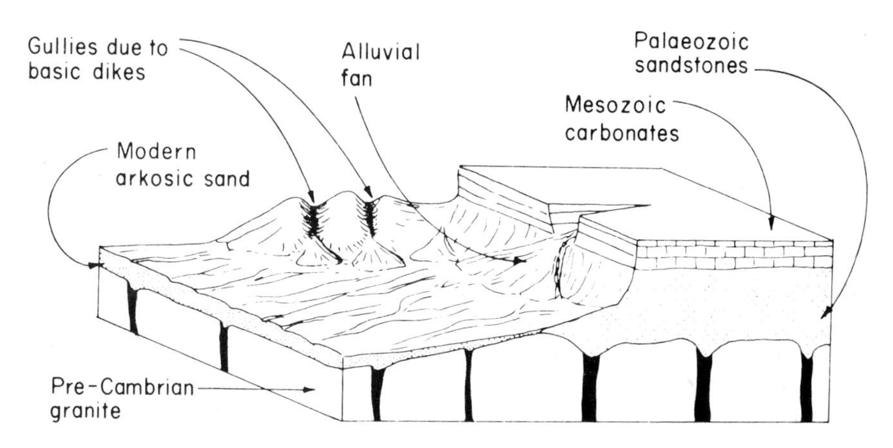

Fig. 23. Field sketch of part of Sinai. Mesozoic limestones overlie Nubian Sandstones which in turn overlie Pre-Cambrian granitoid rocks with basic intrusives. Despite this diversity of rock types the sediment derived from this terrain is essentially a feldspathic quartz sand derived from the granites and Nubian Sandstones. The limestones and basic intrusives contribute little insoluble residue. Their weathering products are largely carried away in solution.

Aside from these generalities, weathering is important because of the way it can concentrate minerals into deposits which are worth exploiting economically. The bauxites and china clays have been already mentioned.

One other important type of weathering is the supergene enrichment of certain ore bodies. This is typically seen on top of sulphide ores such as the Rio Tinto copper deposit of Spain. Four zones can be distinguished in the weathering profile of ores such as this (Fig. 24). At the top is an upper leached zone or "gossan", composed of residual silica and limonite. This grades down into the "supergene" zone, adjacent to the water table.

Immediately above the table oxidized ore minerals are reprecipitated, such as malachite, azurite and native copper. Secondary sulphide ores, principally chalcocite, are precipitated in the lower part of the supergene zone. This grades down into the unweathered hypogene ore, mainly composed of pyrite and chalcopyrite.

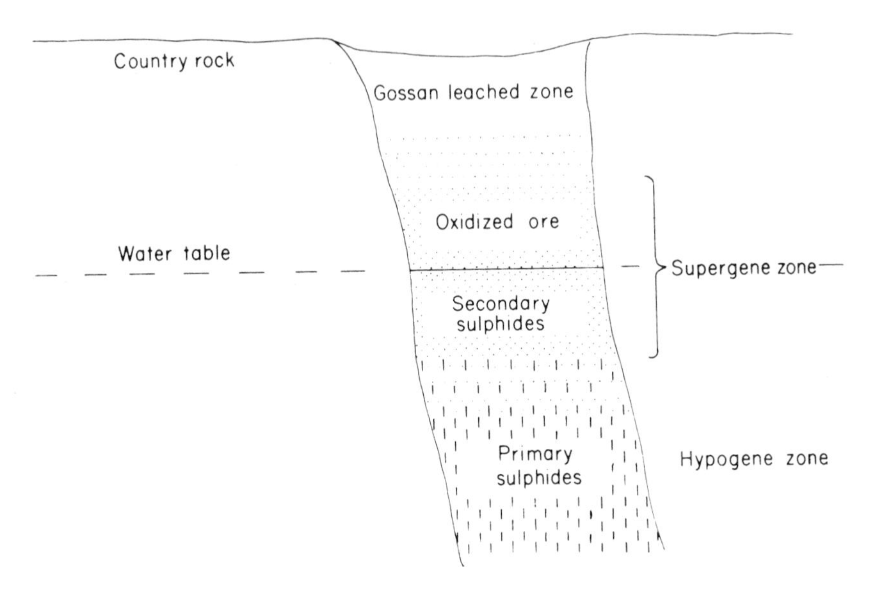

Fig. 24. Diagrammatic section through a weathered sulphide-ore body. Secondary ore minerals are concentrated near the water table by oxidization and leaching in the upper gossan zone.

Other examples of residual ores include placers of gold and other heavy minerals which develop from the weathering to destruction of all the other minerals of the rock within which they were dispersed.

This brief review shows how weathering is a process of considerable direct economic importance as well as being one of fundamental geological significance.

IV. REFERENCES

Bristow, C. M. (1969). Kaolin deposits of the United Kingdom. Rep. 23rd int. geol. Cong. Prague No. 15, 275–288.

Buchanan, F. (1807). "A Journey From Madras Through the Counties of Mysore, Kanara and Malabar" 3 Vols. East India Company, London.

Bunting, B. T. (1967). "The Geography of Soils" (2nd edition). Hutchinson, London. 213pp.

Chapman, R. W. (1974). Calcareous duricrust in Al-Hasa, Saudi Arabia. *Bull. geol. Soc. Am.* **85**, 119–130.

Friend, P. F. and Moody-Stuart, M. (1970). Carbonate deposition on the river flood plains of the Wood Bay Formation (Devonian) of Spitsbergen. *Geol. Mag.* **107**, 181–195.

McFarlane, M. J. (1971). Lateritization and landscape development in Kyagwe, Uganda. *Q. Jl geol. Soc. Lond.* **126**, 501–539.

Ollier, C. C. (1969). "Weathering" Oliver and Boyd, Edinburgh. 304pp.

Parker, A. (1970). An index of weathering for silicate rocks. *Geol. Mag.* **107**, 501–504.

Ruxton, B. P. (1968). Measures of the degree of chemical weathering of rocks. *J. Geol.* **76**, 518–527.

Sivarajasingham, L. T., Alexander, L. T., Cady, J. G. and Cline, M. G. (1962). Laterite. *Adv. Agron.* **14**, 1–60.

Valeton, I. (1972). "Bauxites" Developments in Soil Science, No. 1. Elsevier, Amsterdam. 226pp.

Valeton, I. (1973). Pre-bauxite red sediments and sedimentary relicts in Surinam bauxites. *Geologie Mijnb.* **52**, 317–334.

Williams, G. E. (1969). Characteristics and origin of a PreCambrian pediment. *J. Geol.* **77**, 183–207.

Williams, J. J. (1968). The stratigraphy and igneous reservoirs of the Augila field, Libya. *In* "Geology and Archaeology of Northern Cyrenaica, Libya" (F. T. Barr, Ed.), 197–206. Petrol. Explor. Soc. Libya, Tripoli.

Woolnough, W. G. (1927). The duricrust of Australia. J. Proc. R. Soc. N.S.W. **61**, 24–53.

4 Allochthonous Sediments

I. INTRODUCTION: THE CLASSIFICATION OF SEDIMENTS

Chapter 3 showed how natural processes tend to separate the various products of weathering and, after erosion and transportation, deposit them in discrete entities such as sand, mud and carbonates. Thus sediments are grouped and segregated spontaneously on the earth's surface.

Geologists have attempted to name and classify these different rock types. There are two main reasons for such exercises. First, effective communication requires a uniformity of nomenclature. Secondly, in a particular study it is often necessary to differentiate, compare and contrast rock types. The following section describes first the problems of sediment classification, and then the classification of the allochthonous rocks, their petrology and petrophysics.

Sediment is "what settles at the bottom of a liquid: dregs: a deposit" (Chambers, 1972 edition). This definition is itself unacceptable to most geologists since it would exclude, for example, eolian deposits and biogenic reefs from the realm of the sedimentologist.

Similar dilemmas will be encountered with most other descriptive terms applied to sedimentary petrography.

Essentially, five main genetic classes of sediment may be recognized: chemical, organic, residual, terrigenous and pyroclastic (Hatch *et al.*, 1971).

The chemical sediments are those which form by direct precipitation in a subaqueous environment. Examples may include evaporites such as gypsum and rock salt, as well as tufa and perhaps some lime muds (see p. 125).

The organic sediments are those which are composed of organic matter of both animal and vegetal origin. Examples include skeletal limestones and coal.

The residual sediments are those left in place after weathering, examples include the laterites and bauxites described in Chapter 3.

The terrigenous sediments are those whose particles were originally derived from the earth, and include the mudrocks, siliciclastic (as opposed to carbonate) sands and conglomerates.

Pyroclastic sediments are the product of volcanic activity. Examples include ashes, tuffs, volcaniclastic sands and agglomerates.

Additional terms to introduce at this point are clastic, detrital and fragmental. These all tend to be used in the same way; describing a rock as formed from the lithifaction of discrete particles, a sediment no less.

The five genetic classes previously defined are untenable when closely scrutinized. For example, are the gypsum sand dunes of Abu Dhabi chemical or terrigenous deposits? Are phosphatized bone beds chemical, organic or residual deposits, and so on? These five main genetic classes of sedimentary rocks can be divided into two separate types: the allochthonous and the autochthonous deposits.

The allochthonous sediments are those which are transported into the environment in which they are deposited. They include the terrigenous and pyroclastic classes, together with rare reworked (allodapic) carbonates.

The autochthonous sediments are those which form within the environment in which they are deposited. They include the chemical, organic and residual classes. Table VII shows the relationship between these various terms.

The end-member concept is one of the most useful ways of classifying rocks. Sediments containing three constituents can be classified in a triangular diagram in which each apex represents 100% of one of the three constituents (Fig. 25). Four component sediment systems can be plotted in three dimensions within the faces of a tetrahedron (Fig. 26). Since sedimentary rocks contain many components neither of these systems is entirely satisfactory. Enthusiasts for classificatory schemes will have to

Table VII
A classification of sedimentary rocks

Group	Class	
I. Autochthonous sediments	(a)	*Chemical precipitates*—the evaporites: gypsum, rock salt, etc.
	(b)	*Organic deposits*—coal, limestones, etc.
	(c)	*Residual deposits*—laterites, bauxites, etc.
II. Allochthonous sediments	(d)	*Terrigenous deposits*—clays, siliclastic sands and and conglomerates
	(e)	*Pyroclastic deposits*—ashes, tuffs, volcaniclastic sands and agglomerates

Like most classification of geologic data this one has several inconsistencies. Note particularly that many limestones, though organic in origin, are detrital in texture. Cannel, drift or boghead coals are not truly autochthonous. Many evaporites are diagenetic in origin.

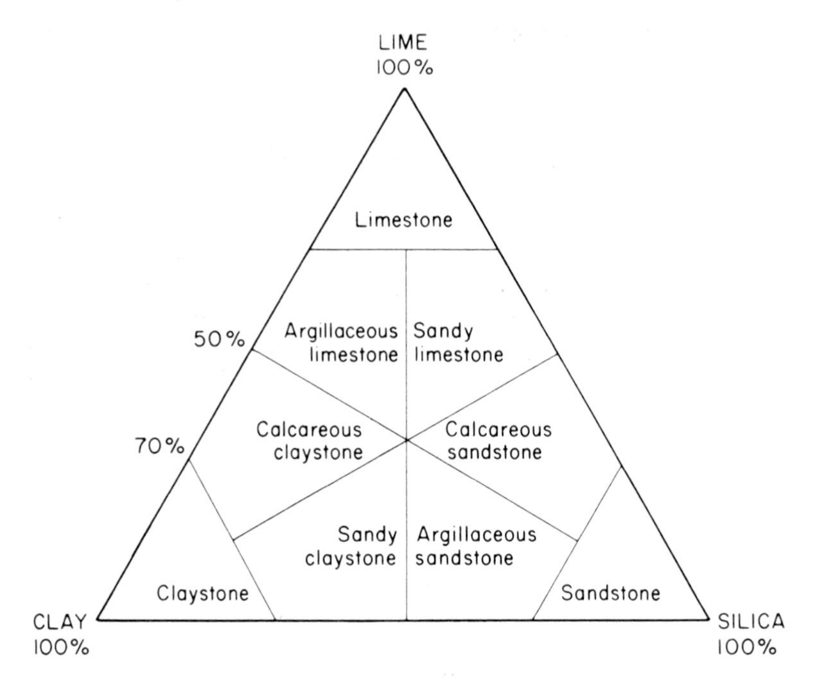

Fig. 25. An example of the end-member classification style. This triangle differentiates sands, claystones and limestones.

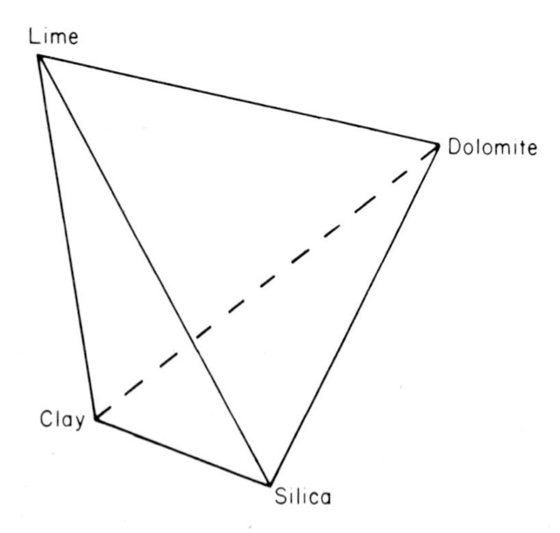

Fig. 26. An example of the end-member classification style. This tetrahedron differentiates four component systems, in this case limestones, dolomites, sandstones and clays.

indulge in statistical gymnastics using factor analysis, etc. (e.g. Harbaugh and Merriam, 1968, p. 157).

II. ALLOCHTHONOUS SEDIMENTS CLASSIFIED

The allochthonous sediments, as previously defined, consist of the terrigenous and pyroclastic classes. The quartzose terrigenous sands are sometimes termed siliciclastic to differentiate them from the bioclastic and pyroclastic detrital sediments. The pyroclastic sediments, those derived from volcanic activity, are included for the sake of completeness. They are not a volumetrically significant part of the present sedimentary cover of the earth, and will not be described in detail.

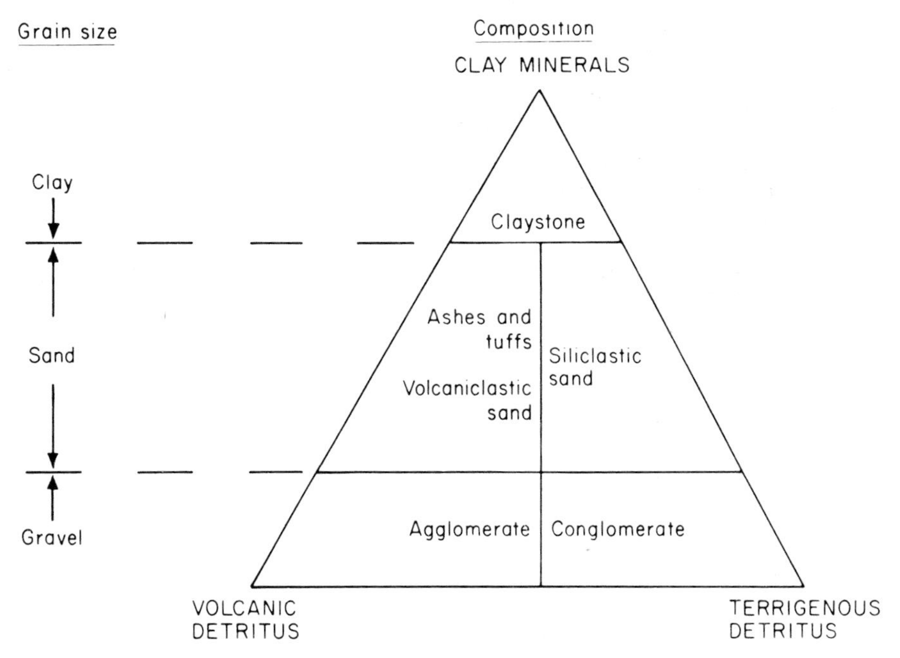

Fig. 27. A classification of the allochthonous sediments based on grain size and composition.

The allochthonous sediments may be conveniently classified using the end-member triangle shown in Fig. 27. The scheme attempts to reconcile the inconsistency that the nomenclature of sedimentary rocks is founded on two immiscible parameters: grain size and composition. It is interesting to see the different emphasis placed on sediment nomenclature in modern and ancient deposits. Studies of modern sediments are primarily concerned with the hydrodynamics of transport and deposition. Sediment description predominantly emphasizes grain size and texture rather than mineral composition. Hence a scheme similar to that in Fig. 28 is generally used. In ancient sediments, however, emphasis is placed much more on composition and this is reflected by the abundance of rock names which stress composition rather than texture (e.g. arkose, tuff, etc.).

The four main types of allochthonous sediments will now be described. These may be grouped conveniently, if illogically, into the mudrocks (composed dominantly of clay minerals), the pyroclastics, the sandstones and the rudaceous rocks.

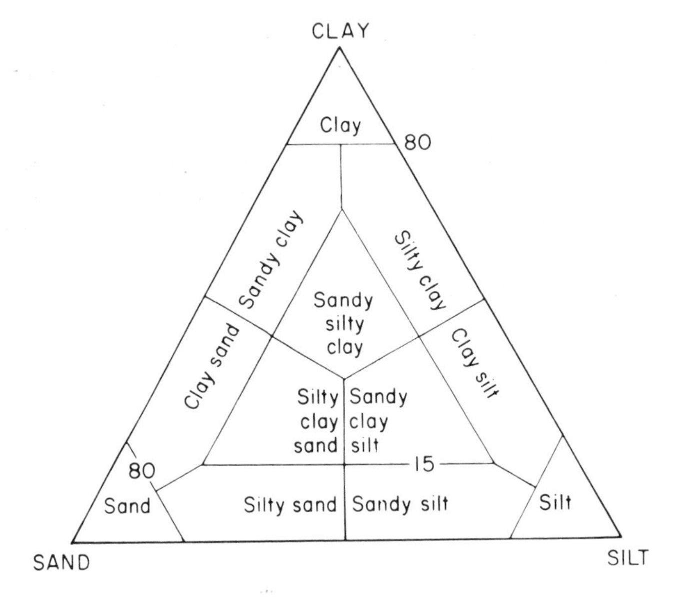

Fig. 28. An end-member triangle for classifying unconsolidated sediment on a basis of grain size. (From Shepard, 1954.)

III. MUDROCKS

The term mud is ill-defined and loosely used. In Recent deposits, sediments referred to as mud are wet clays with a certain amount of silt and sand. Its lithified equivalent is termed "mudstone". More rigidly defined, however, are clay and silt in the Wentworth grade scale (p. 10). Clays are sediments with particles smaller than 0·0039 mm. Silts have a grain size between 0·0039 and 0·0625 mm. Their lithified equivalents are claystone and siltstone respectively.

Shale is another term applied to fine-grained sediment. It is ill-defined and does not differentiate silt from clay-grade sediment. Shaley parting, on the other hand, is a valid description for fissility in a fine sediment. Generally this is due to traces of mica aligned on laminae. The term shale, however, could perhaps usefully be abandoned by geologists, except when communicating to engineers or management.

As discussed later (p. 73), clays are deposited with a primary water-saturated porosity of up to 80%. Most of this is quickly lost, first by dewatering (syneresis, see p. 239) and later by compaction. Other major constituents of mudrocks include clay minerals, detrital grains, organic matter, and carbonates.

The clay minerals will be described in some detail later. The detrital grains consist of fine angular particles of quartz, micas and heavy minerals such as zircon and apatite. Organic matter in mudrocks is chemically very complex and the nomenclature of organic mudrocks is not rigidly defined.

Essentially mudrocks can be named with reference to an end-member triangle whose apices represent pure organic matter, pure lime and pure clay minerals (Fig. 29). Mudrocks composed largely of admixtures of various clay minerals may be termed claystones, or more pedantically, ortho-claystones, to indicate their purity.

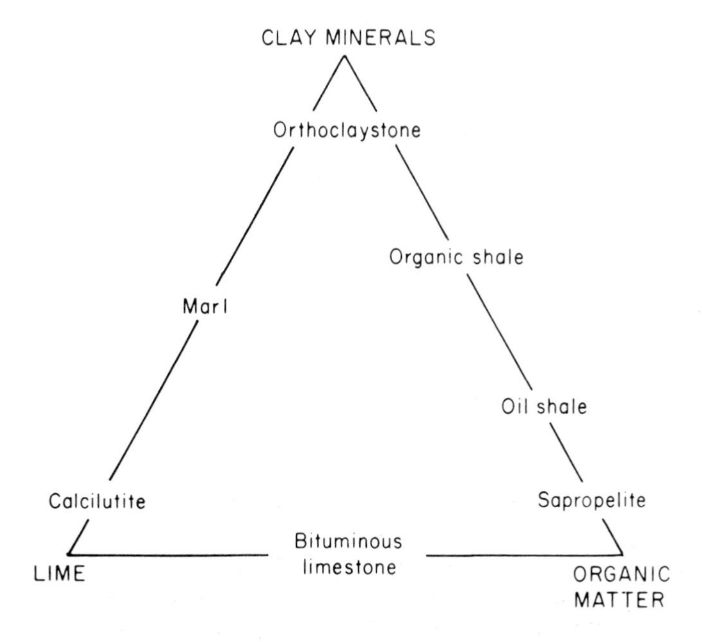

Fig. 29. Triangular diagram to illustrate the nomenclature and composition of the mudrocks.

With increasing lime content, claystones grade into marls and on into micrites which are pure lime mudrocks.

A. Sapropelites and Oil Shales and Oil Source Rocks

Organic matter is present in at least small quantities in all sediments. It is, however, most abundant in the mudrocks and has been intensively studied, both because organic-rich shales are commonly thought to be the source of crude oil, and because certain mudrocks, termed "oil shales", release crude oil on heating (Yen and Chilingarian, 1976). Sapropelite is a name given to organic-rich mudrocks.

Essentially the organic matter in sediments is of four types: kerogen, asphalt, crude oil and natural gas. These are organic compounds of great complexity. Their nomenclature is ill-defined and these four groups span a continuous spectrum of hydrocarbons.

Kerogen is a dark grey–black amorphous solid which is present in varying quantities in mudrocks. It contains between 70–80% carbon, 7–11% hydrogen, 10–15% oxygen and traces of nitrogen and sulphur. Kerogen burns on ignition in free air and is thus sometimes termed pyrobitumen. The exact molecular structure of kerogen is not well known, but, by definition, kerogen includes those hydrocarbon compounds which are not soluble in normal petroleum solvents, such as ether, acetone, benzene or chloroform (Dott and Reynolds, 1969, p. 79). Kerogen is the major constituent of the organic-rich mudrocks (see also Tissot and Welte, 1978, and Hunt, 1979).

Asphalt, or bitumen, is similar to kerogen, but it is soluble in normal petroleum solvents. It contains 80–85% carbon, 9–10% hydrogen and 2–8% sulphur, with traces of nitrogen and negligible oxygen. Asphalt occurs both infilling sediment pores and in fractures.

Hydrocarbons which are liquid at normal temperatures and pressure are termed crude oil. Oil contains 82–87% carbon, 12–15% hydrogen and traces of sulphur, nitrogen and oxygen. Oils are extremely variable in molecular composition but generally consist of varying proportions of four main groups, the paraffins, aromatics, naphthenes and asphalts. Crude oil occurs in pore spaces of many rocks in favourable circumstances.

Table VIII summarizes the properties and composition of these various hydrocarbons and includes natural gas for the sake of completeness.

Mudrocks with substantial traces of organic matter are simply referred to as organic claystones. They are generally dark coloured.

Oil shales are mudrocks which are sufficiently rich in organic matter to yield free oil on heating (Duncan, 1967). They form in restricted anaerobic low-energy environments, both marine and non-marine. A notable oil shale occurs in the Green River Formation of Utah, Colorado, and Wyoming.

Table VIII
Properties and composition of the main groups of organic hydrocarbons

Organic matter	Properties	Average composition (% weight) C	H_2	$S+N+O_2$, etc.
Kerogen	Solid at surface temperatures and pressures. Insoluble in normal petroleum solvents	75	10	15
Asphalt	Solid or plastic at surface temperatures and pressures. Soluble in normal petroleum solvents	83	10	7
Crude oil	Liquid at surface temperatures and pressures	85	13	2
Natural gas	Gaseous at surface temperatures and pressures	70	20	10

This was deposited in a series of Eocene lakes (Donell *et al.*, 1967). Association of the oil shale with desiccation cracks and other shallow-water features together with juxtaposition to trona (sodium carbonate) suggest a continental sabkha depositional environment (Surdam and Wolfbauer, 1973). With rising world oil prices these deposits become increasingly attractive economic sources of distilled crude.

Another notable oil shale occurs in the Carboniferous deltaic facies of the Midland Valley of Scotland (Greensmith, 1968) and was one of the first sources of mineral oil extracted in the world. This oil shale is also lacustrine in origin. Torbanite is a particularly pure variety of these Scottish oil shales consisting almost entirely of the alga *Botryococcus*. This rock is sometimes termed boghead coal since it is closely allied to the sapropelic coals, also termed cannel coals (see p. 148).

It can be seen, therefore, that with increasing organic content, the claystones grade from organic claystone into oil shale (or, strictly, oil claystone) and thence into the dominantly carbonaceous sapropelites, torbanites and coals.

The organic-rich mudrocks have been the subject of intensive research because they are popularly thought to be the source rocks from which liquid hydrocarbons are generated.

It is now realized that a mudrock needs to contain over 1·5% organic carbon to be a significant source rock. The type of kerogen determines the type of hydrocarbon which may be generated. Terrestrial humic kerogen tends to be gas prone. Marine sapropelic kerogen tends to be oil-prone. Temperature is also important however, critical thresholds needing to be crossed for oil and gas generation (p. 373). The subject of oil source rocks is covered in far greater depth in textbooks by Tissot and Welte (1978) and Hunt (1979).

B. Orthoclaystones and Clay Minerals

The orthoclaystones are clay-grade rocks composed mainly of the clay group of minerals.

The clay minerals are an extensive and complex mineral suite which, as pointed out in Chapter 3, form largely by chemical degradation of pre-existing minerals during weathering.

The mineralogy and origin of the main clay minerals will be described before discussing the claystone rocks.

There are three principal groups of clay minerals: the illites, the montmorillonites or smectites and the kaolins. To these three may also be added the chlorites and glauconite. These last two differ somewhat from the other clay minerals in mode of formation, but show similarities in composition and atomic structure.

All five of these mineral groups are hydrous aluminosilicates. Crystallinity is unusual in all but the chlorites and kaolins. Where detectable, it is of the monoclinic system with the possible exception of glauconite, which occurs amorphously almost without exception. For discussion of the complex crystal structure of clay minerals, see Grim (1968). Essentially the clays are arranged in layers of tetrahedra of oxygen and hydroxil radicals. Within the spaces of the tetrahedra lie silica and alumina radicals. In the kaolin clays no other elements are present within this two-layer lattice framework. The illites and smectites are based on a three-layer lattice framework of tetrahedra with substitution of alumina radicals by potassium, calcium and magnesium. The chlorites and glauconites are more complex still, with a structure of mixed two- and three-layer lattices to accommodate their more complex radicals of iron and magnesium.

1. Kaolin

Kaolin is, therefore, the simplest clay mineral in structure and the purest in composition. It is formed from feldspars both by hydrothermal alteration and by superficial weathering. Kaolin is a common detrital clay mineral in sediments derived from granitic and gneissose sources. Unlike the other clay minerals, kaolin often undergoes considerable crystallization during diagenesis to form characteristic kaolin "books". Aggregates of long accordion or worm-like crystals are also found in some sediments. Kaolin may be an important authigenic cement in some sandstones (see p. 101).

In certain circumstances kaolin sedimentation is sufficiently abundant to form a pure kaolin claystone. This rock type is variously termed pipe clay, china clay, and fire clay. The Oligocene Bovey Tracey lake beds of Devon are composed largely of pure kaolin derived from the adjacent Dartmoor granite.

Tonsteins are another distinctive rock type, composed largely of kaolin. These occur intimately associated with coals in the Carboniferous strata of Europe. Individual beds are thin yet laterally very extensive. While it is possible that tonsteins formed from the clays of weathered granite, their regional persistence favours an origin due to alteration of volcanic ash falls (Price and Duff, 1969).

Tonsteins are tough indurated rocks but the younger kaolin claystones are generally white and plastic. Kaolin is extensively used in the ceramic, paper-making and pharmaceutical industries. Pure kaolin rocks are non-marine in origin because kaolin quickly transforms to more complex clays in the presence of seawater.

2. Illite

The illite clays, sometimes termed the hydromicas, are three-layer aluminosilicates with up to 8% K_2O. This potassium may either be present due to the incomplete degradation of potash feldspars to kaolin, or to diagenesis of kaolin within a marine environment. Illite is the most abundant clay mineral in sediments but it is less obvious than kaolin because it is seldom present in crystals that can be seen with an optical microscope. Even under an electron microscope illite crystals are smaller and less well developed than those of kaolin (see p. 102).

3. Montmorillonite

The third group of clay minerals are the smectites, of which montmorillonite is the chief example. These are three-layer lattice types which have the unusual property of expanding and contracting to adsorb or lose water. Montmorillonite can contain up to 20% water as well as calcium and magnesium. Mudrocks composed largely of smectite clays are termed bentonites. Bentonite deposits may be recognized at outcrop by their powdery cauliflower-like surface. Placed in water, montmorillonite lumps expand visibly and fall apart.

Bentonites are formed by the alteration of volcanic ash *in situ*. This may occur in both marine and continental environments. Fragments of detrital glass, frequently devitrified, are generally present, together with microscopic grains of quartz, micas, feldspars and heavy minerals.

The type montmorillonite comes from Montmorillon in France. Montmorillonite is also the chief constituent of the Jurassic and Cretaceous Fuller's Earth in southern England. Bentonites occur in the Upper Cretaceous rocks of Arkansas, Oklahoma, Texas and Wyoming, notably the Mowry Shale of the Big Horn basin (Slaughter and Earley, 1965). Because of their peculiar properties, bentonites are a major constituent of circulating mud systems used in rotary drilling. Conversely, montmorillonite is an unfortunate clay mineral to have as a matrix in an oil reservoir. During production water entering the reservoir may cause the clay matrix to expand and thus destroy permeability.

4. Chlorite

The chlorites are similar to the clay minerals just described in many ways, but also share affinities with the mica mineral group. The chlorites are mixed-layer lattice clays with up to 9% FeO and 30% MgO. Chlorites occur as an alteration product of primary micas and are a common accessory detrital mineral in immature sands and mudrocks. Conversely chlorite replaces illite and other clay minerals at the point where diagenesis merges into metamorphism; it is a characteristic constituent of the microcrystalline matrix of greywackes.

5. Glauconite

The fifth clay mineral to consider is glauconite. This term is used in two ways. It is applied to pretty, rounded green grains commonly seen in marine sediments, and to a particular mineral. Analyses of the former show them to contain a mixture of clays whose lattices are in various stages of ordering. Glauconite proper is a three-layer clay mineral containing magnesium, iron and potash. Unlike the other clay minerals, glauconite does not form from the hydrothermal or terrestrial weathering of pre-existing minerals. Glauconite occurs in dark green amorphous grains seldom larger than fine-sand grade. It is found both in mudrocks and sandstones. Where especially abundant in sands, the rock is commonly named "greensand". Glauconite formation accompanied by greensand sedimentation occurred at particular times on a world-wide basis, notably in the Cambrian and through the Upper Cretaceous and early Tertiary (Pettijohn *et al.*, 1972, p. 229).

Glauconite grains appear to have formed by the replacement of faecal pellets and by infilling foraminiferal tests which have subsequently been destroyed. Glauconite also occurs infilling larger shells and replacing detrital micas. It is a matter of observation that glauconite occurs in ancient sediments of marine origin. It is easily weathered and, with one or two exceptions, does not occur in any abundance as a reworked mineral. There has been considerable debate both as to the nature of the chemical reactions responsible for glauconite formation and of the parameters which control the reaction. Glauconite has generally been considered to form by the transformation of degraded smectitic or illitic-type sheets into more ordered mixed-layer lattices. Recent studies, however, have argued for neoformation of glauconite within voids (Odin, 1972; Bjerkli and Ostmo-Saeter, 1973). No doubt a polygenetic origin for glauconite formation will

Table IX
Summary of the salient features of the main groups of clay minerals and their associates

Mineral	Composition (additional to hydrated alumino-silicate) Ca	Mg	Fe	K	Atomic lattice structure	Source	Rock name
Montmorillonite	│	│			Three-layer	Volcanics	Bentonite
Chlorite		│	│		Four-layer	Mafic minerals	
Glauconite		│	│	│	Three-layer	Submarine diagenesis	
Illite				│	Three-layer	Feldspars	
Kaolin					Two-layer	Feldspars	China clay Fire clay Tonstein

ultimately be clearly proven.

Geochemical evidence suggests that glauconite formation, by whatever process, occurs in seawater at low temperatures and in an environment which is neither strongly oxidizing nor reducing. Optimum depth for glauconite genesis appears to be between 50 m and 1000 m. However, these parameters are hard to define because, once formed, glauconite is stable in seawater and can survive transportation in a marine environment. Thus glauconite is found disseminated in marine mudrocks, in clean well-sorted cross-bedded sand shoals, and as a minor constituent of basinal turbidites. Table IX summarizes the salient features of the main clay minerals.

6. Compaction of clays

Figure 30 shows that clays are deposited with porosities ranging from between 50 to 80%. Water is lost from most clays immediately after deposition by consolidation, the process whereby a clay is changed to claystone. This includes both cementation and dehydration, as well as compaction due to overburden pressure. It is important to note that dewatering of clays at shallow depths is not due solely to overburden pressure. Dewatering also occurs in many clays as a spontaneous process termed "syneresis" (see White, 1961), forming small desiccation cracks at the mud: water interface (see also p. 239).

The compaction of clay at shallow depths has been intensively studied by engineering geologists. This is because it is critical to know the physical properties of a clay if it is to be a foundation for civil engineering projects, such as the building of high-rise blocks, motorways and dam sites.

Furthermore, it is important not only to know the physical properties of the clay at the preliminary stage of site investigation, but also to be able to predict the degree of compaction to be expected if the site is drained.

The compaction of clay down to depths of at least 3000 m may be expressed by Terzaghi's law:

$$p_o = \sigma - u$$

where p_o is the effective overburden pressure, σ is the total vertical pressure exerted by all the materials (solids and liquids), and u is the pore water pressure.

This relationship was formulated by Terzaghi (1936), based on empirical studies of clay compaction at different depths and pressures. Modern work substantiates this equation (Bishop in Skempton, 1970, p. 376).

As Fig. 30 shows, the rate of water expulsion and porosity loss in clays decelerates with increasing depth of burial. Simultaneously the problems of clay compaction move from the field of engineering geology to petroleum exploration. The expulsion of oil from compacting muds has been implicit in most theories of petroleum generation in the western world. Dott and Reynolds (1969, pp. 174–179) have reviewed some of the fundamental papers and concepts in this field.

One of the most interesting ideas on clay compaction to emerge has been the significance of changes in clay mineralogy with increasing depth of burial. Powers (1967) has produced evidence to show how montorillonite converts to illite, releasing water and generating high pore pressures. In the Tertiary oil fields of the Gulf of Mexico, there is a close correlation between the depth of the producing reservoirs and this clay mineral change (Burst, 1969, Figs 12 and 15).

In conclusion it can be seen that primary porosity in clays is lost quickly during early burial, mainly by the physical process of gravitational compaction. Below about 2000 m porosity decreases very slowly with depth and is destroyed largely by chemical processes of recrystallization.

The compaction and consolidation of clays at shallow depths is of significance to the engineering geologist. The changes which occur in clays as they change to claystones at greater depths is of interest to the petroleum geologist since this may be relevant to theories of petroleum genesis and migration. Similarly, mining geologists have considered that low-temperature ore bodies may have been derived from the residual fluids of compacting clays aided by brines acting as transporting media (e.g. Davidson, 1965; Amstutz and Bubinicek, 1967).

In concluding this review of clay minerals, the following points should be noted. The clay mineral suite which is found in a particular rock at a particular time is due to four main variables. The nature of the source rock controls the input of clays minerals. There is a higher probability of kaolin and illite forming from a granite source than from a volcanic hinterland.

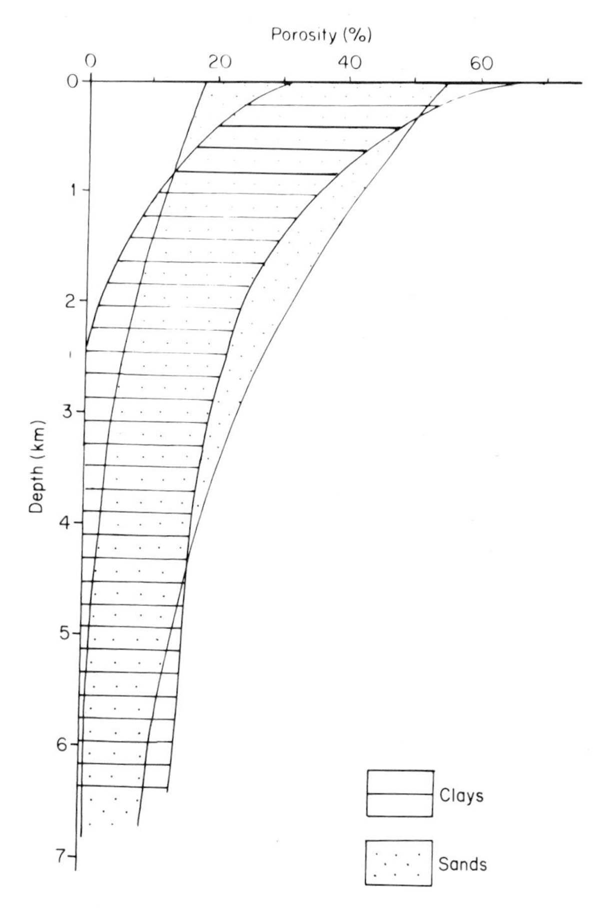

Fig. 30. Burial curves for sands and clays constructed from compilations of data in Selley (1978) and Magara (1980).

Similarly, the more intense the weathering the higher is the probability of kaolin occurring at the expense of illite. Size sorting during transportation may also segregate the various clay mineral species. Kaolin crystals tend to be larger than illites and illite to be larger than montmorillonite. This fact is complicated, however, by the tendency of clay particles to form floccules when they are carried from acidic fresh water to basic seawater. Simultaneously various transformations of clay minerals take place within seawater both during transportation and early burial (Keller, 1970).

Thus while most studies of modern marine coasts show some regular zonation of clay minerals, it is hard to demonstrate whether this results from differential transportation or incipient diagenesis (e.g. Porrenga, 1966). More advanced diagenetic changes occur after burial. These generally lead to the transformation of kaolin to illite and to the dewatering of montmorillonite. The possible importance of these clay mineral changes to hydrocarbon migration has been already mentioned (p. 74). The effect of authigenic clay cements on the permeability of sandstones will be discussed later (p. 101).

More exhaustive accounts of clays and clay minerals will be found in Grim (1968), Millot (1970), Muller (1967), Mortland and Farmer (1978), Shaw (1980) and Potter *et al.* (1980).

IV. PYROCLASTIC SEDIMENTS

Figure 27 showed how the allochthonous sediments are divisible into the claystones (defined by their grain size and clay mineral composition), the siliciclastic sands of terrigenous silica minerals, and the volcaniclastic sediments.

The volcaniclastic sediments are relatively rare by volume in the earth's crust, but deserve mention for the sake of completeness. The minerals of volcanic rock are mostly unstable at temperatures and pressures. For this reason, therefore, detritus derived from volcanic activity is commonly only preserved interbedded with lava flows and can seldom survive bulk transportation far from the volcanic centre from which it came.

The volcaniclastic sediments can be classified into three groups according to their particle size.

Agglomerates are the counterpart to conglomerates. They are formed both by explosive eruptions and by scree movement of volcanic detritus both within a caldera and on the flanks of volcanoes.

Sand-grade volcaniclastic sediment is of two types. Erosional volcaniclastic sands are produced by normal subaerial or subaqueous processes acting on eruptive rocks. The pyroclastic sediments in contrast are ejected into the atmosphere during volcanic eruptions. Pyroclasts include, therefore, both "bombs" which fall close to the vent, sands which fall around the vent for a distance of kilometres, and dust which may be carried into the upper atmosphere and transported around the world.

In many cases it is impossible to distinguish whether an ancient volcaniclastic sediment was produced by normal erosion of lavas or by pyroclastic action. Volcaniclastic sands are generally referred to as tuffs or ashes. They may be subaerial or subaqueous. Volcaniclastic sands are composed essentially of crystals, glass and rock fragments. The crystals are of minerals associated with the eruption, such as olivine and quartz. Glass occurs both as globules and angular irregularly-shaped shards. Rock fragments are composite grains of volcanic minerals and glass.

Volcaniclastic sands are generally poorly sorted because if extensively transported and reworked they are rapidly destroyed. Eolian volcaniclastic sands are an exception to this rule and dunes of basalt sand occur, for example, around the modern volcanic crater of Waw en Namus in the Libyan Sahara.

Fine-grained volcanic ash tends to undergo intensive post-depositional alteration giving rise to the bentonites and (perhaps) tonsteins discussed in the previous section. Palagonite is a mineraloid produced by the alteration of basalt glass. It is sometimes sufficiently abundant in the Pacific to make a sediment termed palagonite mud, and also occurs lithified in Pacific volcanic islands.

Traces of volcanic ash are a common constituent of modern pelagic sediments.

More detailed accounts of the genesis and petrology of volcaniclastic sediments will be found in Ross and Smith (1961) and Pettijohn *et al.* (1972, pp. 261–292).

V. SANDSTONES

This section describes the important group of sedimentary rocks: the sandstones. Specifically it is concerned with the quartzoze or siliciclastic sands as distinct from the volcaniclastic and carbonate sands. About 30% of the land's sedimentary cover is made of terrigenous sand and sandstone. This group of rocks is a fruitful topic for study both for itself and to aid the exploitation of economic materials within sandstones. Because they are often highly porous, sandstones are frequently major aquifers and hydrocarbon reservoirs. They are more uniform in stratigraphy and petrophysical character than carbonates and it is, therefore, easier to predict their geometry and reservoir performance.

The following account discusses the problems of sandstones nomenclature and classification, summarizes the features of the commoner sandstone types, and concludes by discussing the effect of diagenesis on sand porosity and permeability.

A. Nomenclature and Classification of Sandstones

The nomenclature and classification of sandstones has always been a popular academic pursuit. Klein (1963) and Pettijohn *et al.* (1972, pp. 149–174) have written reviews of this topic. It is important that the diverse types of sandstones are named with a terminology that is tolerably familiar to, and agreed on by, practising geologists. Any nomenclatural system has to have arbitrary bounding parameters which separate one rock type from another.

These bounding parameters are most useful when based on some underlying concepts of sand genesis. The basic problem of classifying sands is that they can be grouped according to their physical composition (i.e. grain size and matrix content), or according to the chemical composition (i.e. mineralogy). There are more textural and mineralogical components deemed to be significant in sandstone nomenclature than can be conveniently represented in an end-member triangle or tetrahedron.

Higher, statistically based classificatory schemes, such as factor analysis, may be more logical. On the other hand, they lack the simplicity and visual appeal of end-member classifications.

Thus the majority of sand classifications are based on end-member triangles, the three components generally being chosen from quartz, clay, feldspar or lithic content.

One of the most fruitful concepts on which sandstone nomenclature is based is the idea of maturity. The maturation of a sand takes place in two ways. It matures chemically and it matures physically. Sediments form from the weathering of mineralogically complex source rocks. Throughout weathering and transportation relatively unstable minerals are destroyed and chemically stable minerals thus increase proportionally. Quartz is the most abundant stable mineral and feldspar is a common example of an unstable mineral. An index of the chemical maturity of a rock might, therefore, be the ratio of quartz to feldspar. As sediments are reworked, perhaps through two or more cycles of sediment, they thus tend to mature to pure quartz sands.

Physical maturity, on the other hand, describes the textural changes which a sediment undergoes from the time it is weathered until it is deposited. These changes involve both an increase in the degree of sorting and a decrease in matrix content. Thus an index of the degree of physical maturation might be the ratio of grains to matrix. Total clay content is a useful index of textural maturation, given certain reservations to be discussed shortly.

Both physical and chemical maturation occur during the history of a sand population, but they are not closely related. Thus a chemically mature sand may be physically immature and vice versa. This is because chemical composition is essentially a result of provenance, while textural composition is a result of process.

From the preceding analysis, it would appear that the most appropriate triangular classification to adopt would have stable grains at one apex, matrix at a second and unstable grains at the third (Fig. 31). As a sand population increased in textural maturity it would move away from the matrix apex. As it improved in mineralogical maturity it would move away from the unstable grain apex.

Since both types of maturation occur simultaneously, albeit at different rates, the net tendency is for a sediment to move to the stable grain apex. This may not be achieved in a single sedimentary cycle, but it is the logical destination of any sand sediment.

These concepts of maturity can be used as a basis for sandstone nomenclature (Fig. 32). The total amount of clay material in a sand is obviously the best indicator or matrix content. Feldspar is a common and often volumetrically abundant unstable mineral which may be used as an index of chemical immaturity. Quartz is the obvious choice as the index mineral for the apex of chemical stability.

Using quartz, feldspar and clay as end-members, a sand classification scheme can be drawn up as shown in Fig. 33. This shows that sands may be divided into two broad textural types, the arenites and the wackes. Arenites, with a matrix content of less than 15%, are texturally mature. Wackes, with a matrix content of between 15–75% are immature. Rocks with more than 75% matrix are not sandstones but mudrocks.

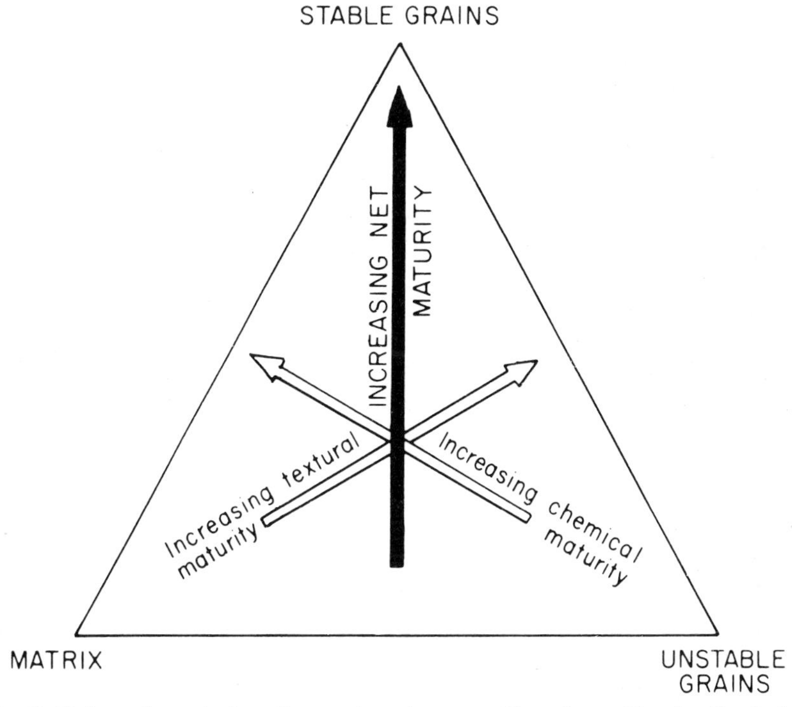

Fig. 31. End-member traingle to illustrate how the composition of a sand is a function both of its textural maturity, expressed as matrix content, and of chemical maturity, expressed as unstable grain content.

Similarly, sands may be divided into chemically mature arenites and chemically mature wackes. These both have less than 25% of feldspar. These two sandstones are designated protoquartzite and quartz-wacke. Logically the mature arenites should be (and have been) called quartz-arenites, but protoquartzite is a long-established name for sands of this general composition. Space is found in the quartz apex for orthoquartzite sands, the purest and most mature of all.

Sands with more than 25% feldspar are chemically immature and are divided into the arkoses and greywackes for the arenites and wackes respectively. Logical synonyms for these two names would be feldspar-arenite and feldspar-wacke, but the terms arkose and grey-wacke are too well established to be omitted.

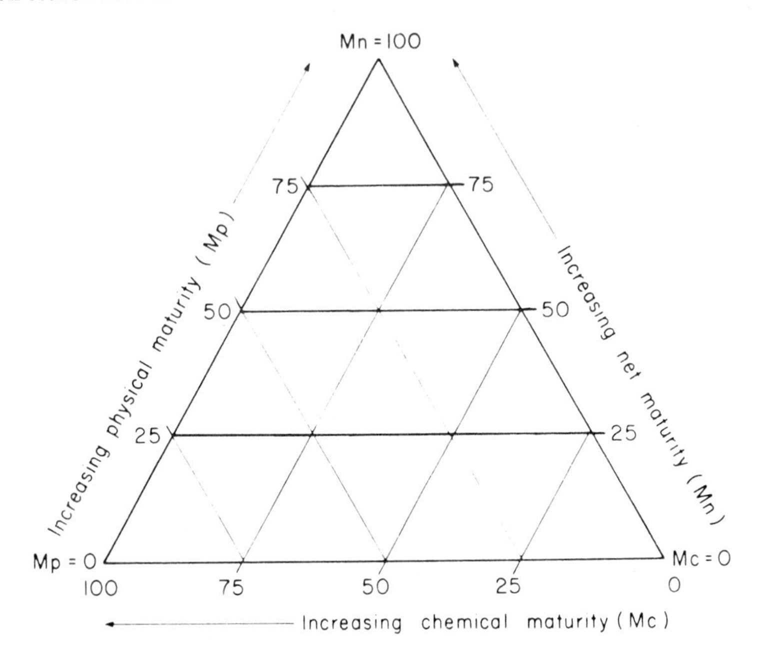

Fig. 32. End-member triangle in which a sand may be plotted to illustrate its maturity indices expressed as the percentage composition of matrix, stable and unstable grains. For full explanation see text. Gs: chemically stable grains; Gu: chemically unstable grains; M: matrix content; Mp: physical maturity index; Mc: chemical maturity index; Mn: net maturity index.

Essentially, therefore, this scheme arbitrarily divides sands into four main groups depending on the degree of physical and chemical maturation. This system is similar to many others which juggle the choice of end-members, rock names and percentage cut-offs within triangles. Probably every geologist has his own private version; not all go public with it in the press, fortunately.

Before describing these four main sandstone clans, a few cautionary remarks are needed concerning matrix and rock fragments (known also as lithic grains).

Clay content has been proposed as an index of the degree of textural maturity of a sediment. This is based on the general observation that in modern sediments clay tends to be winnowed out from a sand during transportation and in environments of persistent high energy such as shallow marine shelves. This is broadly true but the clay content of a lithified sand is not necessarily all syndepositional in origin. Some matrix probably infiltrates pore spaces shortly after deposition. Some clay is transported as silt- and sand-grade particles. On compaction they are squashed to form a matrix between more resistant grains.

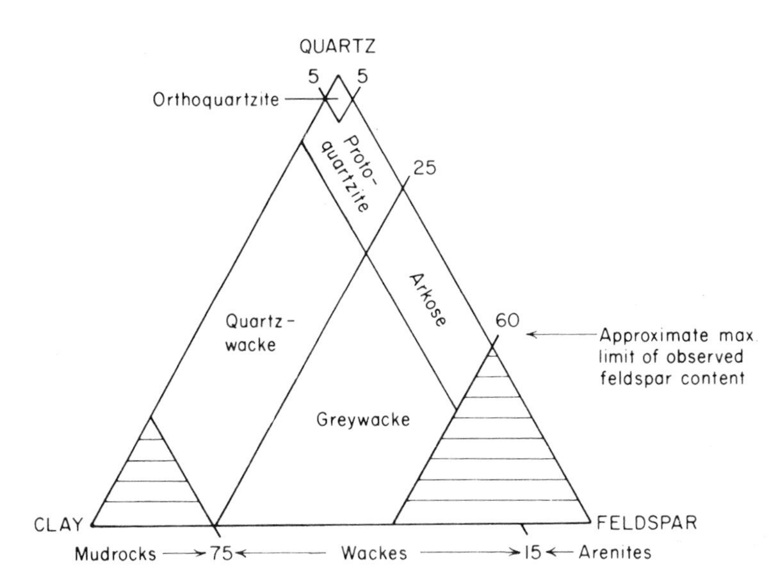

Fig. 33. A classification of sandstones based on the use of clay as an indicator of textural maturity, and feldspar as an index of chemical maturity.

During diagenesis unstable detrital grains break down and form a microcrystalline matrix composed largely of clay minerals. This is a commonly observed feature of greywackes. The question has been asked to what degree do greywackes truly indicate an origin as a muddy sand and to what degree is the matrix due to the decay of labile grains (Cummins, 1962). An analogous problem is often seen in arkoses. Feldspar grains commonly show varying degrees of alteration to kaolin, and individual grains of kaolin can be seen in varying stages of compaction between quartz grains. It is hard to accurately measure feldspar content and clay-matrix content in such specimens.

This discussion shows, therefore, that clay matrix must only be regarded as a rough guide of textural maturity. In lithified sands the clay content is probably rather higher than the original depositional matrix content.

Another important constituent of sandstones which deserves special mention are rock fragments. Many sands contain grains which are not monomineralic, but which are composite grains. These are called lithic grains or rock fragments. Lithic grains are a popular choice for an end-member of many sandstone classifications, and rock names such as lithic greywacke and litharenite have appeared.

There are two important points to note about lithic grains. The first is that the lithic grain content of a sand is likely to be related to the particle size of the source rocks. Lithic grains are unlikely to be common in an arkose derived from a coarsely crystalline granite. Conversely, lithic grains may be abundant in sands derived from microcrystalline volcanics, metamorphics or from well-indurated mudrocks.

The second point to note about lithic grains is that their abundance is dependent on grain size. Figure 34 illustrates this very predictable relationship. Obviously the larger a sand grain is then the larger is the probability of it containing more than a single mineral crystal.

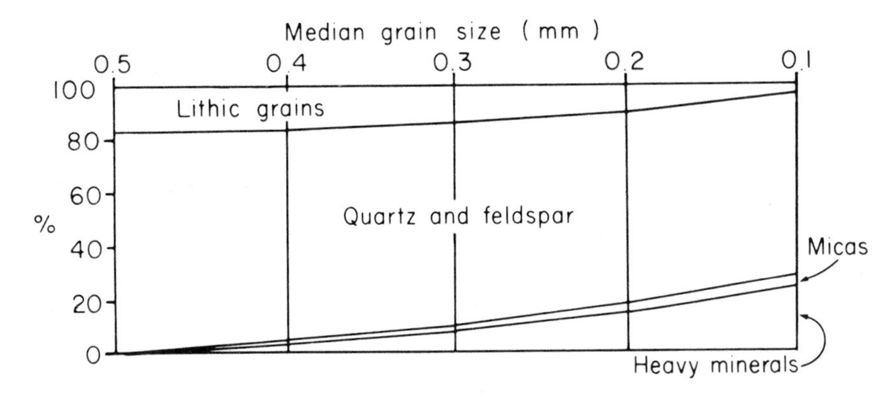

Fig. 34. Relationship between grain size and composition in Torridonian Sandstone, Scotland. (From Selley, 1966.) Lithic grains increase with grain size, conversely mica and heavy minerals decrease.

Similarly, clay content often increases with decreasing grain size (Fig. 35). The anomalous situation that this causes can be seen in some graded turbidites where, within one sedimentation unit of a few inches, an arkose may grade up into a greywacke.

Having pointed to some of the hazards and limitations of sandstone nomenclature, the main groups will be described. For this purpose it is convenient to group them into the wackes (including both quartz-wacke and greywacke), the arkoses, and the quartzites (including both the proto-quartzites and the orthoquartzites).

Beyond the world of the lecture theatre and the learned journals, the problems of sand classification and nomenclature seem to vanish like a hangover. In industry the general practice is either to describe a sand as a sand, and no more. Alternatively, as on a log, a rock may be described more fully using a format similar in content and sequence to this:

"LITHOLOGY, *Colour, hardness, grain size, grain shape, sorting, mineralogy, fossils (if any), porosity, hydrocarbon shows, i.e. stain, fluorescence or cut.*"

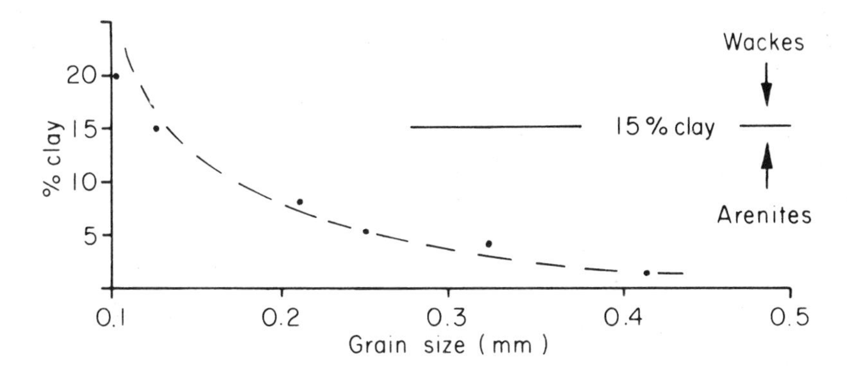

Fig. 35. Relationship between grain size and clay content in Torridonian Sandstone, Scotland. (From Selley, 1966.)

The arkoses, wackes and quartzites will now be described.

B. Sandstones Described

1. Quartzites

Sands which are mature in texture and mineralogy are broadly referable to the quartzites or quartz-arenites. Various authors have proposed the minimum percentage of quartz necessary for a sand to qualify. The term quartzite has been used in metamorphic geology for any sand, regardless of composition, which breaks across grains rather than around grain boundaries. Amongst sedimentary petrographers however, a quartzite generally refers to a quartz-rich sand irrespective of its degree of lithifaction.

As defined in the scheme in Fig. 33 the quartzites have less than 25% feldspar and less than 15% matrix. The quartzites are generally arbitrarily divisible into the protoquartzites which contain some feldspars and matrix, and the orthoquartzites which are almost pure silica.

The quartzites are, therefore, by definition, sands which are mature in texture and mineralogy.

Typical quartzites are white, pale grey or pink in colour. They range from unconsolidated to splintery in their degree of lithifaction. Grain size is variable, but sorting is generally good and individual grains are normally well rounded.

The main detrital grains are quartz, derived from igneous and metamorphic rocks, and cherts (made of quartz, chalcedony or chrystobalite) reworked from sediments. Rare feldspar and mica grains may be present. Heavy minerals are generally the stable residue such as zircon, tourmaline, apatite and garnet. Intraformational autochthonous detrital grains are often quite common in quartzites. Glauconite, phosphate pellets and skeletal debris are typical examples. Probably the majority of glauconitic sands are referrable to the quartzite group (were it not for the presence of that mineral).

Because of their uniform grain size, rounded grain shape and low clay content, quartzites possess high porosities and permeabilities at the time of their deposition. Cementation is generally by calcite or secondary silica, but where it is absent, quartzites make the best aquifers and hydrocarbon reservoirs of all the sandstone types.

It is probable that most, if not all, quartzites are polycyclic in origin. That is to say that they have been through more than one cycle of weathering, erosion, transportation and deposition to achieve the necessary maturity to qualify.

It is a matter of observation that the majority of quartzites were deposited in marine sand-shoal environments. Extensive quartz sand sedimentation occurred in the Lower Palaeozoic shelf seas around the border of the Canadian Shield and the northern margin of the Saharan Shield (e.g. Bennacef *et al.*, 1971). Quartzose sands are also found in eolian facies such as the Permian Rotliegende of the North Sea basin (Glennie, 1972) and the early Mesozoic dune sand formations of the Colorado Plateau (Baars, 1961). Eolian and marine shoal environments provide the optimum conditions for quartzites to be deposited by selective winnowing. Nevertheless, they do also occur in many other environments, including modern deep-sea sands (Hubert, 1964) and ancient turbidites (Sturt, 1961). Figure 36 illustrates a thin section of the Silurian Simpson Group orthoquartzite of Oklahoma.

2. *Arkoses*

The term arkose was first proposed by Brogniart in 1826 for a coarse sand composed of quartz and substantial quantities of feldspar from the Auvergne in France. This name is still used for rocks of that general description. Essentially the arkoses are sands which are relatively mature in texture (i.e. low in clay matrix) and are immature in mineralogy as shown by the abundance of feldspar. The clay content must generally be less than 15% and there must be more than about 25% of feldspar. Few arkoses have more than 60% feldspar because this would presuppose a source in which feldspar was more abundant than quartz. The converse is generally true and the ratio of quartz to feldspar begins to rise as soon as the source rock is submitted to weathering.

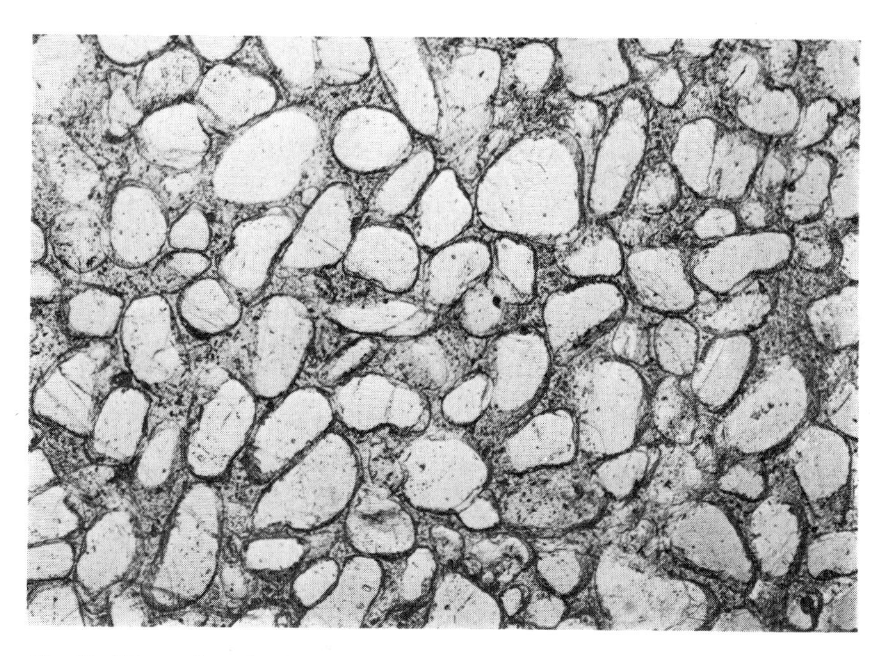

Fig. 36. Thin section of orthoquartzite from the Simpson Group (Silurian) of Oklahoma. Ordinary light. Well-rounded quartz grains with negligible cement and intergranular pores soaked in oil (×30).

Arkoses, therefore, are the product of the incomplete degradation of acid igneous and metamorphic rocks such as granites and gneisses. It was once argued that arkoses were indicators of cold and/or arid climates in which physical weathering processes dominated chemical ones. Krynine (1935) showed, however, that modern alluvial arkoses occur in tropical rain forest climates. These form because rapid run-off and erosion cut gullies through the weathered zone and erode unweathered granite. Nevertheless, some arkoses, such as the Torridonian (Pre-Cambrian) example of Scotland, contain well-rounded feldspar grains (Fig. 37). It is unlikely that these grains are polycyclic. The implication is that rounding occurred within the first and only cycle of sedimentation, and it is known that eolian action is far more effective at this than running water (p. 10). There is a strong presumption that the Torridonian arkoses were deposited fluvially in an arid climate in which wind action rounded feldspar grains.

Fig. 37. Thin section of arkose from the Torridon Group (Pre-Cambrian) of north-west Scotland. Note well-rounded quartz and feldspar grains with negligible matrix. Polarized light ($\times 30$).

The typical arkose is a pink or red coloured rock, less commonly it is grey. Pink coloration is due to the feldspars, but the red coloured examples owe this feature to absorption of red ferric oxide into the clay matrix. Arkoses show a wide range of grain sizes and are often poorly sorted. Arkose often forms *in situ* on granites forming a transitional weathering zone, termed "granite wash", in which it is hard to distinguish sediment from igneous rock. This situation is a well-site geologist's nightmare. It is very hard to distinguish arkose from granite on the basis of drill cuttings, yet the contact can easily be picked subsequently from geophysical logs.

The grains of arkoses are typically angular to subrounded and there is generally a significant amount of clay matrix. For this reason arkoses seldom stay unconsolidated for long like the quartzites. Most ancient arkoses show some degree of lithifaction due to clay bonding. In extreme cases porosity can be completely obliterated by a silica or carbonate cement.

The feldspars in arkose are of various types, depending on the nature of the source rock, but microcline and albite tend to be more abundant than the less stable calcic feldspars. Alteration of the feldspars to kaolin and sericite is a common feature, but it is hard to tell whether this occurred by hydrothermal alteration of the source rock, during weathering, or through diagenesis. Micas are a common accessory mineral and arkoses often contain a diverse suite of heavy minerals. These may give some indication of the source terrain, specifically whether it was igneous or metamorphic. In addition to the ubiquitous stable suite of zircon, apatite, garnet and tourmaline, it is not uncommon to find opaque iron ores as well as many other heavy minerals. Huckenholz (1963) has described the petrography of the type arkoses of the Auvergne.

Most arkoses seem to occur in fluvial facies adjacent to granitoid basement. They are, therefore, the characteristic sediment type of fault-bounded intracratonic basins (see p. 335). Because such sediments are deposited in an oxidizing alluvial fan environment, many arkoses are of what are termed the "red bed" facies assemblage.

Typical examples of arkoses formed in this setting occur on the Pre-Cambrian shields all over the world. Specific examples include the Torridonian (Pre-Cambrian) of Scotland (Selley, 1966) and the adjacent Jotnian, Dala and Sparagmite Pre-Cambrian series of the Scandinavian Shield. The Triassic Newark group of the Connecticut trough, USA, is another classic example of an arkose.

Though probably the bulk of arkoses occur in fluvial environments, they are also found elsewhere. Arkoses mixed with coralgal debris occur on modern beaches of the fault-bounded Red Sea. Ancient analogues occur on the oil-soaked, reef-crowned granite high of the Sirte basin, Libya.

3. The wackes

The wackes are, by definition, texturally immature sands with more than 15% matrix. Chemically mature wackes, with less than 25% feldspar, are termed quartz-wacke. Chemically immature wackes are the classic greywackes.

The name greywacke comes from the German "grauewacke" which was applied to the Palaeozoic sandstones of the Harz Mountain. Petrographic descriptions of examples from the type area have been given by Helmbold (1952) and Mattiatt (1960).

The greywackes are characteristically hard, dark grey-green rocks that break with a hackly fracture. Under the hand lens, greywackes are very poorly sorted with particles ranging from very coarse sand grains down into clay-grade matrix. They have been aptly described as microconglomerates. Grains are commonly angular and of poor sphericity. Quartz is overshadowed by an abundance of other detrital minerals. Feldspar is present, but so also are mafic grains such as horneblende and, occasionally, pyroxenes. Some of the larger grains are lithic rock fragments, and, depending on the source, these may have been derived from volcanics or older metasediments such as quartzite or slate. Micas are abundant and include both muscovite and biotite, as well as microcrystalline diagenetic chlorite and sericite. A diverse suite of unstable heavy minerals is also typical. Figure 38 illustrates a Jurassic greywacke from the northern North Sea.

Fig. 38. Thin section of a greywacke from the Jurassic of the North Sea. Polarized light. Poorly sorted and mineralogically immature (note feldspars and lithic grains), this is not a good oil reservoir rock unlike many Jurassic sands from this area ($\times 40$).

All these detrital grains are set in the abundant matrix. This is a microcrystalline paste of clay minerals, chlorite, sericite, quartz, carbonate (often siderite), pyrite and occasionally carbonaceous matter. Corrosion of detrital grain boundaries is sometimes seen in the form of a characteristic "*chevaux de frise*" of micas. This rims not only the unstable grains but sometimes even quartz.

The quartz-wackes differ from the typical greywacke just described in that they lack the diverse suite of unstable detrital minerals. Their absence is coupled with an increase in the number of quartz and sedimentary lithic grains, though the clay paste is still present and in hand specimen quartz-wackes and greywackes are hard to distinguish. The quartz-wackes correspond in part to the subgreywackes and lithic wackes of some texts (Fig. 39).

Fig. 39. Thin section of a quartz-wacke. Palaeozoic of Chios, eastern Aegean. Poorly sorted, with quartz and chert grains, this rock lacks the unstable mineralogy of a greywacke. Polarized light (×40).

Particular attention has been given to the origin of the matrix in wackes (e.g. Dott, 1964). The poor sorting of these sandstones presupposes that the matrix may be largely syndepositional. On the other hand, the presence of authigenic minerals and the corrosion of detrital grains shows that some of the matrix is diagenetic in origin (Cummins, 1962; Brenchley, 1969). In this context it is important to note that the typical wackes are largely Pre-Cambrian and Palaeozoic in age. This suggests that time, deep burial and/or high geothermal gradients are needed to generate greywackes. These are thus perhaps metamorphic rocks.

Where this argument is of particular importance is in the turbidite problem. Most ancient greywackes occur in flysch facies which are commonly interpreted as turbidites (see p. 185). Is clay a necessary feature of the turbidite process? Hubert (1964) has pointed out that many modern deep-sea sands are well-sorted. These have often been interpreted as turbidites. The problem is threefold: is clay necessary for turbidity flows or not? If it is necessary, then modern deep-sea sands may be due to normal bottom traction currents. Similarly if the clay matrix of ancient flysch greywackes is diagenetic, does this support or detract from a turbidite genesis?

The greywackes, as stated in the previous paragraphs, are commonly found in pre-Mesozoic flysch facies. These typically occur in subductive troughs and it is apparent, both from their petrology and regional setting, that greywackes are often derived from the rising island arcs of volcanic origin. Hence the unstable suite of mafic minerals and the relatively high percentage of iron and magnesia in greywackes.

The Steinmann Trinity is a picturesque name given to a commonly observed association of greywackes, cherts and volcanics (Steinmann, 1926). This is often present in the early phase of the evolution of subductive troughs (see p. 350). Crustal thinning is held responsible for submarine volcanics and the generation of ultrabasic ophiolites and pillow lavas. It has been argued that the cherts form from blooms of radiolaria caused by an abundance of volcanically generated silica, while the greywackes are formed from sediments shed off rising volcanic island arcs (see p. 357).

Quartz-wackes do also occur in geosynclinal settings, as for example in the Palaeozoic flysch of the eastern Aegean. These chemically mature rocks on the other hand are more usually derived from pre-existing sediments. Sands contribute the quartz, shales produce the clay and the lithic fraction comes from the indurated equivalents of both. Thus quartz-wackes are typically found not so much in flysch settings, but in proximal continental deposits. Quartz-wackes occur in fanglomerate and alluvial environments. Examples include those continental Mesozoic sandstone formations of the Sirte basin, Libya, which have been derived from Palaeozoic sediments. Continental quartz-wackes are often red–brown in colour due to impregnation of the clay matrix by red ferric oxide.

With increasing transportation, the quartz-wackes lose some of their clay content and assume a rock type often termed "subgreywacke". This is found in both fluvial and deltaic facies. Examples of subgreywackes occur in the fluvial Devonian and deltaic Pennsylvanian (Upper Carboniferous) sandstones on both sides of the North Atlantic.

This brief review of sandstone petrography shows that the composition of a newly deposited sand is a product of provenance and process. The chemical maturity of a recently eroded sediment will depend on the source rock and the extent of weathering. Chemically mature sediments are generally polycyclic in origin and owe their maturity to derivation from pre-existing sedimentary formations. Chemically immature sands are generally first cycle material derived from igneous and high-grade metamorphic rocks.

Recently eroded sediment is commonly poorly sorted and rich in argillaceous matrix. Eolian and aqueous processes increase the textural maturity of a sand, glacial processes may reverse it.

The final section of this chapter shows how post-depositional changes affect sandstone composition and attempts to relate these to the evolution of porosity and permeability in the terrigenous sands.

After deposition a sand may be buried and turned into sandstone by lithifaction. This consists of physical compaction and chemical diagenesis. These will now be discussed in turn.

D. The Effect of Diagenesis on the Porosity of Sandstones

1. Introduction

The term diagenesis has been applied in varying ways to the post-depositional, yet pre-metamorphic processes which affect a sediment. Dunoyer de Segonzac (1968) has reviewed the history and semantics of this term.

For the purposes of the following account the definition of Pettijohn is used:

> Diagenesis refers primarily to the reactions which take place within a sediment between one mineral and another or between one or several minerals and the interstitial or supernatant fluids (Pettijohn, 1957, p. 648).

This definition limits diagenesis to essentially chemical processes distinct from physical processes, such as compaction, described in the previous section (p. 73).

The study of the diagenesis of sandstones is described in detail elsewhere in texts such as those by Pettijohn (1957), Larsen and Chilingar (1962), Folk (1968) and Pettijohn *et al.* (1972). The following account concerns only those aspects of sandstone diagenesis which effect porosity.

Before proceeding to the details of sandstone cementation, it is necessary to consider the chemistry of the fluids which move through the pores and their effect on cementation and solution.

Pore fluids may be placed in three groups according to their origin. Meteoric water is that of the oceans, rivers and rain. It is commonly present in the pores within the top hundred metres or so of the earth's surface, but may also be preserved fossilized below unconformities. Meteoric water contains relatively low concentrations of dissolved salts. Meteoric ground water tends to have a positive Eh, due to dissolved oxygen, and a low pH, due to carbonic and humic acids.

Connate water was originally defined as residual sea water within the pores of marine sediment. Sea water is oxidizing and neutral. It is now realized that sea water trapped in pores undergoes considerable modification as chemicals are precipitated or dissolved. The term "connate water" is retained, however, for such deeply buried fluids. The Eh and pH of connate fluids vary widely, extending from acidic and oxidizing, where they mingle with meteoric waters, to alkaline and reducing, where they are associated with oil and gas accumulations.

The salinity of pore fluids gradually increases with depth, going from the meteoric to the connate zone.

Juvenile waters are those of hydrothermal origin, characterized by high temperature and bizarre chemistry. Figure 40 shows the Eh:pH ranges of these fluids.

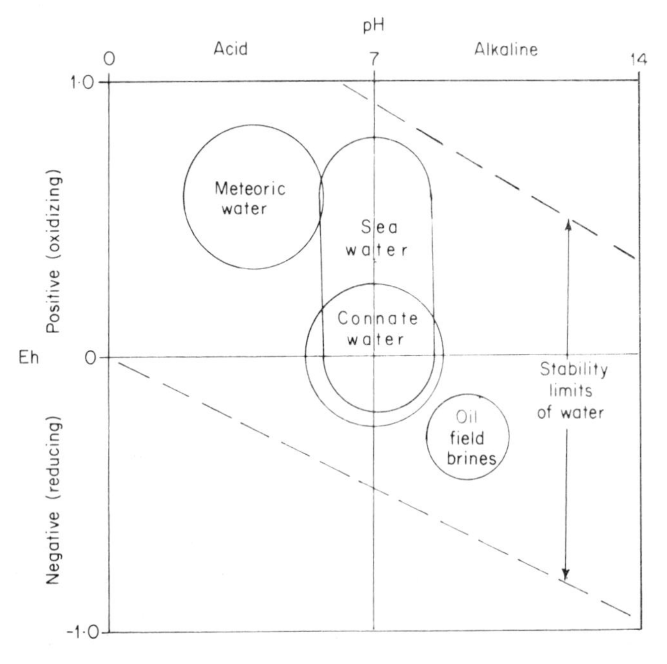

Fig. 40. Diagram showing the Eh and pH of the various types of water which move through pores and thus affect diagenesis.

The diagenetic history of a sandstone will obviously be controlled by the chemistry of the fluids which have moved through its pore system. The factors which determine mineral precipitation or solution include the chemistry of the sediment and the composition, concentration, Eh and pH of the pore fluids.

There are many chemical reactions which occur during sandstone diagenesis. Only about half a dozen of these are of major significance in sandstone cementation and porosity evolution. These are those which control the precipitation and solution of silica, calcite and the clay minerals. These will now be considered in turn.

The solubilty of calcite and silica are unaffected by Eh, but are strongly affected in opposing ways by pH. Silica solubility increases with pH, while calcite solubility decreases. Thus in acid pore fluids, such as meteoric waters, calcite tends to dissolve and quarts overgrowths to form, while in alkaline waters calcite cements develop and may even replace quartz. For mildly alkaline fluids (pH between 7 and 10) quartz and calcite cements may both develop (Fig. 41).

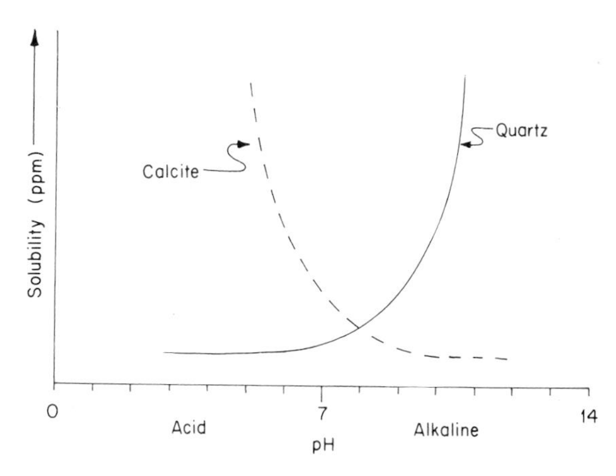

Fig. 41. Diagram showing the way in which the solubility of calcite and quartz and related to pH. Simplified from Blatt (1966).

Clay minerals are similarly sensitive. Kaolin tends to form in acid pore fluids, while illite develops in alkaline conditions. This is a very important point which significantly controls the permeability of a sand.

Siderite, glauconite and pyrite are all stable in reducing conditions. Figure 42 summarizes the stability fields of these various minerals.

Dapples (1967) has classified sandstone diagenesis into a continuous one-way spectrum of three phases termed redoxomorphic, locomorphic and phyllomorphic. At any point in this sequence a sandstone may be uplifted and submitted to weathering. This leads to a fourth phase, termed epidiagenesis; the development of secondary porosity beneath a potential unconformity. These four phases of sandstone diagenesis will now be discussed in relation to the way in which they effect porosity.

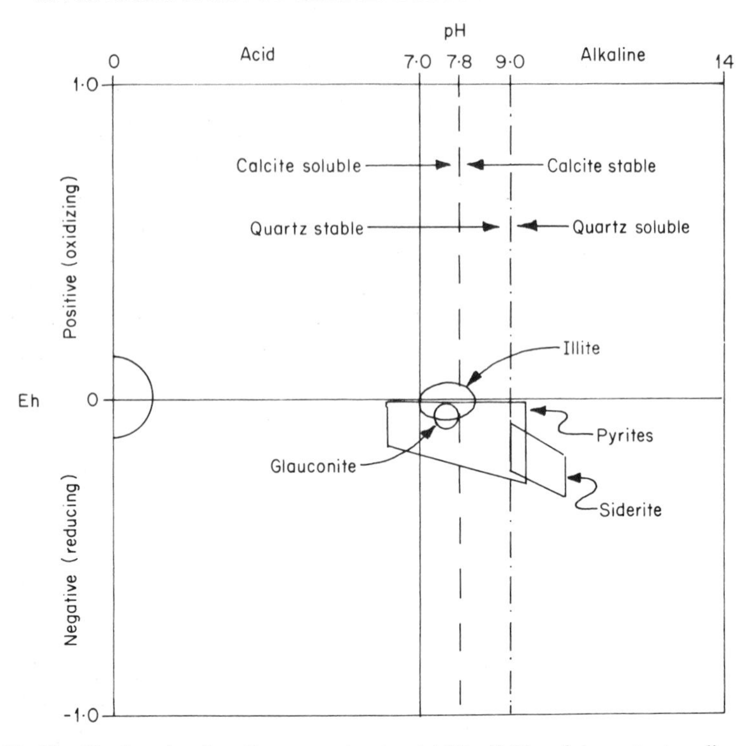

Fig. 42. Eh:pH plot showing the approximate stability fields of important sedimentary minerals. Compare with Fig. 40.

2. *The redoxomorphic phase*

When a sand has been deposited it first goes through the essentially physical processes of compaction and dewatering. The early chemical diagenetic changes are dominated by reduction or oxidation, hence the term redoxomorphic phase. These reactions primarily concern oxygen, naturally; iron, sulphur and organic matter. They are largely the consequence of bacterial action.

Essentially a sand which has high permeability and is deposited above the water table will be subjected to oxidizing reactions. This is because the pore system will be subjected to both free air and oxygenated ground water. Organic matter is oxidized, sulphur compounds are oxidized and carried off as soluble sulphate ions. Iron tends to be preserved as ferric oxide. This is red in colour and forms a pellicle around detrital grains and is mixed in with such clay matrix as may be present. This is why the majority (but not all) of red sandstones are of continental origin, both eolian and fluvial.

By contrast, low permeability argillaceous sands and those deposited below the water table, tend to be dominated by reducing reactions due to the relative shortage of free oxygen. This is largely because of bacterial action. Organic matter may be preserved and iron and sulphur combine to form pyrites. This combination of substances together with the lack of red ferric oxide, imparts an overall drab grey-green colour to the sediment.

A considerable amount of work has been done on the mineralogy of these oxidizing and reducing reactions (Van Houten, 1973; Folk, 1976; Walker, 1976). The consensus of opinion is that they are truly diagenetic and that the original oxidized or reduced state of the iron soon achieves equilibrium with the diagenetic environment. Field relationships and petrographic studies show that these reactions are completed at an early stage of diagenesis.

A clear example of this occurs in the Pre-Cambrian Torridon Group of northwest Scotland. Here there is a sequence of three deposits: scree, lacustrine (or shallow marine) and braided alluvium. These deposits are banked against an irregular basement topography with marginal fanglomerate development. The three facies are red, grey-green and red in vertical sequence. These colours extend into the marginal fanglomerate, indicating that the iron coloration is a primary feature. The reducing water-logged conditions of the lake or shallow sea allowed grey-green chloritic cement to develop both in it and the adjacent fanglomerate. The oxidizing environment of the earlier scree and later braided deposits caused a red ferric oxide to develop in them and their lateral conglomeratic equivalents (Fig. 43).

Other examples are described by Van Houten (1968), Walker (1967) and Friend (1966).

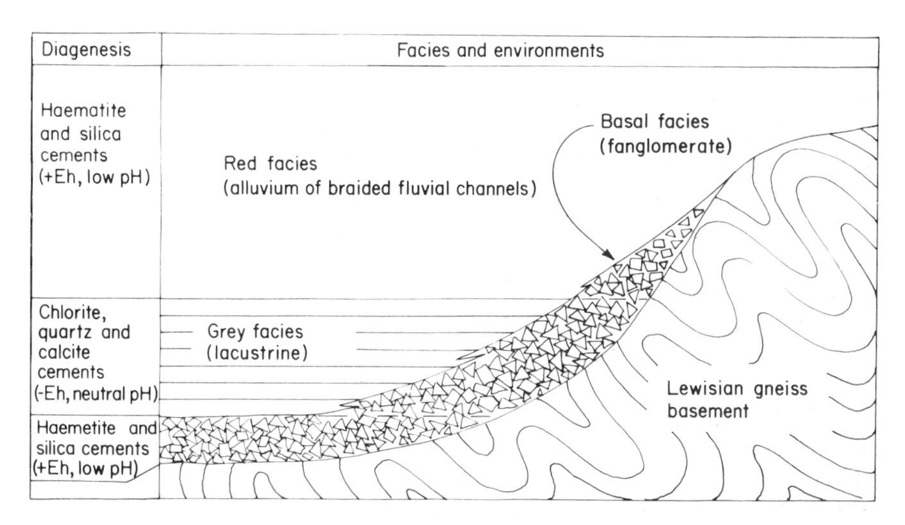

Fig. 43. Diagram showing the correlation between facies and diagenesis in Torridonian (Pre-Cambrian) sediments of north-west Scotland. This suggests that cementation was early and directly related to the chemistry of syndepositional pore fluids. (From Selley, 1966.)

Early red ferric oxide cements can change to grey-green ferrous iron during later burial. This is very characteristic around oil fields in red sandstone reservoirs. The strongly reducing oil field brines often form a grey-green halo in the sandstones of the water zone around the field (e.g. the Lyons Sand of the Denver Basin, Levandowski *et al.*, 1973).

Throughout the redoxomorphic phase of diagenesis porosity of a sand is lost slowly, but this is due primarily to the effects of compaction and dehydration rather than to the chemical effects of diagenesis. This is in contrast to the second phase of sandstone diagenesis.

3. The locomorphic phase

The second phase of sandstone diagenesis primarily involves cementation and is termed the locomorphic stage.

A cement is a crystalline substance which is precipitated within the pores of a sediment after it is deposited. It should be distinguished from the matrix which is microgranular material which occurs in pores and is of syndepositional origin.

The two most common cements of sandstones are silica and carbonates. A host of other authigenic minerals are found in sandstones but they are seldom sufficiently abundant to form the dominant cementing mineral. These minerals include barytes, celestite, anhydrite (gypsum at outcrop), halite, haematite and feldspar.

The net effect of all these mineral cements is to diminish or completely destroy the primary intergranular porosity and the permeability of the sandstone. Because of their abundance and significance, the origin of carbonate, silica and clay cements will now be described.

(i) Carbonate cements

Carbonate cements in sandstones consist of both calcite and dolomite.

Modern sediments have been found with cements of both aragonite and calcite (Allen *et al.*, 1969; Garrison *et al.*, 1969, respectively). These reports indicate that carbonate cements can form at surface temperatures and pressures. Their genesis does not require high temperatures or pressures. These cements are precipitated from solutions which gained their calcium carbonate both from connate water expelled by compaction and from the dissolution of shells. In ancient sandstones aragonite is unknown as a cement due to reversion to calcite, the stable form of calcium carbonate.

Carbonate cements range from fringes of small crystals rimming detrital grains through sparite-filled pores, to single crystals centimetres across which completely envelop the sand fabric. This latter type of texture is termed poikilitic or poikiloblastic (Fig. 44). It is easily identified in hand specimens because the sandstone tends to break along cleavage fractures which twinkle in the sunlight. This is known as "lustre mottling". Calcite cements can, therefore, be present in a sandstone in sufficient quantities to infill all primary intergranular porosity.

Fig. 44. Thin section of Jurassic sandstone from the North Sea showing poikilotopic cement. Polarized light. Note how large calcite crystals, at various shades of extinction, enclose quartz grains. Negligible porosity and permeability ($\times 40$).

Dolomite is the second common type of carbonate cement found in sandstones. It occurs typically in rhomb-shaped crystals which, by themselves, seldom completely destroy porosity.

In argillaceous sandstones microcrystalline calcite, dolomite and siderite are often present within the clay matrix. As previously discussed, the presence of a carbonate cement indicates that the sand has been bathed in alkaline pore fluids.

(ii) Silica cements

Sandstones are commonly cemented to varying degrees by silica. Rarely this is in the form of amorphous colloidal hydrated silica, opal. This occurs in younger rocks at low pressures but sometimes at high temperatures, as in some hot springs. Opal dehydrates with age to microcrystalline quartz, termed chalcedony. This is quite a common cement in sandstones of various ages.

By far the commonest type of silica cement, however, is quartz overgrown in optical continuity on detrital quartz grains. These authigenic overgrowths develop in a variety of styles (Fig. 45). In some sands with high primary porosities the quartz overgrowths form beautiful euhedral faces; extreme cases show bipyramidal terminations over the original detrital grain. Sorby (1880) drew attention to this phenomenon in the Permian Penrith sandstone of England (see also Waugh, 1970a, b).

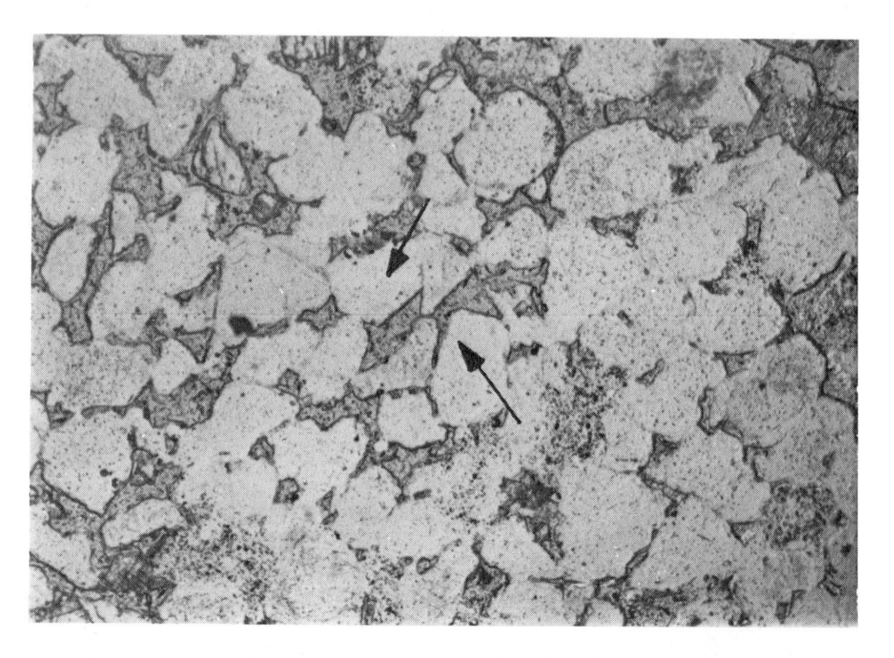

Fig. 45. Thin section of Lossiemouth Sandstone (Permian) Scotland. Intergranular pores partially occluded by angular crystal faces of overgrowths on quartz grains (arrowed) ($\times 40$).

More typically, however, secondary silica overgrows detrital quartz grains in optically continuous jackets which generally conform to the shape of the adjacent pore space.

The genesis of secondary silica cement has been extensively studied because this is the most common type of porosity destroyer in sandstones. Particular attention has been paid to finding the depth below which effective porosity is absent in a particular sedimentary basin. This may be used to predict the "economic basement" below which it would be futile to search for aquifers or hydrocarbon reservoirs.

Attention has been directed towards the source of silica, the physico-chemical conditions which govern its precipitation, and the relationship between silica cementation and pressure solution (Ireland, 1959).

There is no doubt that silica cements may have been precipitated from solutions which derived their silica from organic debris such as radiolaria, diatom tests and siliceous sponge spicules. Likewise some silica-rich solutions must have been expelled from compacting clays. A number of successful attempts have been made to grow silica overgrowths artificially. These have been achieved at high temperatures and pressures (Heald and Renton, 1966; Paraguassu, 1972), and also at normal temperatures and pressures too (Mackenzie and Gees, 1971).

Study of the relationship between secondary silica, porosity and depth of burial is inextricably linked with the phenomenon of pressure solution, or pressure welding. Many thin sections give the impression that quartz grains have been squashed into one another and that, adjacent to the points of contacts, secondary silica has been precipitated. These observations have suggested to many workers that as sand is compacted, silica is dissolved at the points of grain contacts and reprecipitated instantaneously (Fig. 46). Recently, Rittenhouse (1971) has given a quantitative analysis of porosity loss which integrates pressure solution with grain shape and packing.

Taylor (1950) showed how, with increasing depth of burial, the number

Fig. 46. Thin section of sandstone under polarized light showing apparent pressure solution of quartz grains (arrowed) (×50).

of grain contacts per grain increased from about one or two near the surface up to five or more at great depth. Simultaneously Taylor showed how the nature of the grain contacts changed with increasing depth of burial. At shallow depths tangential or point contacts are typical. These grade down into long contacts where grain margins lie snugly side-by-side together. At greater depths still, concavo-convex and sutured grain boundaries prevail where there has been extensive pressure solution. These changes in the number and nature of grain contacts are accompanied by a gradual decrease in porosity (Fig. 47).

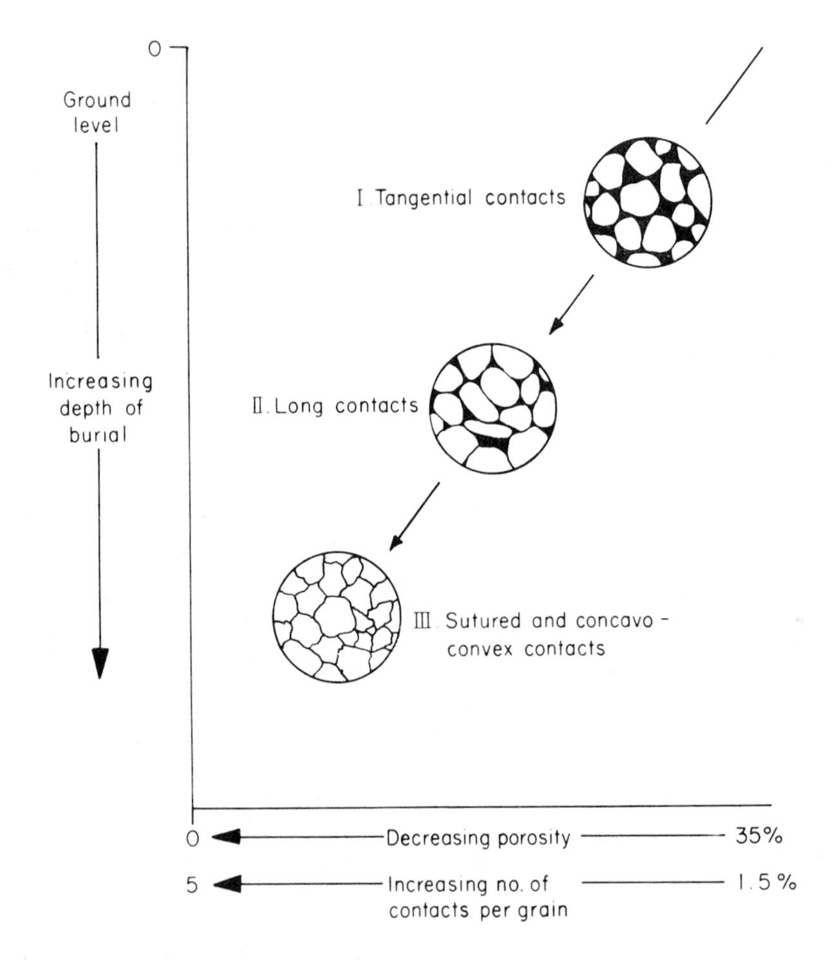

Fig. 47. The destruction of porosity in sandstones with increasing depth. Previously interpreted as largely due to pressure solution, cathodoluminescence shows that diagenetic silica cementation is the main process. (From Sippel, 1968.)

Great care must be taken in studying the relationship of secondary silica to pressure solution. Sippel (1968) and Sibley and Blatt (1976) have shown that cathodoluminescence examination of sands reveals far greater amounts of secondary quartz than examination under a polarizing microscope. Many sands which appear to have lost all porosity by extensive pressure solution have, in fact, lost it by extensive secondary quartz cementation. This calls into question the accuracy of all modal analysis of sandstone composition and most studies of pressure solution and silica cementation.

(iii) Clay cements

Clay mineral reactions also take place during the locomorphic phase. As previously discussed (p. 70), there are three types to consider: kaolin, illite and montmorillonite. The type of clay in a sand has a strong effect on its permeability (Wilson and Pittman, 1977; Guvan *et al.*, 1980). Montmorillonite has the worst effect because its lattice structure can absorb water, expand and destroy permeability (Fig. 48). Illite is the next most destructive clay. It tends to form fibrous crystals which grow radially from the sand grains. These furry illite jackets entwine across the throat passages, thus diminishing permeability (Fig. 49). Kaolin, by contrast, grows as discrete crystals which, though they may diminish porosity by the same amount, volume for volume, as illite, are far less destructive of permeability

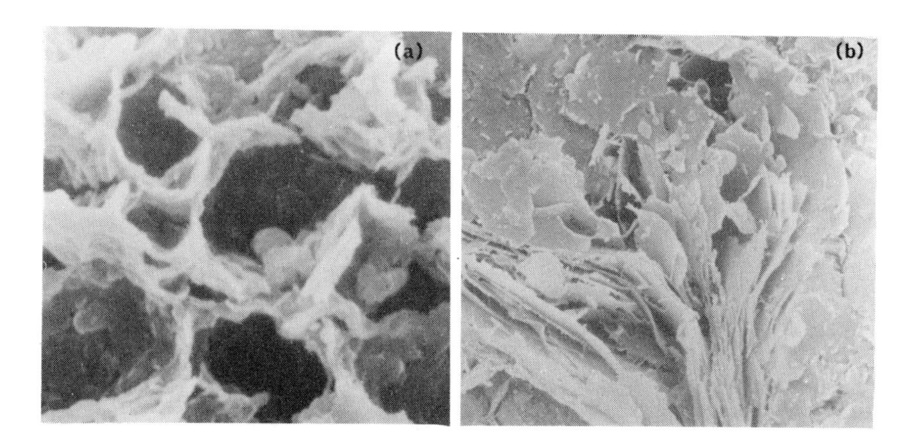

Fig. 48. Scanning electron microscope photographs of montmorillonite clay cements from Niger delta Tertiary sandstones. (a) shows typical honeycomb structure (× 5500). (b) shows swollen clay laminae due to reaction to drilling fluid (× 2400). The expansion of montmorillonite clays destroys permeability. Photos by courtesy of Lambert-Aikhionbare.

(Fig. 50). Whether an oil or gas reservoir may be produced economically may be controlled by its clay mineralogy. In the Lower Permian (Rotliegende) gas fields of the North Sea, illitic cemented sands tend to be subcommercial, and kaolin-cemented ones economic (Stadler, 1973). Thus it is important to be able to predict the distribution of authigenic kaolin and illite.

Fig. 49 (left). Scanning electron microscope photograph of fibrous authigenic illite infilling pore in Westphalian sandstone (× 5400). Photo by courtesy of J. Huggett.
Fig. 50 (right). Scanning electron microscope photograph of kaolin crystals, showing chunky book-like habit, within pore of Westphalian sandstone (× 2000). Photo by courtesy of J. Huggett.

Because of its stability in low pH waters, kaolin tends to be deposited in continental environments; it may be converted to illite during diagenesis in the presence of alkaline connate waters.

Illite, by contrast, is more commonly the detrital clay found in marine deposits. Recrystallization of illite from detrital flakes to furry coatings may take place in alkaline connate waters.

Conversely, kaolin may form authigenically either from the breakdown of detrital grains, such as feldspar, or by the recrystallization of illite fibres to kaolin crystals. These changes occur in low pH conditions in several ways. Meteoric water may flush recently deposited sands during a marine regression, or lithified rocks undergoing weathering. The permeable kaolinitic zone may then be preserved beneath an unconformity.

Kaolinization may also be caused by carbonic-acid rich waters produced by carbonic-acid rich waters produced by the devolatolization of coal beds.

4. The phyllomorphic phase

The third diagenetic phase defined by Dapples (1967), termed phyllomorphic, lies on the boundary between diagenesis and low-grade metamorphism. By the end of the previous locomorphic stage all primary porosity has been lost by cementation. In the phyllomorphic phase of argillaceous sandstones, clays and labile minerals are recrystallized to form muscovite, biotite and chlorite micas leading to the development of incipient schistose textures. In the purer sandstones detrital grains bond tighter together as they begin to change into the metamorphic "quartzites".

5. Porosity gradients in sandstones

As a general statement it is true to say that the porosity of sandstones increases with depth. The porosity of a sandstone at a given depth may be expressed thus (Selley, 1978):

$$\phi^{d} = \phi^{p} - G.D$$

where ϕ^{d} is the porosity at depth d, ϕ^{p} is the original porosity at the surface, G is the porosity gradient, and D is the depth below surface. The original porosity at the surface will depend on depositional process and provenance (as discussed on p. 34). The porosity gradient will be a function of: sandstone composition, pore fluid composition and history, temperature and time.

There are now a number of studies of porosity gradients in sandstone basins, notably by Fuchtbauer (1967), Nagtegaal (1978), Wolf and Chilingarian (1976), Selley (1978), and Magara (1980). Figure 30 shows the envelope of sandstone porosity gradients which have been recorded. It is interesting to note that porosity loss is largely linear with depth and the rapid loss of porosity with shallow burial shown by clays is largely absent. This is because compaction is of less importance in clean sands (see p. 74).

The less mature the mineralogy or texture of a sand the faster it will lose porosity during burial; similarly, porosity is lost faster in basins with higher geothermal gradients than with lower ones. Overpressure can preserve porosity to greater depths than anticipated and the early invasion of hydrocarbons inhibits porosity loss by excluding connate fluids from the pores of a sand (e.g. Philipp *et al.*, 1963; Fuchtbauer, 1967).

It can be seen, therefore, that it is extremely difficult to predict the porosity gradient in a given situation. General statements can be made to the effect that porosity will be preserved to greater depths in cool basins of quartzose sands than in hot volcaniclastic ones. Accurate gradients for a particular case can only be established from well data in which all the variables (mineralogy, facies, geothermal and pressure gradients) are known.

6. Secondary porosity in sandstones

While a sediment is going through the sequence of redoxomorphic, locomorphic and phyllomorphic stages, it may be uplifted at any time and subjected to weathering. In some instances this weathering can be deep and intensive, causing significant increases in the porosity and permeability of the rock (Fig. 51). The term "epidiagenesis" is used for this process which runs counter to the usual tendency for diagenesis to reduce porosity in sandstones (Fairbridge, 1967). The normal processes of weathering are described in Chapter 3. The following account discusses only those aspects which affect the generation of secondary porosity.

First, physical weathering, the release of overburden pressure and mass movement on sloping ground, can generate fracture porosity. At the same time diverse chemical changes can generate secondary porosity by leaching. In carbonate cemented sandstones, ground water rich in humic and other acids will leach out the cement and carry the carbonate away in solution.

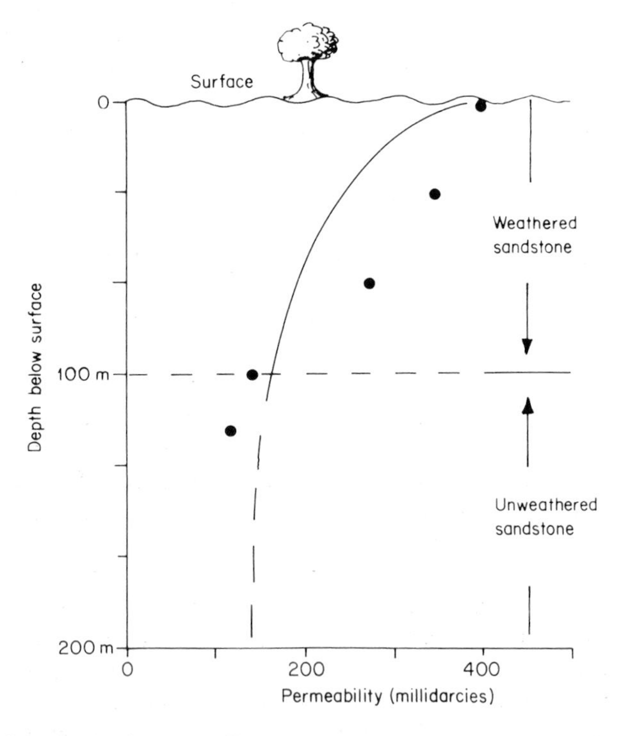

Fig. 51. Diagram showing the effect of solution weathering on the permeability of Mississippian sandstones of Indiana. (From Hrarber and Potter, 1969.) This demonstrates how epidiagenetically produced secondary porosity may be preserved beneath unconformities.

This may leave an unconsolidated sand with a porosity approaching that which it had when it was first deposited.

Weathering profiles in modern deserts provide excellent examples of secondary porosity development in sandstones. Unfortunately due to the great age and, often, long exposure of the rocks, it is not possible to prove the time or climatic conditions which produced the weathering profile seen at the present time. On the intensely weathered bare rock surfaces of the Saharan hamadas, sandstones typically show a dark brown ferruginous layer a few millimetres thick. Beneath this crust two zones may be distinguished. The upper zone is one of increased porosity. The lower zone is one of decreased porosity. In the upper zone iron and carbonate are removed in solution; micas, feldspars and illitic clays are altered to kaolin and the total clay content is reduced by leaching. Detrital silica grains can be corroded in the more porous sand, though silica cementation may occur in the less permeable sands. This upper zone of leaching, and hence increased secondary porosity, varies from decimetres to hundreds of metres in thickness. The second lower zone is one where porosity is decreased by precipitation of the minerals percolating down from the zone above. Silica is the dominant cement. This is often opaline hydrous silica in modern weathering profiles; though this ages to chalcedony in ancient examples. Iron may also be precipitated in this zone, particularly in the form of ferruginous crusts at the contacts between permeable sands and impermeable shales. The overall effect of these reactions is to decrease the porosity and permeability of this zone.

The previous account, based largely on a detailed petrographic study by Hea (1971), is summarized in Fig. 52.

The diagnostic petrographic criteria of solution porosity in sandstones have been reviewed by Schmidt *et al.* (1977).

An understanding of the processes of epidiagenesis, briefly reviewed above, is of some significance in the search for porous sand bodies such as aquifers and hydrocarbon reservoirs. Every modern weathering profile is essentially a potential unconformity and, because of epidiagenesis, pegrographic studies of rocks based on outcrop samples are unrepresentative of the formation as a whole.

Epidiagenetically induced porosity is generally destroyed after burial by compaction and cementation. It can be preserved however by the early invasion of hydrocarbons and/or overpressure. Epidiagenesis is thus one of several factors which make unconformity zones favoured sites for hydrocarbon accumulation (Levorsen, 1967, pp. 333–339).

The Sarir and similar fields in Libya, and many of the Jurassic oil fields of the northern North Sea owe their good reservoir qualities, in part, to epidiagenetic porosity enhancement.

An alternative process to epidiagenesis for the development of secondary porosity has been suggested by Rowsell and De Swardt (1974, 1976). They postulate that the acidic fluid necessary for the leaching may have been produced by the decarboxylation of organic matter, both coal and kerogen dispersed in shales.

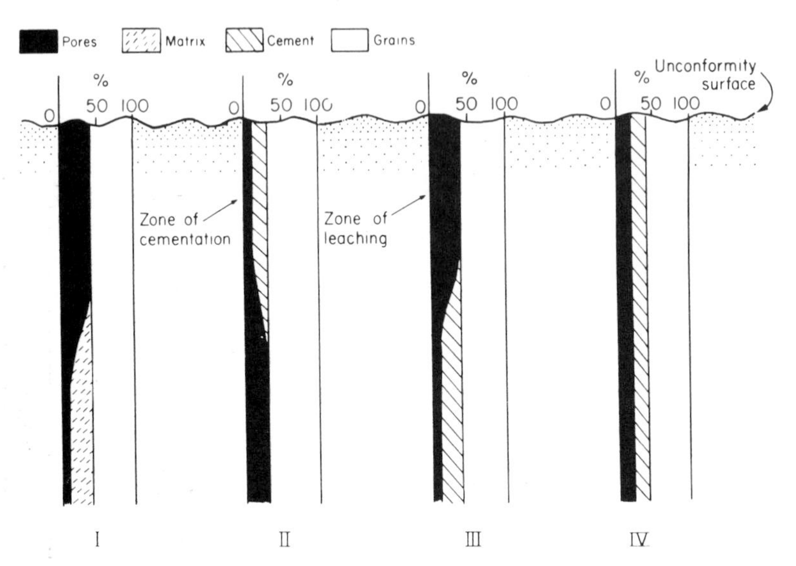

Fig. 52. Varieties of epidiagenesis and porosity development beneath weathering profiles. I. Extensive porosity may form from the weathering of an argillaceous sand. II. Porosity may be destroyed by cementation in a clean friable sand. III. Porosity may form by the solution of calcite cement. IV. Silica cemented sand may undergo little modification when subjected to weathering.

6. Sandstone porosity: Summary

The preceding pages have reviewed the factors which control the petrophysical characteristics of sandstones.

The evolution of porosity in sandstones is much simpler than in carbonates because of the greater chemical stability of silica.

The porosity of a sand is a reflection of its texture, mode of deposition and extent of diagenesis. The grain size, grain shape, sorting and packing of a sediment play an important role in determining primary intergranular porosity (p. 34). Pryor (1973) has shown how these vary for different environments and has documented the spatial variation of porosity and the vectorial variation of permeability in different types of sand body.

The effect of compaction appears to be negligible in sands and most of the observed decrease of porosity with increasing depth is attributable to cementation. Of the various sandstone cements, silica is the most common and is generally the irreversible end of the spectrum of sandstone diagenesis. At any point in its diagenetic evolution, however, a sandstone may be subjected to epidiagenesis. This intensive weathering beneath a potential unconformity generates secondary fracture porosity and solution porosity by leaching of the cement and unstable detrital grains.

Figure 53 summarizes the evolution of porosity in sandstones.

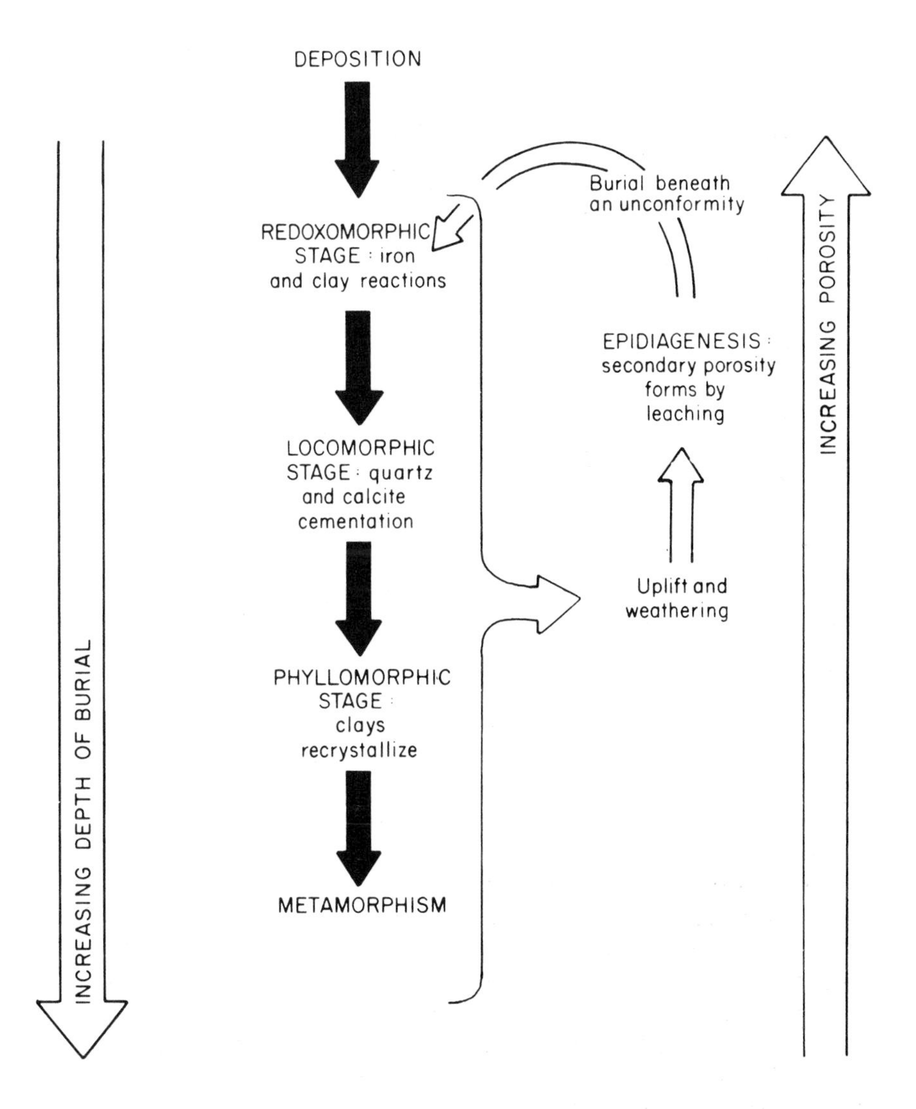

Fig. 53. The relationship between sandstone diagenesis and porosity.

VI. THE RUDACEOUS ROCKS

The rudaceous rocks are sediments at least a quarter of whose volume is made of particles larger than 2 mm in diameter. They grade down through the granulestones into the very coarse sandstones. Traditionally the rudaceous deposits are divided into breccias, whose particles are angular, and conglomerates, whose particles are rounded.

Breccias are rare rocks which occur principally in faults (tectonic breccias) and in some screes. The majority of rudaceous rocks are conglomerates.

The conglomerates can be divided into three groups by their composition. The volcaniclastic conglomerates are termed agglomerates, and have already been briefly described (p. 76). The other two groups are the carbonate conglomerates (known sometimes as the calcirudites) and the terrigenous conglomerates (silici rudites).

Continental carbonate conglomerates are rare because of their solubility in acid ground waters. Marine carbonate conglomerates are more common. The best known examples are probably the "coral rock" boulder beds which form submarine screes around reef fronts. Thin layers of carbonate clasts, often phosphatized, overlie chalk "hardground" horizons (p. 140).

The following account is concerned primarily with the terrigenous conglomerates.

Texturally conglomerates may be conveniently divided into two types: grain-supported and mud-supported conglomerates. The mud-supported rudaceous rocks are properly termed the diamictites. In these the pebbles are seldom in contact with one another, but are dispersed through a finer-grained matrix like currants in a bun. The origin of these rocks are discussed on pp. 199 and 201. It is apparent that the diamictites, or pebbly mudstones, originate from various processes which are not well understood. Some are due to mud flows which occur both in subaerial environments and underwater, where they are sometimes termed "fluxoturbidites". Other diamictites are glacial in origin. Notable examples are the Pleistocene boulder clays.

The second class of conglomerates defined by texture, are grain-supported. In these the individual pebbles touch one another, but the intervening spaces are generally infilled by a matrix of poorly sorted sand and clay. In fluvial conglomerates it is quite likely that this was a primary depositional feature. In marine conglomerates on the other hand, such as those beach gravels that mark marine transgressions, it is more likely that there was no depositional matrix. The matrix that is commonly present is probably due to infiltration from overlying sediment. Conglomerates are deposited with high porosities but, because of their large throat passages, they have excellent permeability. This high permeability enables porosity to be quickly destroyed by matrix infiltration even before any cementation can occur.

Two further classes of conglomerate are defined according to the compositions of their pebbles.

Polymictic conglomerates are composed of pebbles of more than one type. Oligomictic conglomerates have pebbles of only one rock type. Whereas the polymictic conglomerates are of diverse composition, the oligomictic conglomerates are generally quartzose. This is because of the chemical stability of silica. Thus the conglomerate of polycyclic sediments are commonly made of pebbles of vein quartz, quartzite and chert.

Polymictic conglomerates are generally the product of aggradation where tectonically active source areas shed off wedges of fanglomerates.

Oligomictic conglomerates by contrast are generally the product of degradation where tectonic stability allows extensive reworking to produce the laterally extensive basal conglomerates which characterize major unconformities.

A third bipartite division of conglomerates is according to the source of the pebbles. Extraformational or exotic conglomerates are composed of pebbles which originated outside the depositional basin.

Infraformational conglomerates contain pebbles of sediment which originated within the depositional basin. The majority of limestone conglomerates are intraformational in origin for the reason already stated. Infraformational sand conglomerates are rare because unconsolidated sand lacks cohesion and disaggregates on erosion. Mud intraformational conglomerates are commonly called "shale flake" or "shale pellet"

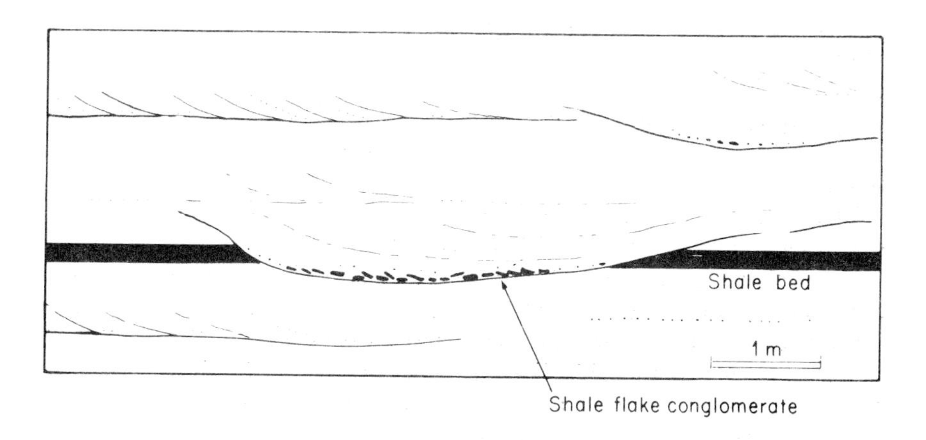

Fig. 54. Intraformational conglomerate of claystone pellets. The channel cuts through a clay unit demonstrating the penecontemporaneous origin of the conglomerate.

Table X
Nomenclature of conglomerates

I. Texture	*Orthoconglomerate*—grain supported *Paraconglomerate* (syn. diamictite)—mud supported
I. Composition	*Polymictic*—pebbles composed of several rock types *Oligomictic*—pebbles composed of one rock type
III. Source	*Intraformational*—pebbles originate within the basin *Extraformational* (syn. exotic)—pebbles extrabasinal in origin

There is no satisfactory classification of conglomerates. Nomenclature and description is based on the criteria of texture, composition and source.

conglomerates. These are volumetrically insignificant occurring in beds generally only one or two clasts thick. Sedimentologically they are significant, however, because they indicate penecontemporaneous erosion close to the site of deposition. Intraformational shale pellet conglomerates are often present at the base of turbidite units and of channels (Fig. 54).

This brief review of conglomerates shows that there is no consistent scheme of nomenclature and classification. Table X shows how descriptive terminology is based on bipartate divisions of texture, composition and source.

VII. REFERENCES

Allen, R. C., Gavish, E., Friedman, G. M. and Sanders, J. E. (1969). Aragonite-cemented sandstone from outer continental shelf off Delaware Bay. *J. sedim. Petrol.* **39**, 136–149.

Amstutz, G. C. and Bubinicek, L. (1967). Diagenesis in sedimentary mineral deposits. *In* "Diagenesis in Sediments" (G. Larsen and G. V. Chilingar, Eds), 417–475. Elsevier, Amsterdam.

Baars, D. L. (1961). Permian blanket sandstones of Colorado Plateau. *In* "Geometry of Sandstone Bodies" (J. A. Peterson, and J. C. Osmond, Eds), 79–207. Am. Ass. Petrol. Geol.

Bennacef, A., Beuf, S., Biju-Duval, B., de Charpal, O., Gariel, O. and Rognon, P. (1971). Example of cratonic sedimentation: Lower Paleozoic of Algerian Sahara. *Bull. Am. Ass. Petrol. Geol.* **55**, 25–245.

Bjerkli, K. and Ostmo-Saeter, J. S. (1973). Formation of glauconite in foraminiferal shells on the continental shelf off Norway. *Mar. geol.* **14**, 169–178.

Brenchley, P. J. (1969). Origin of matrix in Ordovician greywackes, Berwyn Hills, North Wales. *J. sedim. Petrol.* **39**, 1297–1301.

Brogniart, A. (1826). L'arkose, caractères mineralogiques et histoire géognostique de cette roche. *Annls Sci. nat.* Paris, **8**, 113–163.

Burst, J. F. (1969). Diagenesis of Gulf Coast clay sediments and its possible relation to petroleum migration. *Bull. Am. Ass. Petrol. Geol.* **53**, 73–93.

Cummins, W. A. (1962). The greywacke problem. *Lpool Manchr geol. J.* **3**, 51–72.
Dapples, E. C. (1967). Diagenesis of sandstones. *In* "Diagenesis in Sediments" (G. Larsen and C. V. Chilingar, Eds), 91–125. Elsevier, Amsterdam.
Davidson, C. F. (1965). A possible mode of origin of strata-bound copper ores. *Econ. Geol.* **60**, 942–954.
Donell, J. R., Culbertson, W. C. and Cashion, W. B. (1967). Oil shale in the Green River formation. Proc. 7th Wld Petrol. Cong. Mexico, No. 3, 699–702.
Dott, R. H. (1964). Wacke, greywacke and matrix—what approach to immature sandstone classification. *J. sedim. Petrol.* **34**, 625.
Dott, R. H. and Reynolds, M. J. (1969). "Sourcebook for Petroleum Geology". Am. Ass. Petrol. Geol. Mem. No. 5, 471pp.
Duncan, D. C. (1967). Geologic setting of oil shale deposits and world prospects. Proc. 7th Wld Petrol. Cong. Mexico, No. 3, 659–667.
Dunoyer de Segonzac, G. (1968). The birth and development of the concept of diagenesis (1866–1966). *Earth Sci. Rev.* **4**, 153–201.
Fairbridge, R. W. (1967). Phases of diagenesis and authigenesis. *In* "Diagenesis in Sediments" (G. Larsen and G. V. Chilingar, Eds), 19–28. Elsevier, Amsterdam.
Folk, R. L. (1968). "Petrology of Sedimentary Rocks" Hemphill's Book Store, Austin Texas. 170pp.
Folk, R. L. (1976). Reddening of desert sands. *J. sedim. Petrol.* **46**, 604–613.
Friend, P. F. (1966). Clay fractions and colours of some Devonian red beds in the Catskill Mountains, U.S.A. *Q. Jl geol. Soc. Lond.* **122**, 273–292.
Fuchtbauer, H. (1967). Influence of different types of diagenesis on sandstone porosity. Proc. 7th Wld Petrol. Cong. Mexico, 353–369. Elsevier, Amsterdam.
Garrison, R. E., Luternauer, J. L., Grill, E. V., MacDonald, R. D. and Murray, J. W. (1969). Early diagenetic cementation of Recent sands, Fraser River delta, British Columbia. *Sedimentology* **12**, 27–46.
Glennie, K. W. (1972). Permian Rotliegendes of Northwest Europe interpreted in light of modern desert sedimentation studies. *Bull. Am. Ass. Petrol. Geol.* **56**, 1048–1071.
Greensmith, J. T. (1968). Palaeogeography and rhythmic deposition in the Scottish oil-shale group. Proc. UN Symp. Dev. Util. Oil Shale Res. Tallin. 16pp.
Grim, R. E. (1968). "Clay Mineralogy" (2nd Edition). McGraw-Hill, New York. 596pp.
Guvan, N., Hower, W. F. and D. K. Davies (1980). Nature of authigenic illites in Sandstone Reservoirs. *J. sedim. Petrol.* **50**, 761–766.
Harbaugh, J. W. and Merriam, D. F. (1968). "Computer Applications in Stratigraphic Analysis" John Wiley, Chichester. 282pp.
Hatch, F. H., Rastall, R. H. and Greensmith, J. T. (1971). "Petrology of the Sedimentary Rocks" (Revised 5th Edition). Murby, London. 502pp.
Hea, J. P. (1971). Petrography of the Paleozoic-Mesozoic sandstones of the southern Sirte basin, Libya. *In* "The Geology of Libya" (C. Grey, Ed.) 107–125. University of Libya.
Heald, M. T. and Renton, J. J. (1966). Experimental study of sandstone cementation. *J. sedim. Petrol.* **36**, 977–991.
Helmbold, R. (1952). Beitrag zur Petrographie der Tanner Grauwacken. *Heidelb. Beitr. Miner. Petrogr.* **3**, 253–280.
Hraber, S. V. and Potter, P. E. (1969). Lower West Baden (Mississippian) sandstone body of Owen and Greene counties, Indiana. *Bull. Am. Ass. Petrol. Geol.* **53**, 2150–2160.

Hubert, J. F. (1964). Textural evidence for deposition of many western North Atlantic deep-sea sands by ocean-bottom currents rather than turbidity currents. *J. Geol.* **72**, 757–785.

Huckenholz, H. G. (1963). Mineral composition and textures in greywackes from the Harz Mountains (Germany) and in arkoses from the Auvergne (France). *J. sedim. Petrol.* **33**, 914–918.

Hunt, J. M. (1979). "Petroleum Geochemistry and Geology." Freeman, San Francisco. 617pp.

Ireland, H. A. (Ed.) (1959). Silica in Sediments. *Spec. Publs Soc. econ. Palaeont. Miner., Tulsa.* **7**, 110pp.

Keller, W. D. (1970). Environmental aspects of clay minerals. *J. sedim. Petrol.* **40**, 788–813.

Klein, G. de Vries (1963). Analysis and review of sandstone classifications in the North American geological literature. *Bull. geol. Soc. Am.* **74**, 555–576.

Krynine, P. D. (1935). Arkose deposits in the humid tropics, a study of sedimentation in southern Mexico. *Am. J. Sci.* **29**, 353–363.

Larsen, G. and Chilingar, G. V. (Eds) (1962). "Diagenesis in Sediments" Elsevier, Amsterdam.

Levandowski, D., Kaley, M. E., Silverman, S. R. and Smalley, R. G. (1973). Cementation in Lyons sandstone and its role in oil accumulation, Denver Basin, Colorado. *Bull. Am. Ass. Petrol. Geol.* **57**, 2217–2244.

Levorsen, A. I. (1967). "Geology of Petroleum" W. H. Freeman, San Francisco. 724pp.

Mackenzie, F. T. and Gees, R. (1971). Quartz: synthesis at earth-surface conditions. *Science, N.Y.* **3996**, 533–535.

Magara, K. (1980). Comparison of porosity–depth relationships of shale and sandstone. *Jl Petrol. Geol.* **3**, 175–185.

Mattiatt, B. (1960). Beitrag zur Petrographie der Oberharzer Kulmgrauwacke. *Heidelb. Beitr. Miner. Petrogr.* **7**, 242–280.

Millot, G. (1970). "Geology of Clays" Springer-Verlag, Berlin. 429pp.

Mortland, M. M. and Farmer, V. C. (1978). "International Clay Conference." Developments in Sedimentology Vol. 27. Elsevier, Amsterdam. 662pp.

Muller, G. (1967). Diagenesis in argillaceous sediments. *In* "Diagenesis in Sediments" (G. Larsen and G. V. Chilingar, Eds), 127–178. Elsevier, Amsterdam.

Nagtegaal, P. J. C. (1978). Sandstone-framework instability as a function of burial diagenesis. *J. geol. Soc. Lond.* **135**, 101–106.

Odin, G. S. (1972). Observations on the structure of glauconite vermicular pellets: a description of the genesis of these granules by neoformation. *Sedimentology* **19**, 285–294.

Paraguassu, A. B. (1972). Experimental silicification of sandstone. *Bull. geol. Soc. Am.* **83**, 2853–2858.

Pettijohn, F. J. (1957). "Sedimentary Rocks" Harper Bros, New York. 718pp.

Pettijohn, F. J., Potter, P. E. and Siever, R. (1972). "Sand and Sandstone" Springer-Verlag, Berlin. 618pp.

Philipp, W., Drong, H. J., Fuchtbauer, H., Haddenhorst, H. G. and Jankowsky, W. (1963). The history of migration in the Gifhorn trough (N.W. Germany). Proc. 6th Wld Petrol. Cong. Frankfurt. Sect. I, paper 19, 457–481.

Porrenga, D. H. (1966). Clay minerals in Recent sediments of the Niger Delta. Clays Clay Miner. Proc. 14th Natl Conf. 221–233.

Potter, P. E., Maynard, B., and Pryor, W. A. (1980). "Sedimentology of Shales." Springer-Verlag, New York. 306pp.

Powers, M. C. (1967). Fluid release mechanisms in compacting marine mudrocks and their importance in oil exploration. *Bull. Am. Ass. Petrol. Geol.* **51**, 162–189.

Price, N. B. and Duff, P. McL. D. (1969). Mineralogy and chemistry of tonsteins from Carboniferous sequences in Great Britain. *Sedimentology* **13**, 45–69.

Pryor, W. A. (1973). Permeability-porosity patterns and variations in some Holocene sand bodies. *Bull. Am. Ass. Petrol. Geol.* **57**, 162–189.

Rittenhouse, G. (1971). Pore space reduction by solution and cementation. *Bull. Am. Ass. Petrol. Geol.* **55**, 80–91.

Ross, C. S., and Smith, L. R. (1961). Ash-flow tuffs—their origin, geologic relationship and identification. *Prof. Pap. U.S. geol. Surv.* **366**, 81pp.

Rowsell, D. M. and De Swardt, A. M. J. (1974). Secondary leaching porosity in Middle Ecca Sandstones. *Trans. geol. Soc. S. Afr.* **77**, 131–140.

Rowsell, D. M. and De Swardt, A. M. J. (1976). Diagenesis in Cape and Karoo sediments, South Africa, and its bearing on their hydrocarbon potential. *Trans. geol. Soc. S. Afr.* **79**, 81–153.

Schmidt, V., McDonald, D. A., and R. L. Platt (1977). Pore geometry and reservoir aspects of secondary porosity in sandstones. *Bull. Can. Petrol. Geol.* **25**, 271–290.

Selley, R. C. (1966). Petrography of the Torridonian rocks of Raasay and Scalpay, Inverness-shire. *Proc. geol. Ass. Lond.* **77**, 293–314.

Selley, R. C. (1978). Porosity gradients in North Sea oil-bearing sandstones. *J. geol. Soc. Lond.* **135**, 119–131.

Shaw, H. F. (1980). Clay minerals in sediments and sedimentary rocks. *In* "Developments in Petroleum Geology" Vol. 2 (G. D. Hobson, Ed.). Applied Science, Barking. 53–85.

Shepard, F. P. (1954). Nomenclature based on sand-silt-clay ratios. *J. sedim. Petrol.* **24**, 151–158.

Sibley, D. F., and Blatt, H. (1976). Intergranular pressure solution and cementation of the Tuscarora Quartzite. *J. sedim. Petrol.* **46**, 881–896.

Sippel, R. F. (1968). Sandstone petrology, evidence from luminescence petrography. *J. sedim. Petrol.* **28**, 530–554.

Skempton, A. W. (1970). The consolidation of clays by gravitational compaction. *J. geol. Soc. Lond.* **25**, 375–412.

Slaughter, M. and Earley, J. W. (1965). Mineralogy and geological significance of the Mowry Bentonite, Wyoming. *Spec. Pap. geol. Soc. Am.* **83**, 116pp.

Sorby, H. C. (1880). On the structure and origin of non-calcareous stratified rocks. *Proc. geol. Soc. Lond.* **36**, 46–92.

Stadler, P. J. (1973). Influence of crystallographic habit and aggregate structure of authigenic clay minerals on sandstone permeability. *Geologie Mijnb.* **52**, 217–220.

Steinmann, G. (1926). Die ophiolitischen Zoren in den Mediterranen Kettengebirgen. C. r. XIV int. Geol. Cong. Madrid.

Sturt, B. A. (1961). Discussion in: some aspects of sedimentation in orogenic belts. *Proc. geol. Soc. Lond.* **1587**, 78.

Surdam, R. C. and Wolfbauer, C. A. (1973). Depositional environment of oil shale in the Green River formation, Wyoming (Abs.). *Bull. Am. Ass. Petrol. Geol.* **57**, 808.

Taylor, J. M. (1950). Pore space reduction in sandstones. *Bull. Am. Ass. Petrol. Geol.* **34**, 701–716.

Terzaghi, K. (1936). The shearing resistance of saturated soils. Proc. 1st Int. Conf. Soil Mechanics pp. 54–56.

Tissot, B. P. and Welte, D. H. (1978). "Petroleum Formation and Occurrence." Springer-Verlag, Berlin. 538pp.

Van Houten, F. B. (1968). Iron oxides in red beds. *Bull. geol. Soc. Am.* **79**, 399–416.

Van Houten, F. B. (1973). Origin of red beds, a review—1961–1972. *A. Rev. Earth planet. Sci.* **1**, 39–61.

Walker, T. R. (1967). Formation of red beds in modern and ancient deserts. *Bull. geol. Soc. Am.* **78**, 353–368.

Walker, T. R. (1976). Diagenetic origin of Continental red beds. *In* "The Continental Permian in Central, West and Southern Europe", (H. Falke, Ed.) Reidel, Dordrecht-Holland. 240–282.

Waugh, B. (1970a.). Petrology, provenance and silica diagenesis of the Penrith Sandstone (Lower Permian) of northwest England. *J. sedim. Petrol.* **40**, 1226–40.

Waugh, B. (1970b). Formation of quartz overgrowths revealed by scanning electron microscopy. *Sedimentology* **14**, 309–320.

White, G. (1961). Colloid phenomena in sedimentation of argillaceous rocks. *J. sedim. Petrol.* **31**, 560–565.

Wilson, M. D. and Pittman, E. D. (1977). Authigenic clays in sandstones: recognition and influence on reservoir properties and paleoenvironmental analysis. *J. sedim. Petrol.* **47**, 3–31.

Wolf, K. H. and Chilingarian, G. V. (1976). "Compaction of Coarse-grained Sediments" Vol. 2. Elsevier, Amsterdam. 808pp.

Yen, T. F. and Chilingarian, G. V. (Eds) (1976). "Oil Shale" Elsevier, Amsterdam. 292pp.

5 Autochthonous Sediments

I. INTRODUCTION

The second great group of sedimentary rocks are variously referred to as the chemical or autochthonous sediments. These are rocks which form within a depositional basin, as opposed to the terrigenous sands and muds which originate outside the basin.

The chemical rocks are sometimes divided into organic and inorganic groups. Carbonate skeletal sands are a good example of the first type; evaporites of the second. Biochemical research shows, however, that there is no clearly defined boundary between these two groups. For example, apparently spontaneous precipitates of lime mud occur in response to chemical changes in seawater due to the activity of plankton and bacteria (see p. 125). It is a matter of semantics whether such sediments are organic or inorganic in origin.

Though the chemical rocks are directly precipitated within a sedimentary basin they can be subjected to minor reworking. There are, therefore, examples of detrital (i.e. clastic) chemical sediments, but these must be carefully distinguished from the detrital terrigenous sediments of extra-basinal origin.

Table XI shows the main chemical sedimentary rocks. Of these the carbonates are volumetrically the most important. The carbonate rocks include the limestones, made of calcium carbonate ($CaCO_3$), and the dolomites, composed of the mineral dolomite ($CaMg(CO_3)_2$). Some pedants prefer to restrict the term dolomite to the mineral and refer to the rock as "dolostone". The carbonate rocks form by organic processes, by direct inorganic precipitation and by diagenesis. Carbonates are important aquifers and hydrocarbon reservoirs, because of the high porosity which they sometimes contain. Porosity distribution is complex, however, and has merited considerable research. For these reasons carbonate rocks will subsequently be described and discussed in some detail.

A second important group of the chemical sediments are the evaporites. These form both by inorganic crystallization and by diagenesis. The most

Table XI
The major types of chemical rocks

Carbonates	Dolomites Limestones
Evaporites	Anhydrite/gypsum Halite/rock salt Potash salts, etc.
Siliceous rocks	— Chert, radiolarite, novaculite
Carbonaceous rocks	Humic group—coal series Sapropelitic group—oil shales and cannel coals
Sedimentary ironstones	
Phosphates	

The chemical rocks are those which form within the depositional environment. They include direct chemical precipitátes, such as some evaporites, and formation by organic processes, such as coal and shell limestones. Not all chemical sediments are syndepositional. Diagenetic processes are important in the genesis of some evaporites, dolomites, cherts, ironstones and phosphates.

common evaporite mineral is anhydrite, calcium sulphate ($CaSO_4$), and its hydrated product gypsum ($CaSO_4.2H_2O$). Less common evaporites are rocksalt or halite (NaCl) and a whole host of potassium and other salts.

The sedimentary ironstones are a much less abundant chemical sedimentary rock. They also form both as precipitates and by diagenesis. The common ferrugenous sedimentary minerals are pyrite (FeS_2) and siderite ($FeCO_3$). The sedimentary iron ores, however, include the oxides goethite and haematite, and chamosite, a complex ferrugenous alumino-hydrosilicate. These rocks are discussed in the section on synsedimentary ores (p.391).

Phosphorite is a sedimentary rock composed dominantly of phosphates. They form largely during early diagenesis in sediment, immediately beneath the sediment/water interface, aided by reworking and concentration of incipient pellets and concretions. The phosphate minerals, like the ironstones, are chemically very complex. Examples include collophane, dahllite, francolite and fluorapatite. These are calcium phosphates combined with various other radicals. Phosphorites are discussed on p. 150.

Coal is a sedimentary rock formed entirely by biochemical processes. It originates from the accumulation of vegetable detritus as peat, under anaerobic conditions, in swamps, marshes, meres and pools. Extensive coal beds are especially characteristic of ancient deltaic deposits (p. 290).

The last of the chemical sediments to be considered are the siliceous rocks, termed chert. These are composed largely of microcrystalline quartz and chalcedony, a variety of silica with a spherulitic habit. Hydrous silica, opal, occurs in Tertiary rocks, but appears to dehydrate on burial to silica (Ernst and Calvert, 1969).

Cherts occur in several situations. Bedded cherts are typical of basinal facies, often interbedded with dark shales, turbidites and pillow lavas. These cherts are commonly composed largely of the tests of radiolaria.

The occurrence of siliceous oozes of radiolarian and diatom tests on modern ocean floors has lead to the assumption of a bathyal origin for ancient radiolarian cherts (see Grunau, 1965 for a review). One of the best-known examples of such a deposit is the Palaeozoic Caballos Novaculite of the Marathon basin, Texas. While there is general agreement that this formed as a primary chert there is debate as to its environment of deposition. McBride and Thomson (in Folk and McBride, 1976) argue for a bathyal depth, while Folk (1973) and in Folk and McBride (1976) argues for an intertidal origin. Bedded cherts are also known from lacustrine deposits. For example, silicified limestones with the freshwater gastropod *Hydrobia* occur in the Continental Mesozoic beds of the Kufra basin, Libya. Bedded cherts of uncertain origin occur in the Pre-Cambrian cherty ironstone formations (see p. 149).

It has often been remarked that cherts tend to be associated with contemporaneous volcanic activity. It has been argued that vulcanism releases large amounts of silica into the environment. This encourages "blooms" of silica-secreting organisms—such as radiolaria, diatoms and sponges—and thus generates chert formation (e.g. Wenk, 1949; Khvorova, 1968).

Another distinctive type of silica rock is nodular chert. This is especially characteristic of fine-grained limestone, but also occurs in sandstones. Nodular chert beds are commonly found in the late Cretaceous and Tertiary chalks of the Middle East and Europe. This type of chert occurs in irregular rounded nodules of eccentric shape and low sphericity. They are generally concentrated along beds and sometimes plug burrows or replace fossils. Less commonly they infill joints and fractures.

The genesis of chert, both bedded and nodular, has attracted considerable speculation. The main point at issue is whether chert originates as a colloidal silica gel on the sediment/water interface, or whether it occurs by replacement.

Undoubtedly both processes can take place in different situations. Primary precipitation of opal has been described in ephemeral Australian lakes (Peterson and Von der Borch, 1965). Some ancient cherts are also undoubtedly penecontemporaneous with sedimentation, as, for example, the synsedimentary chert breccia described by Carozzi and Gerber (1978). Conversely, the presence of chert in fractures and of chert-replaced carbonate skeletal debris provide conclusive evidence of a secondary origin. Clearly cherts are polygenetic.

II. CARBONATES

A. Introduction

The carbonate rocks are extremely complex in their genesis, diagenesis and petrophysics. There are a number of reasons for this (Ham and Pray, 1962). Carbonate rocks are intrabasinal in origin. Unlike terrigenous sediments they are easily weathered and their weathering products are transported as solutes. Carbonate rocks are, therefore, deposited at or close to their point of origin. Most carbonate rocks are organic in origin. They contain a wide spectrum of particle sizes, ranging from whole shells to lime mud of diverse origin. These sediments are deposited with a high primary porosity. The carbonate minerals are chemically unstable, however. This combination of high primary porosity and permeability, coupled with chemical instability, is responsible for the complicated diagenesis of carbonate rocks, and hence for the problems of locating aquifers and hydrocarbon reservoirs within them.

The following brief account of carbonates first defines their mineralogy, then describes their petrography and classification and concludes by showing the relationship between diagenesis and porosity development.

The carbonate rocks have generated a vast literature. Key works to which the reader is directed for detailed accounts include Chilingar *et al.* (1967a), Chilingar *et al.* (1967b), Bathurst (1975) and Milliman (1974).

B. The Carbonate Minerals

It is necessary to be familiar with the common carbonate minerals to understand the complex diagenetic changes of carbonate rocks.

Calcium carbonate ($CaCO_3$) is the dominant constituent of modern carbonates and ancient limestones. It occurs as two minerals, *aragonite* and *calcite*. Aragonite crystallizes in the orthorhombic crystal system, while calcite is rhombohedral. Calcite forms an isomorphous series with *magnesite* ($MgCO_3$). A distinction is made between high and low-magnesium calcite, with the boundary being arbitrarily set at 10 mol per cent.

Ancient limestones are composed largely of low magnesium calcite, while modern carbonate sediments are made mainly of aragonite and high-magnesian calcite. Aragonite is found in many algae, lamellibranchs and bryozoa. High magnesian calcite occurs in echinoids, crinoids, many foraminifera, and some algae, lamellibranchs and gastropods. An isomorphous series exists between calcite and magnesite ($MgCO_3$). A distinction is generally made between high and low-magnesian calcite with an arbitrary division at about 10 mol per cent $MgCO_3$.

Skeletal aragonite and calcite also contain minor amounts of strontium, iron and other trace elements. The relationship between carbonate secreting organisms, the mineralogy of their shells and their contained trace elements has been studied in detail (e.g. Lowenstam, 1963; Milliman, 1974). These factors are important because their variation and distribution play a controlling part in the early cementation of skeletal sands (Friedman, 1964).

Dolomite is another important carbonate mineral, giving its name also to the rock. Dolomite is calcium magnesium carbonate $Ca.Mg(CO_3)_2$. Isomorphous substitution of some magnesium for iron is found in the mineral termed ferroan dolomite or ankerite $Ca(MgFe)(CO_3)_2$.

Unlike calcite and aragonite, dolomite does not originate as skeletal material. Dolomite is generally found either crystalline, as an obvious secondary replacement of other carbonates or as a primary or penecontemporaneous replacement mineral in cryptocrystalline form. The problem of dolomite genesis will be elaborated later.

Siderite, iron carbonate ($FeCO_3$), is one of the rarer carbonate minerals. It occurs, apparently as a primary precipitate, in ooliths. These sphaerosiderites, as they are termed, are found in rare restricted marine and freshwater environments. Sphaerosiderite is often associated with the hydrated ferrous aluminosilicate, chamosite, in sedimentary iron ores (see p. 392). Siderite also occurs as thin bands and horizons of concretions in argillaceous deposits, especially in deltaic facies. It is not uncommon to find siderite bands contorted and fractured by slumping. Siderite clasts are also found in intraformational conglomerates. These facts suggest that siderite forms diagenetically during early burial while the sediments are still uncompacted. Its formation is favoured by alkaline reducing conditions (Fig. 42).

Table XII summarizes the salient features of the main carbonate minerals.

To conclude this brief summary of a large topic, the following points are important. The principal carbonate minerals are the calcium carbonates, calcite and its unstable polymorph aragonite; and dolomite, calcium magnesium carbonate. Modern carbonate sediments are composed of both aragonite and calcite. Only calcite, the more stable variety, occurs in lithified limestones. Dolomite does not occur as a biogenic skeletal mineral. It forms as a secondary replacement mineral or, rarely, by primary precipitation or penecontemporaneous replacement of other carbonates.

C. The Physical Components of Carbonate Rocks

Carbonate rocks, like sandstones, have four main components: grains, matrix, cement and pores. Unlike sandstones though, the grains of

Table XII
Summary of the common carbonate minerals

Mineral	Formula	Crystal system	Occurrence
Aragonite	$CaCO_3$	Orthorhombic	Present in certain carbonate skeletons. Unstable and reverts to stable polymorph calcite
Calcite	$CaCO_3$	Hexagonal	Present in certain carbonate skeletons, as mud (micrite) and cement (sparite)
Magnesite	$MgCO_3$	Hexagonal	Present in minor amounts within the lattices of skeletal aragonites and calcites
Dolomite	$CaMg(CO_3)_2$	Hexagonal	Largely as a crystalline diagenetic rock, also penecontemporaneously associated with evaporites
Ankerite (ferroan dolomite)	$Ca(MgFe)(CO_3)_2$	Hexagonal	A minor variety of dolomite
Siderite	$FeCO_3$	Hexagonal	Found as concretions and ooliths (sphaerosiderites)

carbonate rocks, though commonly monominerallic, are texturally diverse and polygenetic. The various grain types, matrix and cement will now be described. They are tabulated in Table XIII.

1. Grains

Grains are the particles which support the framework of a sediment. They are thus generally of sand grade or larger. As Table XIII and Fig. 55 show, carbonate grains are of many types. These will now be briefly described.

First are the detrital grains. These are of two types. They include rock fragments, or *lithoclasts*. These are grains of non-carbonate material which originated outside the depositional basin. Quartz grains are a typical example of lithoclasts and, as they increase in abundance, limestones grade into sandy limestones and thence to calcareous sandstones.

The second type of detrital grains are *intraclasts*. These are fragments of reworked carbonate rock which originated within the depositional basin. Early cementation followed by penecontemporaneous erosion is a common feature of carbonate rocks, and is responsible for the generation of intraclasts. Intraclasts may range from sand size, via intraformational

Table XIII
The main components of carbonate rocks

I.	Grains	(a) Detrital grains	lithoclasts; intraclasts
		(b) Skeletal grains	
		(c) Peloids (including faecal pellets)	
		(d) Lumps	composite grains; algal lumps
		(e) Coated grains	ooliths; pisoliths; algally encrusted grains
II.	Matrix	micrite; clay	
III.	Cement —	sparite	
IV.	Pores		

After Leighton and Pendexter (1962).

conglomerates of penecontemporaneously cemented "beach rock", to, arguably, fore-reef slump blocks.

The most important of all grain types is *skeletal detritus*. As pointed out in the previous section, this is composed of aragonite or calcite with varying amounts of trace elements. The actual crystal habit of skeletal matter is varied too, ranging from the acicular aragonite crystals of lamellibranch shells to the single calcite crystals of echinoid plates. The size of skeletal particles is naturally very variable, ranging down from the largest shell to individual disaggregated microscopic crystals (Fig. 56). Continued abrasion of skeletal debris by wave and current action and by biological processes, such as boring, is responsible for the very poor textural sorting characteristic of carbonate sediments.

Peloids are a third major grain type (Fig. 56). These are grains of structureless cryptocrystalline carbonate (McKee and Gutschick, 1969). This is a useful non-genetic term. Studies of modern carbonate sediments show that peloids form by a variety of processes. Many invertebrates excrete lime mud faecal pellets and this is probably the most significant process. Others include the micritization of skeletal grains by endolithic algae. In certain environments, notably lagoons and sheltered embayments, peloids are sufficiently abundant to be a dominant rock builder.

Lumps are another grain type. These are botryoidal grains which are composed of several peloids held together. They are sometimes termed composite grains or "grapestone". Grains such as these are probably formed by the reworking of peloidal sediment that has already undergone some lithifaction. They are thus nascent intraclasts.

Last of the grain types to consider are the coated grains. These are grains

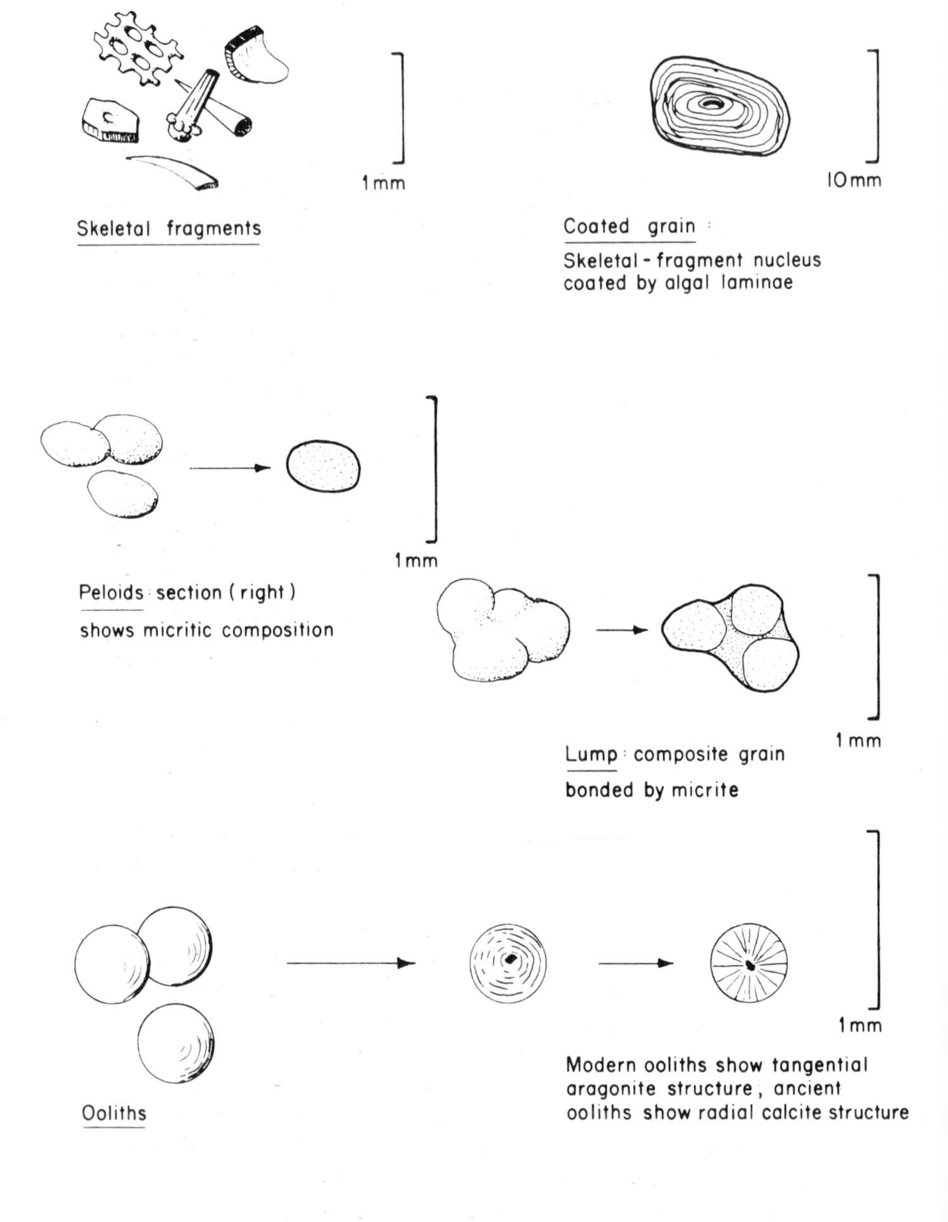

Fig. 55. Some of the principal carbonate grain types.

which show a concentric or radial arrangement of crystals about a nucleus. The commonest coated grains are ooids or *ooliths*. These are rounded grains of medium to fine grain size which generally occur gregariously in sediments termed oolites, devoid of other grain types or matrix. Modern

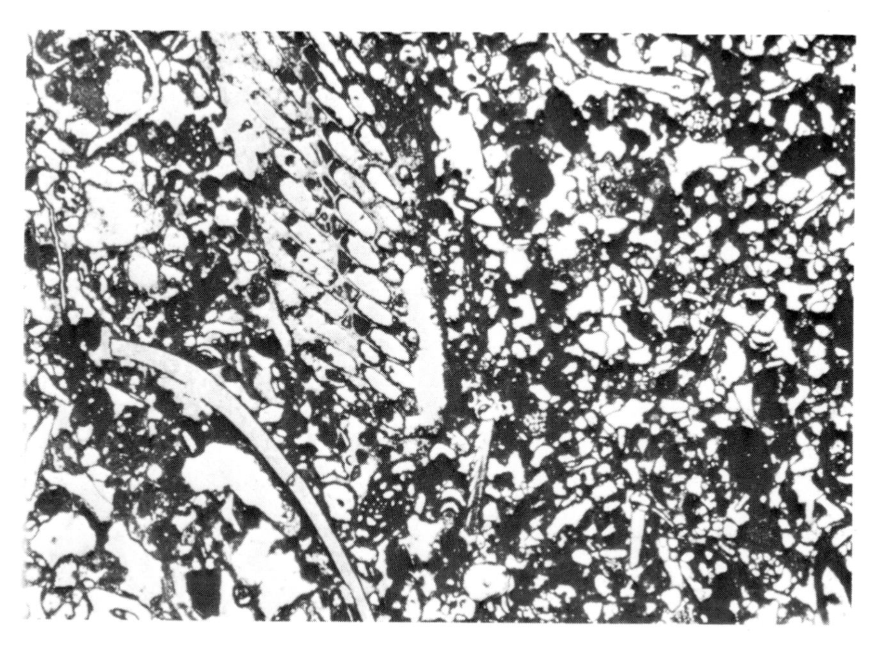

Fig. 56. Thin section of packstone from the Marada Formation (Miocene) of Libya. This is composed largely of skeletal grains (including large lamellibranch and bryozoan fragments), with minor rounded cryptocrystalline peloids (faecal?) and micrite matrix (×40).

ooliths are generally composed of concentric layers of tangentially arranged aragonite. In ancient ooliths this has generally been modified to a radiating arrangement of acicular calcite.

Modern oolite sands occur in high-energy environments, such as sand banks and tidal deltas. Like their ancient analogues they are generally well sorted, matrix-free and cross-bedded. These data all suggest that ooliths form by the bonding of aragonite crystals around nuclei, such as quartz or skeletal grains, in a high-energy environment (Fig. 57).

The physicochemical processes which cause oolith formation are unclear, but it is noted that they generally tend to form where cool dilute sea water mixes with warm concentrated waters of lagoons and restricted shelves (e.g. the modern Bahamas platform). Recent ooliths are covered with a mucilage jacket of blue-green algae which presumably play some role in aragonite precipitation.

Pisoliths are coated grains several millimetres in diameter. They form in caverns (cave pearls). Vadose pisoliths form in caliche crusts beneath weathered zones (Dunham, 1969). A third type of coated grain are oncoliths. These are up to five or six centimetres in diameter and irregularly shaped. The laminae are discontinuous around the grain. Oncoliths are formed by primitive blue-green algae growing on a grain and attracting carbonate mud to their sticky surface. Intermittant rolling of the grain allows the formation of discontinuous lime-mud films. In contrast to ooliths, therefore, pisoliths and oncoliths are indicators of low-energy environments.

2. *Matrix*

Carbonate mud is termed micrite. The upper size-limit of micrite is variously taken as 0·03–0·04 mm diam. Micrite may be present in small quantities as a matrix within a grain-supported carbonate sand, or may be so abundant that it forms a carbonate mudrock, termed micrite or calcilutite (Fig. 58). Modern lime muds are composed of aragonite; their lithified fossil analogues are made of calcite.

Several processes appear to generate lime mud. The action of wind, waves

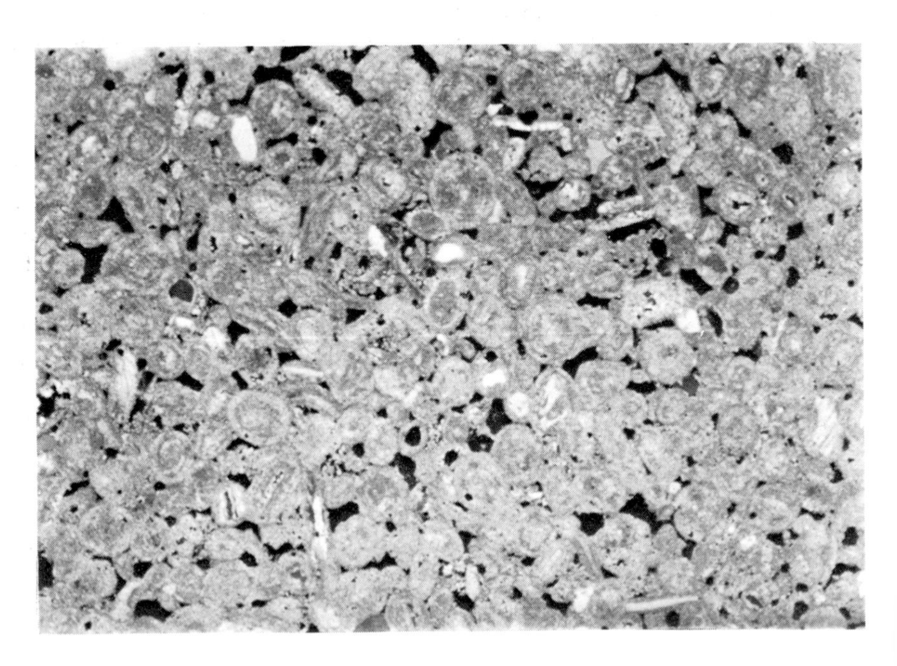

Fig. 57. Thin section of oolite grainstone showing concentric laminae and nuclei within individual ooliths. Note preserved intergranular porosity. Portland Group, Jurassic, Dorset (×40).

and tides will smash up shell debris and ultimately may abrade them into their constituent crystals. Faecal pellets may be similarly disaggregated.

Biological action is also efficient at breaking up carbonate particles to form crystal muds. This includes the parrot fish which eats coral, shell-munching benthonic and burrowing invertebrates, and especially the humble blue-green algae. These form pits within skeletal grains and lead to the micritization of the grain surface. Grains so softened will tend to break up and release the micrite. The calcareous algae, such as *Halimeda*, also secrete aragonite needles within their mucilagenous tissues. On death the mucilage rots to release the aragonite needles.

Some evidence suggests that direct inorganic precipitation of aragonite muds may sometimes take place. In modern carbonate environments such as the Bahamas platform and Persian Gulf, "whitings" have been seen (e.g. Wells and Illing, 1964). These are temporary cloudy patches of lime mud disseminated in sea water. These can be attributed to spontaneous precipitation of aragonite from sea water. Some geochemical data are inconsistent with this interpretation, however, and there are other explanations for these phenomena. Micrite can also form as a crypto-crystalline cement in certain circumstances. Because of this it must be used with care as an index of depositional energy.

In conclusion, it would appear that lime mud, micrite, can well form by a variety of processes.

3. Cement

The third component of carbonate rocks is cement. By definition this applies to crystalline material which grows within the sediment fabric during diagenesis. The commonest cement in limestones is calcite, termed spar, or sparite (Fig. 59). Other cements in carbonate rocks include dolomite, anhydrite and silica.

At the present time it is the custom to restrict the term cement to the growth of crystals within a pore space (e.g. Bathurst, 1971, p. 416). This has also been termed drusy crystallization.

This type of spar is genetically distinguishable from neomorphic spar which grows by replacement of pre-existing carbonate (Folk, 1965). Neomorphism is itself divisible into two varieties. Polymorphic transformations are those which involve a mineral change, as in the reversion of aragonite to calcite. Recrystallization is the development of calcite spar by the enlargement of pre-existing calcite crystals.

Additional discussion of cement is contained in the section on diagenesis.

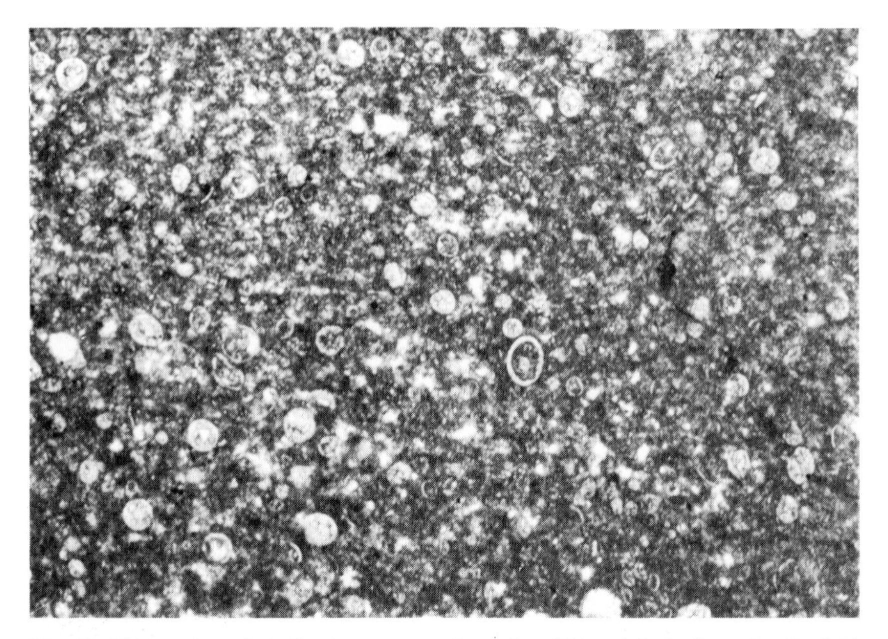

Fig. 58. Thin section of chalk (Cretaceous), North Sea. This calcilutite is made of calcitic coccoliths and is generally porous but impermeable (×40).

D. Nomenclature and Classification

The nomenclature and classification of carbonate rocks has been a topic of great confusion and debate. There are several good reasons for this. There are so many parameters which may be used to define carbonate rock types. These include chemical composition (e.g. limestone, dolomite), grain size, particle type, type and amount of porosity, degree of crystallinity, quantity of mud, and so on.

The concepts on which present-day carbonate nomenclature is based are contained in a volume of papers edited by Ham (1962). Two of these articles deserve special mention because they proposed a series of terms and groupings which are widely used today. Folk (ibid.) divided limestones into four main classes (Table XIV). The first two classes include rocks composed largely of grains (allochems); these he jointly termed the allochemical limestones. One class is dominated by sparite cement, the other by micrite matrix. The third class is for rocks lacking grains, termed the orthochemical limestones. This group includes micrite lime-mud carbonates. The fourth group is for rocks made of *in situ* skeletal fabrics. This group, the authochthonous reef rocks, includes the biolithites (Fig. 59).

Table XIV
Major grouping of the carbonate rocks

Allochemical rocks	—	composed largely of detrital grains
	I	Sparite cement dominant
	II	Micrite matrix dominant
Orthochemical rocks	III	Composed dominantly of micrite
Autochthonous reef rock	IV	Biolithite

After Folk (1962).

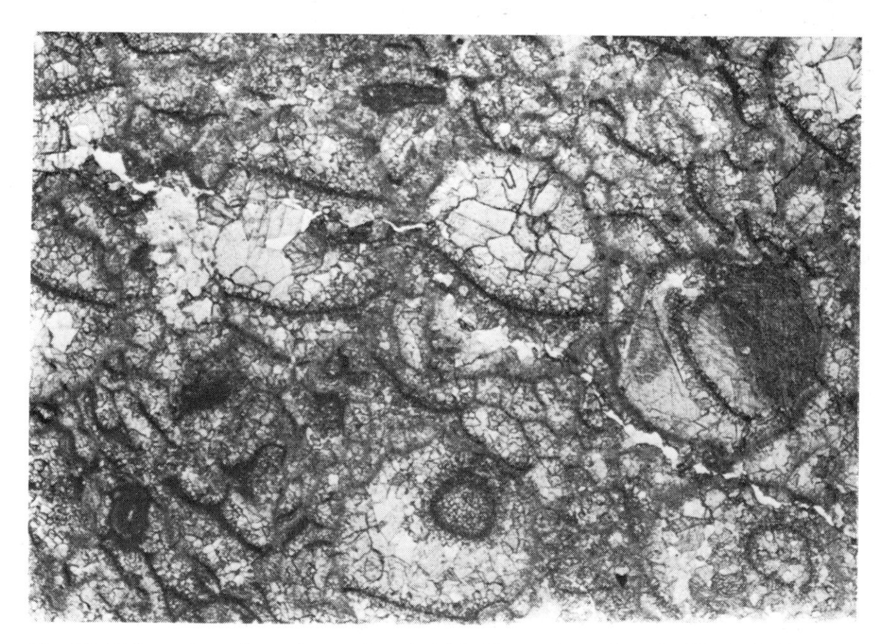

Fig. 59. Thin section of coraline reef rock (biolithite, or boundstone), Wenlock Limestone (Silurian) England. The original high porosity has been destroyed by a sparite cement (× 10).

Folk proposed a bipartite nomenclature for the allochemical rocks. The prefix defined the grain type and the suffix denoted whether sparite or micrite predominated. Thus were born words like "oosparite" and "pelmicrite".

Where more than one allochem type was present, two could be used with the major one first, as in "biooosparite". When both sparite and micrite were present, they could both be compounded with the dominant constitutent first: as in "pelbiomicsparite". Joking apart, this is a logical and flexible scheme for classifying and naming carbonate rocks.

Dunham's (ibid.) approach to the problem was quite different, but equally instructive. Like Folk, he placed the *in situ* reef rocks in a class of their own: the boundstones. A second class was erected for the crystalline carbonates whose primary depositional fabric could not be determined. Dunham divided the rest of the carbonates into four groups according to whether their fabric was grain-supported or mud-supported (Table XV).

Table XV
Classification of carbonate rocks after Dunham (1962)

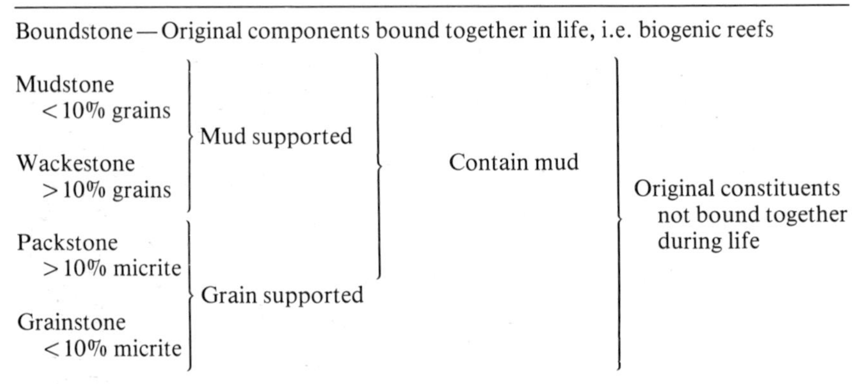

Boundstone—Original components bound together in life, i.e. biogenic reefs			
Mudstone <10% grains	Mud supported	Contain mud	Original constituents not bound together during life
Wackestone >10% grains	Mud supported	Contain mud	Original constituents not bound together during life
Packstone >10% micrite	Grain supported	Contain mud	Original constituents not bound together during life
Grainstone <10% micrite	Grain supported		Original constituents not bound together during life
Crystalline carbonate—Primary depositional fabric destroyed by recrystallization			

Grainstones are grain-supported sands with no micrite matrix. Packstones are grain-supported sands with minor amounts of matrix. Wackestones are mud-supported rocks with a significant but dispersed number of grains (Fig. 60), and mudstones are carbonate muds.

This scheme too has much to commend it. The nomenclature is simple and identification can be made with only a hand lens. Furthermore, this system, by drawing attention to fabric and matrix content, gives an index of the depositional energy. Thus the mud-supported limestones may be taken to indicate deposition in a low-energy environment. By contrast, the matrix-free grain-supported rocks suggest deposition in a high-energy environment in which no mud could come to rest.

These concepts may be usefully applied to siliciclastic rocks, but must be applied to carbonates with certain reservations. The polygenetic origin of micrite has been mentioned already. It is possible for a clean carbonate sand to be deposited in a high-energy environment. Subsequently, however, micrite may develop by bioturbation, algal micritization, cementation and by infiltration due to high permeability. Falls and Textoris (1972) analysed the granulometry of some Recent carbonate beach sands using the same statistical techniques as those applied to siliciclastic sediment, yet they reported difficulties in analysing the fine fraction due to micritization.

Similarly, the large initial size of carbonate skeletal material makes it dangerous to use grain size as an energy index in the same way as it can be used in terrigenous deposits. Consider, for example, the oyster reefs of modern lagoons. In terms of particle size these are essentially low-energy conglomerates.

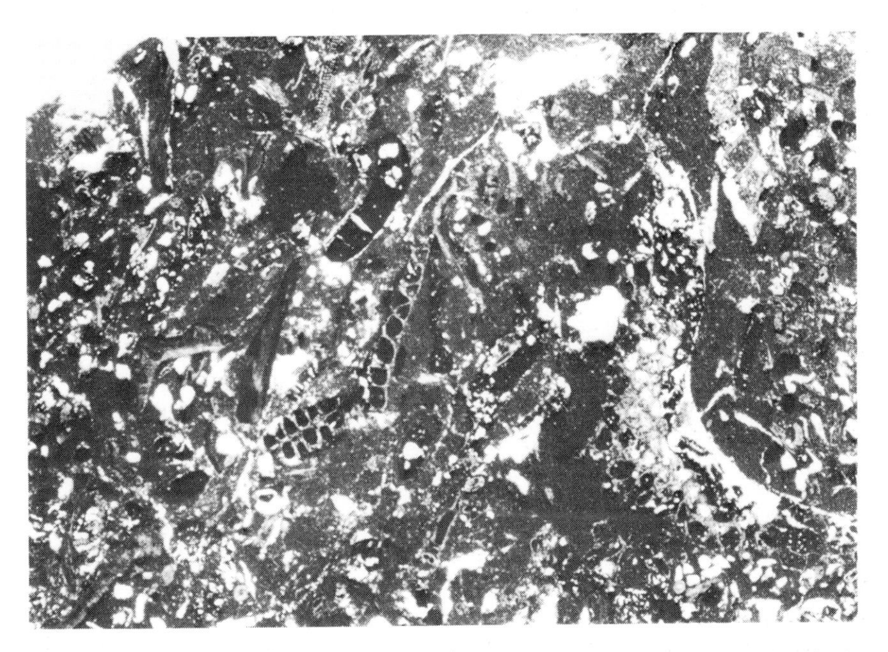

Fig. 60. Thin section of wackestone, Marada Formation (Miocene) Libya. Skeletal and other grains float in a micrite matrix. The fabric of the rock is essentially mud-supported (×30).

Nevertheless, some studies do show that skeletal debris can be comminuted, size-graded and winnowed to produce granulometric characteristics which mimic those of terrigenous sands (e.g. Hoskin, 1971).

To conclude this discussion, it may be stated that grain size, sorting and matrix content can only be used with reservations as indicators of hydrodynamic environment in carbonate rocks.

Nevertheless, the classifications and nomenclature of carbonates proposed by Folk and Dunham are extremely useful and, used in conjunction, encompass most varieties of limestones with flexibility and finesse (Fig. 61).

	GRAIN TYPES			
MUDSTONE > 10% grains	Lime mud, micrite, calcilutite, chalk			
	PELLETS	SHELL DEBRIS	OOLITHS	INTRACLASTS
WACKSTONE > 10% grains, mud supported	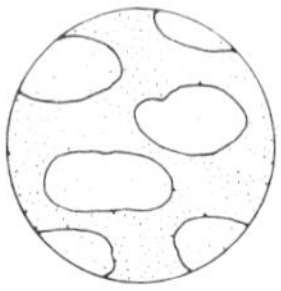 Pelmicrite	Biomicrite	Oomicrite	Intramicrite
PACKSTONE > 5% mud, grain supported	Pelmicsparite	 Biomicsparite	Oomicsparite	Intramicsparite
GRAINSTONE < 5% mud	Pelsparite	Biosparite	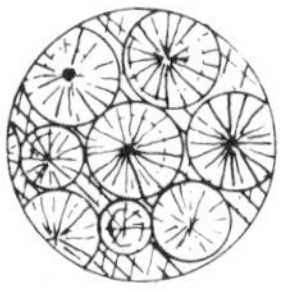 Oosparite	Intrasparite
BOUNDSTONE original components bound together	Reef rock, biolithite			

Fig. 61. Nomenclature of carbonate rocks. (After Dunham, 1962 and Folk, 1962.)

E. Diagenesis and Porosity Evolution

1. Diagenesis and petrophysics

It has already been pointed out that the diagenesis of carbonate rocks is very complex. This is basically because of their unstable mineralogy and because their high initial permeability make them susceptible to percolating reactive fluids. Choquette and Pray (1970) have pointed out the salient petrophysical characteristics of carbonate rocks: primary porosity is generally higher than in terrigenous sands, running at between 40 and 70%. It consists of both interparticle and intraparticle porosity. The ultimate porosity in a carbonate reservoir, however, may be only 5–15%. Little of this is primary porosity. Most of it is of the diverse secondary porosity varieties described in Chapter 2; namely moldic, vuggy and intercrystalline. The size and shape of individual pores is extremely variable within any one rock and, unlike sandstones, there is little correlation between pore volume, pore geometry and grain size, shape and sorting. Because of the erratic petrophysical variations within a small volume of rock, it is necessary to measure porosity and permeability from whole cores rather than from small plugs. Extensive coring of hydrocarbon reservoirs is necessary for accurate calculations of reserves and for effective production.

For detailed accounts of the petrophysics of carbonate oil reservoirs see Chilingar *et al.* (1972) and Langres *et al.* (1972).

The following examination of the relationship between porosity and diagenesis of carbonate rocks is drawn freely from the writings of Murray (1960), Larsen and Chilingar (1967), Choquette and Pray (1970) and Bathurst (1976). An account of the main diagenetic processes and their petrophysical effects is followed by examination of the diagenetic characteristics of some of the major carbonate rock types.

The preceding brief account of cement drew attention to the different modes of formation of sparite—crystalline calcite. Crystallization, used in its restricted sense, means the infilling of primary inter- and intraparticle porosity by the drusy growth of sparite out from the pore walls. This naturally results in a decrease in porosity.

Neomorphism is the term applied to describe the recrystallization or replacement of a mineral (Folk, 1965). Neomorphism can lead to both increasing, or unaltered porosities. Recrystallization, defined as neomorphism in which the mineralogy is unchanged, does not significantly alter the amount or type of porosity. However polymorphic transformations, in which one mineral replaces another, can have large effects on rock porosity. As already mentioned, one of the earliest diagenetic changes is the transformation of aragonite to calcite. This results in an increase in total rock volume of 8%. Note: this is not a decrease of total porosity of 8%, but a decrease of 8% of the primary porosity. Another important diagenetic process is the replacement of one mineral by another, as for example calcite by dolomite. Dolomitization can cause an overall contraction of the rock by as much as 13% (Chilingar and Terry, 1964). It is the intercrystalline porosity caused by this replacement that makes secondary dolomites such attractive reservoir rocks. Conversely, a decrease in porosity is caused by the transformation of dolomite to calcite;

de-dolomitization or calcitization as it is called (Shearman *et al.*, 1961).

Leaching is one of the most important processes giving rise to secondary porosity. Solution porosity may be due to the selective solution of either matrix, cement or specific grain types. Vuggy porosity results from the solution of pores whose boundaries cross-cut the fabric. Moldic porosity describes the selective solution of one particular grain type, e.g. oomoldic porosity, biomoldic porosity, and so on.

Lastly, silicification is another characteristic diagenetic process in carbonate rocks. The silicification of lime muds is described elsewhere (p. 117) and is not of petrophysical significance. In calcarenites and reef rocks, however, silicification can be an important destroyer of primary porosity when it develops as a chalcedonic cement. Silica also occurs either as a whole-sale replacement of the rock or selectively to produce silicified fossils to delight palaeontologists. This change is generally a one-way process.

Table XVI summarizes the major diagenetic processes in carbonate rocks and illustrates their various effects on porosity.

Table XVI
A summary of the main diagenetic processes in carbonate rocks and their effects on the amount and type of porosity

Diagenetic process			Porosity response
1. Drusy crystallization			Decrease of primary porosity
2. Neomorphism	recrystallization	Calcite—Calcite	No change
	polymorphism	Aragonite—Calcite	8% decrease
	replacement	Calcite—Dolomite	13% increase
		Dolomite—Calcite	13% decrease
3. Leaching			Increase in porosity of moldic and vuggy types
4. Silicification	chalcedonic pore filling		Decrease in primary porosity
	replacement		No change

The various diagenetic processes which have just been described take place in response to changes in pore fluid chemistry. These reactions are still imperfectly understood, but this section will attempt to outline the major ones. The various pore fluids, their pH and Eh have already been described in the section dealing with sandstone diagenesis (p. 91). Reference back to Figs 41 and 42 shows that calcite is soluble in acid meteoric waters but stable, and may be precipitated, in alkaline connate waters. It is also necessary however, to consider the stability fields of aragonite and dolomite. Their

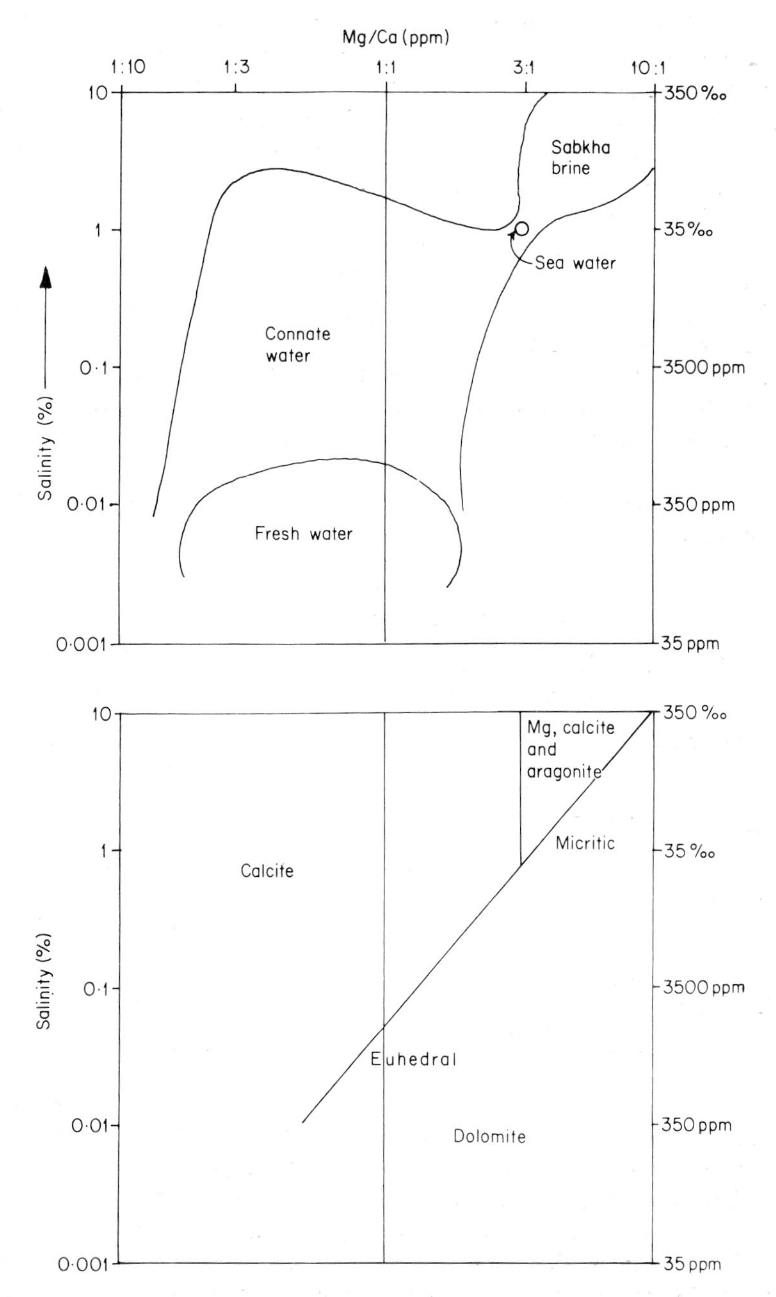

Fig. 62. Diagrams to show the salinity and Mg:Ca ratios encountered in various environments (upper), and the stability fields of carbonate minerals (lower). (From Folk and Land, 1974.)

solubility is also controlled by pH, but salinity and Ca:Mg ratio are also important parameters. Calcite:aragonite:dolomite stability fields are shown in Fig. 62. This shows that aragonite is the stable polymorph of calcium carbonate for saline waters with relatively high Mg:Ca ratios. Dolomite can form in a wide range of Mg:Ca ratios and salinities.

The diagenesis and petrophysical evolution of calcarenites, calcilutites and dolomites will now be described and discussed in turn.

2. Diagenesis of calcarenites

For the purpose of this discussion calcarenites will be deemed to include not only lime sands, but also reefs, i.e. those carbonate sediments which originally contain both high primary porosity and permeability. From the preceding discussion it will be remembered that these sediments are predominantly aragonitic, and that this reverts to calcite, during diagenesis, while at the same time the original pore space will tend to be infilled with a calcite cement. The exact way in which these changes take place depends on the fluids with which the pores are bathed. The following account attempts to synthesize a diffuse literature and is based particularly on papers by Purser (1978) and Longman (1980).

If one considers a cross-section through a carbonate island there are five major pore fluid environments to consider (Fig. 63).

If the topography is sufficient there may be raised ground where carbonate sediment, or previously lithified limestone occurs in the vadose zone above the water table. The pores of the vadose zone will be full of air, and thus chemically inert. When rain falls however, the pores will be flushed with acidic meteoric water. This will tend to corrode the carbonate minerals, generating moldic and vuggy porosity and enlarging pre-existing fractures in lithified rock. This is of course epidiagenesis analogous to that discussed earlier for sandstones (p. 104). If this process continues uninterrupted then cavernous porosity may develop leading ultimately to karstic topography, such as the "cockpit" country of parts of the Caribbean.

A second type of vadose diagenesis occurs near the shore in the intertidal and supratidal spray zone, where sediment pores are intermittantly flushed, not by fresh water, but by sea water. As this evaporates in the pores the salinity increases and generates a characteristic type of cement. Because the pore fluids are only filled intermittantly the cement tends to be irregular. Uniform isopachous rim coats are unusual. Cement is often restricted to throat passages (meniscus fill), or depends from the upper surface of pores like stalactites. Cement fabrics are generally fibrous or micritic and are of aragonitic or high magnesium calcite composition. This is referred to as beachrock cementation.

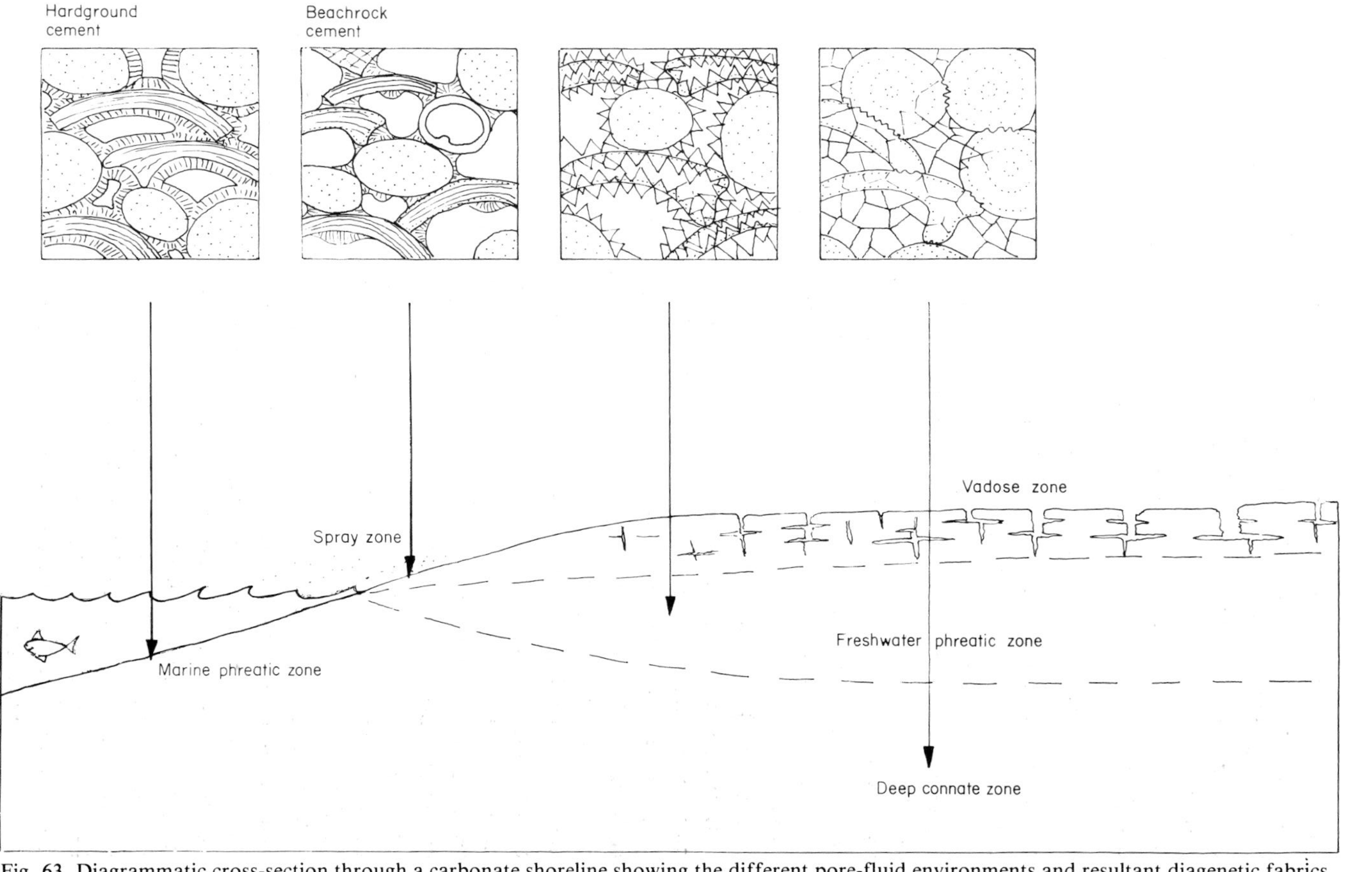

Fig. 63. Diagrammatic cross-section through a carbonate shoreline showing the different pore-fluid environments and resultant diagenetic fabrics. For explanation see text.

Below the water table, in the phreatic zone, the pores are soaked in water. For a carbonate bank or island a biconvex lens of meteoric fresh water overlies denser sea water (near the coast) which merges inland to connate water of modified marine and meteoric origin.

In the non-marine, continental fresh water phreatic zone the downward percolating rain water becomes neutralized and saturated with calcium and carbonate ions from the dissolved carbonates above. Thus precipitation of calcite takes place, first as a dog-tooth rim cement on the grains, which may ultimately form a coarse sparite pore-fill. This may grow, not only outwards, from grains into intergranular pores, but also inwards into moldic pores. This back-to-back dog-tooth cement grows from algally bored micritic films on the original grain surfaces.

Dedolomitization may also occur in the fresh water phreatic zone, leaving dolomoldic pores which may later be infilled by calcite pseudomorphs (Evamy, 1967).

Immediately beneath the sea floor, where the pores are full of sea water a second phreatic zone occurs with its own distinctive diagenetic fabric. Here an aragonitic rim cement develops around the grains. The aragonite develops either as micrite (i.e. cryptocrystalline aragonite, not a depositional matrix), or as a radial acicular fabric. In either case the cement is isopachous, rimming the grains evenly (Fig. 63).

This "hardground" type of cement generally develops as a layer only a metre or so deep below the sea floor. In modern examples the carbonate sand beneath is unconsolidated.

Below the sea-floor "hardground" and beneath the freshwater phreatic zone is the deep connate environment. Here sediment which has escaped the various surface diagenetic processes will still be unconsolidated and highly porous to begin with. Because these sands have not been lithified already they will be very susceptible to porosity loss by compaction, whereas those sediments which have already lost some porosity during early shallow diagenesis will not undergo compaction to the same degree.

The effects of compaction are manifest by signs of pressure solution at the point of contact of grains, and by stylolites due to wholesale solution of rock (Park and Schot, 1968; Wanless, 1979). Simultaneously the pores become infilled with a coarse sparite mosaic. Unlike the spar fill of freshwater cementation this does not grow inwards from dog-tooth crystals, but develops crystals of uniform size, though these may sometimes be seen cross-cutting "ghosts" of grains and earlier shallow diagenetic fabrics.

Particular attention has been directed towards finding the source of the calcium carbonate for this deep connate type of cement. Two major ones have been proposed: internal and external. The internal source is provided by pressure solution. If this is so then the inhibition of compaction is significant in preserving porosity. As already mentioned early diagenesis may be important in this role.

Calcium carbonate may also be brought into a sediment by groundwater movement. The ions may have been derived from the solution of shallower limestones, as previously discussed, or from fluids squeezed from compacting clays. The problem with invoking an external source for the cement is that we know how much calcium carbonate is required and we know the concentrations of calcium and carbonate ions in pore fluids. It has been calculated that whole oceans of water must pass through a carbonate sediment before it is completely cemented (Weyl, 1958).

Finally it is important to enquire what the foregoing has to do with predicting porosity distribution in carbonates. The following general guide lines can be proposed. Early shallow diagenesis is good news. It causes some porosity loss, but by lithifying the sediment, prevents compaction when deep burial occurs.

Porosity loss may also be inhibited by cutting off the sediment from migrating waters. This may occur where a porous carbonate lens is sealed within impermeable carbonates, shales or evaporites. Alternatively early hydrocarbon invasion is quite delightful. Once oil or gas fill pores they keep the calcium and carbonate ions out.

When all primary porosity has been destroyed secondary porosity may yet develop. This may be due to tectonic fracturing, dolomitization or solution where low pH, generally meteoric, waters invade the rock. Thus

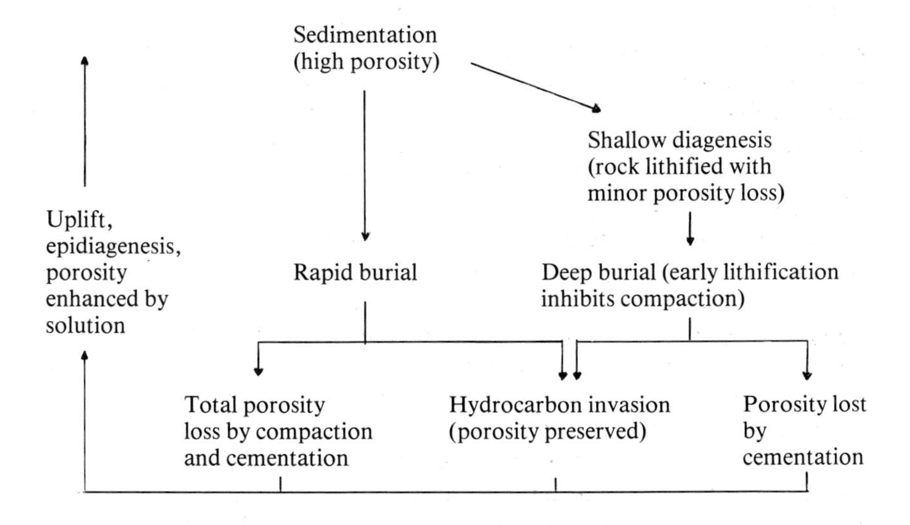

Fig. 64. Flow chart illustrating possible diagenetic pathways for carbonates.

secondary porosity may be anticipated beneath unconformities. Figure 64 summarizes these various diagenetic pathways.

The end result of carbonate diagenesis is that the distribution of porosity and permeability in a limestone may be unrelated to the primary porosity and permeability distribution with which it was deposited. In such cases facies analysis is of little value in predicting the distribution of porosity and permeability. This is illustrated by the Intisar "A" oil field of Libya (Fig. 65).

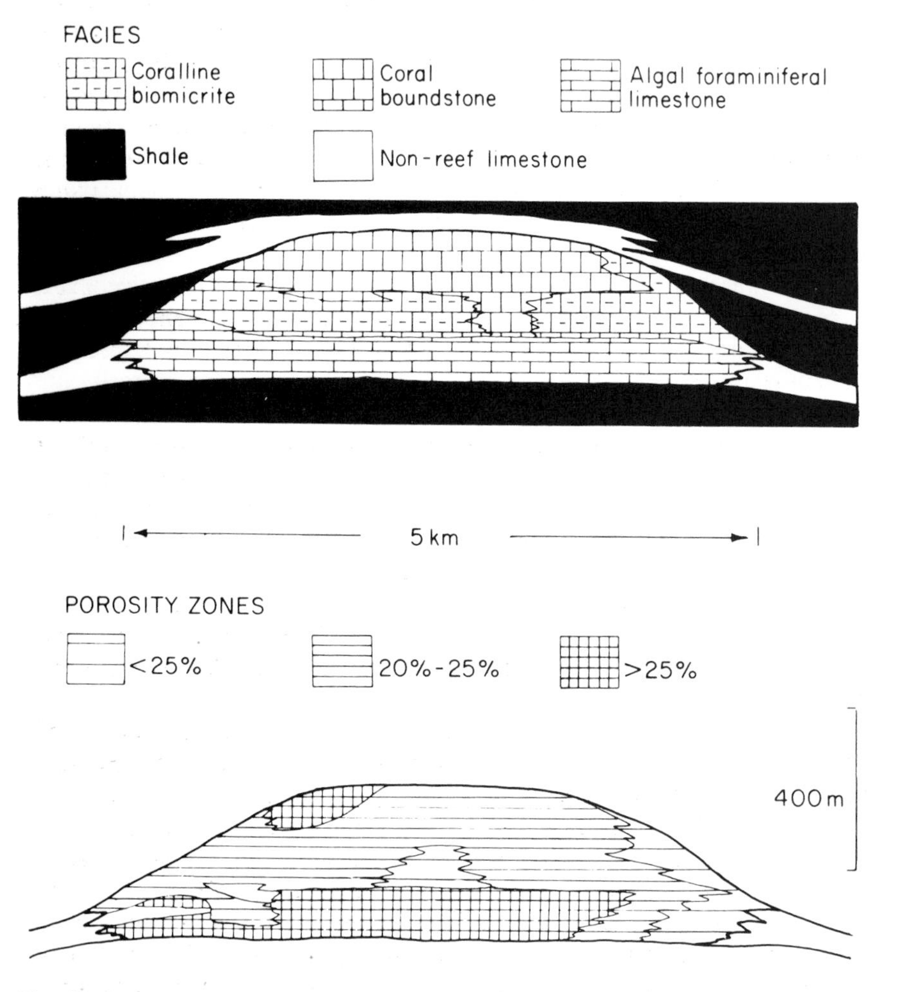

Fig. 65. Cross-section of the Intisar Reef, Libya, from Terry and Williams (1969). This demonstrates that in carbonates, unlike clastics, facies (top) are not correlative with petrophysical properties (lower).

3. Diagenesis of lime muds

The diagenesis and petrophysics of lime muds are far simpler than for calcarenites. The main reason for this is that though lime muds are often as porous as calcarenites, they are far less permeable due to the small size of the throat passages. They are thus not nearly as susceptible to flushing by fluids of diverse chemistry. An important distinction must be made at the outset between lime muds made of aragonite, and those made of calcite.

From the Pre-Cambrian until the Cretaceous lime muds were made almost entirely of aragonite. When buried the aragonite reverts to calcite. As already seen this reaction results in an 8% increase in volume, and a corresponding decrease in porosity (Note: this is not an 8% decrease in porosity, but an increase of 8% by volume of rock, i.e. an aragonite mud with 40% porosity and 60% grains, would lose 4·8% porosity by the aragonite: calcite reversion). The rearrangement of crystals, coupled with compaction, however, generally leads to a total loss of porosity and permeability.

Thus it is a matter of observation that most lime mudstones are hard, tight splintery rocks of negligible porosity and permeability. They can only become reservoirs if secondary porosity has been generated by fracturing, dolomitization or solution.

An exception to this general rule is provided by a particular type of lime mud known as chalk. Chalk is a fine-grained limestone composed largely of coccoliths, the calcitic plates of coccospheres. These are the skeletal remains of a group of nannoplanktonic golden-brown algae (Black, 1953). Coccolithic limestones first became important towards the end of the Jurassic Period. By the middle of the Cretaceous Period vast quantities of coccolith muds began to be deposited across the continental shelves and ocean basins of the world. These gave rise to the Chalk Group of northwest Europe, the Austin Chalk of Texas and analogous formations (Hakansson *et al.*, 1974).

Chalk is deposited with porosities and permeabilities similar to those of aragonitic lime muds. Because chalks are of calcitic composition however, they do not undergo the early diagenetic recrystallization of aragonite to calcite. Thus chalks generally remain as "chalky" friable rock, retaining porosities in the order of 20–30%. They have considerable storage capacity and may act as aquifers or reservoirs for oil and gas (Hancock, 1976; Scholle, 1977). To actually yield up their contained fluids chalks must have permeability. This may occur due to fracturing. Alternatively chalks may still retain some original intergranular permeability. This is best preserved where compaction has been inhibited by abnormally high pore pressures (relieving the stress at grain contacts) or hydrocarbon invasion. These conditions are responsible for the productivity of the chalk reservoirs of the Ekofisk group of fields in the Norwegian North Sea (Byrd, 1975; Heur, 1980). Here the Cretaceous chalk has been domed over salt structures and fractured. Simultaneously rapid burial and high heat flow favoured overpressuring and hydrocarbon generation.

Early diagenesis of chalks is minimal, but worthy of comment. Horizons which show evidence of early cementation are widespread stratigraphically though not volumetrically important. These penecontemporaneously cemented layers are termed "hardgrounds". Evidence for the early cementation is provided by the fact that the upper part of the beds are often extensively affected by the activities of boring organisms, or were colonized by organisms such as oysters that required a rigid substrate on which to attach themselves. These bored surfaces are frequently immediately overlain by thin intraformational conglomerates of chalk, often phosphatic and glauconitic, in a marly matrix (Fig. 66). Examples of such hardground horizons have been described from the Upper Cretaceous chalk of north-west Europe (Jefferies, 1963; Bromley, 1967).

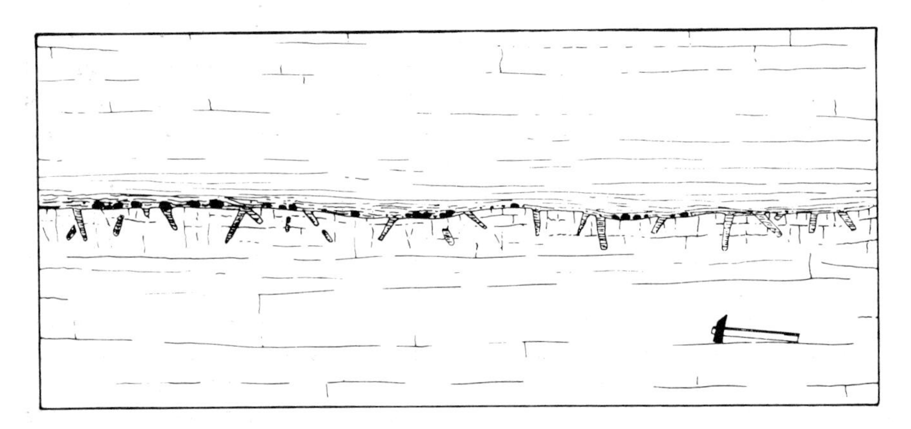

Fig. 66. A hardground horizon in a chalk limestone. A pure micrite is overlain by an argillaceous micrite. The contact is bored and immediately overlain by an intraformational conglomerate of phosphatized clasts with some glauconite. Eocene, Jordan.

Hardgrounds are found at intervals throughout many limestone successions (e.g. Lindstrom, 1979). It may be useful to speculate on the possible influence that they may have on the diagenetic evolution of the interbedded sediments. The hardgrounds effectively seal off the underlying sediment, and render them an essentially closed system if the succession is progressively buried. Thus constraints are set on the diagenetic processes that can take place in the interbedded sediments. There is evidence to suggest that in some instances diagenesis follows the simple pattern described in the previous section. Thus hardgrounds may play a role in preserving porosity, by restricting the extent of early cementation in the underlying sediment. The presence of hardgrounds within a limestone formation that is otherwise porous and permeable, may cause vertical permeability barriers within the succession. As such they may influence the migration and entrapment of oil, and affect the reservoir engineering characteristics of the formation as a whole.

F. Dolomite

1. Introduction: chemical constraints on dolomite formation

The term dolomite is applied both to the mineral $Ca.Mg(CO_3)_2$ and to rock of this mineralogical composition. The term dolostone is sometimes used for the latter. The exact genesis of dolomite is still a fruitful field for research despite many years of field observation and laboratory research (see Zenger, 1980, for recent papers reviewing the dolomite problem). Figure 62 showed that a high ratio of magnesium to calcium is not necessarily a prerequisite for dolomite precipitation. Normal sea water has an Mg:Ca ratio of about 3:1. Dolomite forms in supersaline environment where Mg:Ca ratios exceed this value. It is noteworthy however, that dolomite may form at the expense of calcite for Mg:Ca ratios less than 1:1 if the salinity is very low (Folk and Land, 1975).

Conditions necessary for dolomite formation appear to include initial permeability within the host sediment, coupled with sufficient pressure differential to permit pore fluid movement, an adequate and continuous supply of magnesium ions, and a fluid which is undersaturated with respect to calcium ions. The last two of these conditions seem to be fulfilled in the so-called Dorag model of Badiozamani (1973). Sea water and fresh water may both be saturated with respect to dolomite and calcite, but mixtures with between 5 and 50% sea water are undersaturated with respect to calcite and supersaturated with dolomite (Fig. 67). This suggests that dolomitization may be expected where marine and fresh waters mix. Having now examined the theoretical constraints for dolomite formation it is now pertinent to examine the observational evidence for dolomite formation. It has long been known that there are two main types of dolomite, primary or syngenetic, and secondary, or diagenetic. These will now be considered in turn.

2. Primary dolomites

Primary dolomites are defined as those which formed at the time of deposition. There is discussion as to whether genuine direct precipitation of dolomite occurs, or as to whether it is in fact a replacement of previously formed minerals, i.e. penecontemporaneous rather than strictly primary.

Recent dolomite deposits have been described from many arid hypersaline coasts (termed "sabkha" from the Arabic for saltmarsh). In some

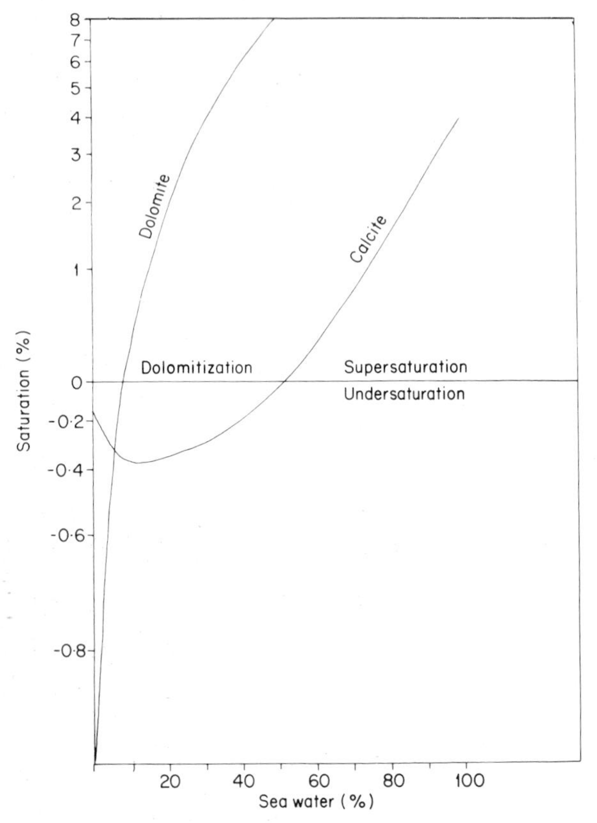

Fig. 67. Graph of saturation curves for calcite and dolomite for various mixes of sea and fresh water. Dolomitization occurs where the percentage of sea water varies between 5 and 50%. (From Badiozamani, 1973.)

examples these are believed to be direct precipitates of dolomite within the pore spaces of aragonite mud. Friedman (1979) has described such a case from marginal pools of the Red Sea, and von der Borch (1976) has cited another from the Coorong Lagoon of Australia. In other instances, however, the modern dolomite has been interpreted as a replacement of pre-existing aragonite or calcite; as for example the account by Butler (1969) of the sabkha dolomite of Abu Dhabi.

In all these cases there is general agreement that the dolomites are primary, or penecontemporaneous. These modern examples are all micritic and cryptocrystalline with a grain size of less than 10–20 μm. Petrophysically they are like chalk: porous, but of low permeability.

Analogues at these Recent primary dolomites occur in ancient carbonate sequences. They too are characterized by a cryptocrystalline texture and low permeability. Evidence for their primary origin is provided by their bedded concordant nature, as opposed to the irregular discordant occurrence of secondary dolomites. They occur in "sabkha" facies, interbedded with faecal pellet muds, stromatolitic algal limestones and evaporites. Associated sedimentary, structures include desiccation cracks and "tepee" structure (a wigwam-like buckling of bedding due to penecontemporaneous hydration and expansion of anhydrite (see, for example, Kendall, 1969).

3. Secondary dolomites

Secondary dolomites are defined as those which are obviously of post-depositional origin. This is clearly shown by the way in which such dolomites have an irregular distribution, discordant to bedding and cross-cutting sedimentary structures. Unlike primary dolomite this type has crystals of over 20 μm diameter, which are occasionally euhedral or idiomorphic and cross-cut relic microfabrics of the original limestone

Fig. 68. Thin section of coarsely crystalline secondary dolomite. Zechstein (Upper Permian) North Sea. Unlike primary dolomites, secondary dolomites are generally both porous and permeable (×15).

(Fig. 68). These secondary dolomites have a characteristic sugary texture (sometimes referred to as sucrosic or saccharoidal). This has resulted in part from the bulk volume shrinkage, as calcite is replaced by dolomite; and in part from the dissolution of residual calcite during the final stages of dolomitization. Thus secondary dolomites are frequently delightfully porous with intercrystalline pores connected to one another by planar throat passages. Unlike primary dolomites these secondary dolomites can act as excellent hydrocarbon reservoirs.

III. COAL

There is no doubt that coal is of vegetable origin because coals not only contain recognizable plant remains, but transitions can be found between obvious accumulations of vegetable matter, e.g. peat, through lignites or brown coals, into true coals and on into anthracite. This series, from peat, through lignite, into the humic coals and finally anthracite is called the "coal series". The position of a coal in the series is termed its "rank". Thus lignite is a very low-rank coal, while at the other extreme, anthracite is a very high-rank coal. Physical appearance, physical properties and chemical composition of coals change with rank as do their utilization characteristics, e.g. calorific value, coking properties and gas generation potential. Thus knowledge of the rank of a coal can be a guide to its utilization.

The changes that vegetable matter undergoes in the course of alteration to coal are termed maturation or coalification. Maturation takes place in two stages: the peat stage and the burial stage. In the peat stage the plant material suffers a measure of biochemical degradation, and, when it is buried, progressive increase in both overburden load and temperature bring about dynamothermal maturation that slowly turns the peat into coal. The peat stage is an essential prerequisite for the formation of coal.

Under normal circumstances when plants die, they are exposed to air and are broken down primarily by oxidation and also by various organisms, particularly the fungi and aerobic bacteria. Where plant remains accumulate in swamp or bog environments, however, they become water saturated. Aerobic decay soon depletes the water of oxygen, the aerobic organisms die off and anaerobic bacteria take over. The anaerobic bacteria operate without oxygen but they are equally as capable of breaking down organic matter as the aerobic forms. Because of the stagnant nature of swamps and bogs, however, the waste products of the bacteria are not flushed away, but build up in interstitial waters and ultimately render the environment sterile. Bacterial activity is thus curtailed and the partially decomposed plant material remains in a state of arrested decay. In this state the material is peat. If the peat is drained the toxic materials are flushed out, decomposition sets in again and the peat may ultimately be destroyed. If the peat is not drained, however, but is buried under relatively impermeable sediments, its geological preservation becomes possible.

These conditions are generally met with in certain continental environments. Coal formation is especially characteristic of deltas, but may also take place in lakes, on alluvial flood plains and in coastal lagoons. Detailed environmental analysis is essential to the economic exploitation of coals (Dapples and Hopkins, 1969).

Chemically, coals comprise the three elements carbon, hydrogen and oxygen with minor proportions of sulphur and nitrogen and mineral impurities. The latter remain as ash after the coal has been burnt and, clearly, high ash content is undesirable. Inherent sulphur is normally present in a very small amount and was probably derived from sulphur proteins in the original plants. Some coals have a high sulphur content due to the presence of disseminated pyrites: this is deleterious because it generates sulphurous fumes on combustion of the coal. Although the nitrogen content of coal is small, it was economically important in the production of ammonia as a by-product of the coal-gas industry.

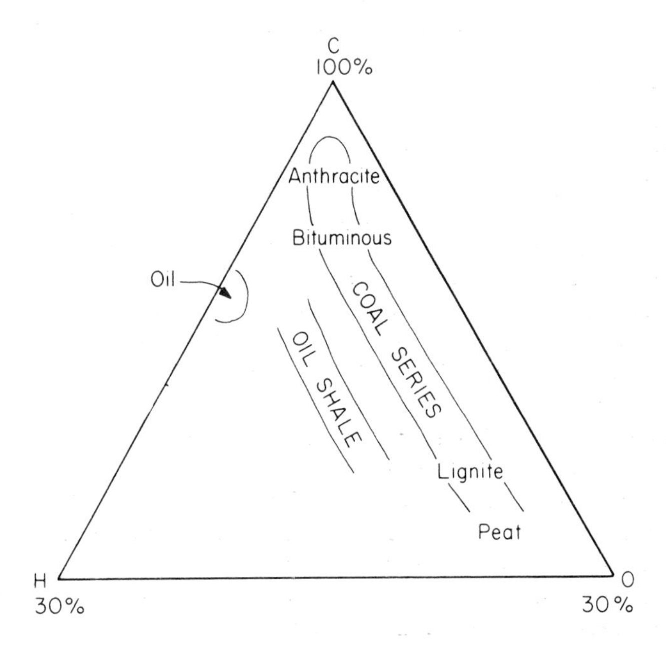

Fig. 69. Triangular diagram showing the coal series and the relationship between coal, oil shale and crude oil. (From Forsman and Hunt, 1958.)

Dynamothermal diagenesis of coal on burial is expressed by the changes in carbon, hydrogen and oxygen contents that accompany maturation. If the elemental composition of coal is plotted in terms of these three elements (Fig. 69) it will be seen that the coals lie within a narrow zone, called the coal belt. Carbon content increases progressively with increase in rank, but the proportion of hydrogen remains fairly constant at between 5 and 7% in the humic coals before it falls rapidly in the semianthracites and anthracites. Oxygen decreases with increase in rank.

The chemical changes that accompany maturation or coalification are not well understood but they no doubt involve production of gaseous carbon dioxide and methane. Some occurrences of natural gas, e.g. that in the Permian Rotliegendes of Holland and the southern parts of the North Sea appear to have been derived from devolatilization of underlying Westphalian coals (Patijn, 1964).

In general terms the rank of coals tends to increase with age, in the sense that most lignites or brown coals are Tertiary or Mesozoic in age whereas the Upper Palaeozoic occurrences are true coals. Age is only coincidental, however, and the prime control appears to be thermal, which is generally related to depth of burial. In any normal vertical succession of humic coals the carbon content, i.e. the rank, increases with depth. This relationship is called Hilt's law (Hilt, 1873), and it is likely that increase in temperature with burial is an important factor in maturation.

There is debate about the causes of the change from coal into anthracite. In some instances it may be the natural continuation of thermal maturation, but in many occurrences coals change laterally into anthracites as they approach zones of tectonic deformation. The latter relationship is found in the South Wales coalfield where the coals pass westwards into anthracite as they enter the zone of shearing associated with the Ammonford compression. Similarly in North America the Pennsylvanian coals pass laterally into anthracites as they enter the Appalachian fold belt. Coals may be devolatized by thermal metamorphism where they are cut by igneous intrusions. Examples of this are seen where Tertiary dykes cut Westphalian coal measures in north-east England. Although the physical appearance of the coals is only affected for a few feet on either side of the intrusions, devolatization as expressed by increase in carbon content spreads out regionally.

Most brown coals or lignites are obviously woody, lustreless, and not well-jointed; humic coals on the other hand do not normally show a woody appearance in hand specimen. They are well-bedded and display various degrees of lustre from bright to dull. On the basis of lustre and physical appearance, the humic coals are divided into four "rock" types: vitrain, clarain, durain and fusain (Stopes, 1919). Vitrains are the bright coals which have a vitreous lustre. Clarains are less bright with a silky appearance, while the durains are dull, and matt. Fusain is sooty black and very friable so that, unlike vitrain, clarain and durain, it readily soils the hands like charcoal. It is important to distinguish between coal type and rank. Coal type is the expression of the composition of the original plant materials, whereas rank is a measure of the extent of maturation or diagenesis that the vegetable matter has undergone. Some coal seams consist mainly of one rock type, but many are composite and comprise alternations of layers of vitrain, clarain, durain and fusain of varying thicknesses.

Coal normally breaks in three directions into roughly prismatic blocks. Two of the surfaces are joints and are termed "cleat" and "end" respectively. The cleat is the more strongly developed joint and in the days of manual extraction its direction often determined the layout of the underground workings. The third surface is parallel to bedding and usually coincides with a fusain layer. It is these bedding surfaces of fusain on a block of coal that soil the fingers.

As with most other rocks, the "rock" types of coal are each composed of a series of discrete components, termed macerals (the prefix "mac-" implying that they are macerated plant remains, and the termination "-erals" indicating that they are analogous to minerals). Macerals are named with the suffix "-inite", to distinguish them from the rock type of coal, termination "-ain". The macerals were all plant tissues or plant degradation products that have been modified chemically by diagenetic processes, i.e. by maturation. The common ones include: vitrinite, the main constituent of the coal type vitrain, and there are two varieties: tellinite and collinite. Under the microscope, tellinite shows compressed cellular structure of what was formerly woody tissue or xylem, whereas collinite which has similar optical properties is structureless. Suberinite, cutinite and exinite are macerals formed by coalification of bark, leaf cuticles and spore jackets respectively. Fusinite stands apart from other macerals in that it is nearly pure carbon. It often shows cellular structure under the microscope. The cell walls are usually preserved as a mass of broken fragments "*bogenstrukture*". This contrasts markedly with cellular tissues that are preserved as tellinite, where the cell walls have been deformed but not broken by compaction. It appears therefore that where cellular tissues are preserved as fusinite, the alteration must have taken place very early in the maturation history, certainly before significant compaction due to overburden load. It has been suggested that fusinite may have formed during the peat stage by a process analogous to "dry rot", or by forest fires. Micrinite is the general term applied to aggregates of very fine-grained macerated plant materials most of which are too small to be identified.

The coal type vitrain is composed mainly of the maceral vitrinite and with increase in proportion of other macerals, vitrain grades into clarains and durains. The durains are characterized by micrinite and/or exinite, and some of the grey durains are very rich in exinite.

The degree of maturation of the macerals vitrinite and exinite can be assessed optically under the microscope by change in their colour under ordinary transmitted light. In the humic coals, vitrinite changes from translucent yellow, through orange and red to deep red with increase in rank of coal. Exinite retains shades of yellow throughout much of the range of humic coals and only darken to orange and red in the higher ranks, i.e. semianthracite. The colour transmission of the macerals is affected by thickness of the thin section and determinations have to be made on very thin sections, with strict control of thickness. Vitrinite can also be studied by reflected light and, using the techniques of ore microscopy, its reflectance can be measured. The reflectance of vitrinite changes with increase in rank, and can be used as an index of maturation. Spores and scraps of vitrinite are present in minor traces in many sediments other than coal, and where they occur determination of their degree of maturation can provide a guide to the thermal history of the rocks that contain them.

This knowledge is important in the subsurface exploration of potential oil-bearing formation because maturation of liquid hydrocarbon runs parallel to that of coal.

Most coal seams appear to have been of *in situ* origin: that is to say the coal-forming peat accumulated where the plants lived and died. Some coal appears to be of "drift" origin, and the plant remains were transported, e.g. as log rafts, to the site of deposition. An *in situ* origin is demonstrated by the presence of roots and rootlets that lead down from the coal into an underlying fossil soil or "seat earth". Seat earths are of interest in their own right, by virtue of their refractory properties. Clayey seat earths, termed fire clays, consist mainly of kaolin minerals (disordered kaolinite) and are of value for manufacture of refractory bricks. Sandy seat earths, "ganisters", are almost pure quartz rocks and are used for producing silica fire bricks. There can be little doubt that the seat earths were deposited as normal clays and sands, but that they were leached of alkali metals and reconstituted by humic acids from the overlying peats (see also p. 71).

IV. SEDIMENTARY IRONSTONES

Iron sulphide (pyrite) and iron carbonate (siderite) are common diagenetic minerals present in uneconomic quantities in many sedimentary rocks.

In some places, however, iron is sufficiently concentrated in sedimentary rocks to become an ore body. Three main types of sedimentary iron ore may be distinguished, the Pre-Cambrian iron formations, the Palaeozoic Clinton ironstones, and the Jurassic minette ores.

The Pre-Cambrian iron formations occur in three remarkably similar settings: the Transvaal basin of South Africa, the Hammersly basin of Western Australia and the Animikie series of the Lake Superior region, Canada. These rocks have been extensively documented (Gross, 1972; Goodwin, 1973; Unesco, 1973), and Trendall (1968) has made a comparative study of the main Pre-Cambrian iron formation basins.

The iron ores consist of the oxides magnetite and haematite and of the silicate greenalite. The ore occurs thinly interbedded with chert, hence the common name "banded ironstone". Individual ironstone formations are between 50 and 500 m thick and often have basinwide continuity of hundreds of miles. The genesis of these Pre-Cambrian ironstones has been debated. Though there is general agreement that the ores are syngenetic, there has been considerable discussion as to the source and mode of precipitation of the iron.

It has been suggested that the iron originated from intrabasinal volcanism or from the residual concentration of iron on an extensively lateritized hinterland. Theories of the mode of precipitation of the iron frequently invoke atmosphere and sea water different in composition from those of the present day (e.g. Holland, 1972).

The second main group of iron ores are those referred to as of "Clinton type". These occur in the Silurian limestones of the Appalachians, notably at Birmingham in Alabama. The main ore mineral is haematite, which occurs both as a matrix, as ooliths, and replacing shells. Evidence cited in support of a primary origin includes the presence of reworked ironstone clasts in supradjacent formations, the presence of unreplaced carbonate shells in haematite matrix, and the converse, namely scattered haematite ooliths in normal carbonate beds. Furthermore, the distribution of Clinton-type ironstones is not restricted to modern outcrops and unconformities, as one would expect if they were replacement deposits. The fauna in such ironstones is also commonly dwarfed, suggesting a hostile environment.

Not all haematite oolite deposits are necessarily syngenetic. A strong case for replacement of primary carbonate ooliths by iron-rich ground water can be argued for those which immediately underlie unconformities or modern land surfaces. Examples of this situation include the Carboniferous haematite oolites of the Forest of Dean, in England, and of the Chatti Valley, Libya (Goudarzi, 1971).

The third main type of sedimentary iron ore is the minette variety. This is developed in the Jurassic rocks of Europe, notably in Lorraine and northern England. The minette ore minerals are generally chamosite, siderite and limonite. These occur as ooliths, matrix and replaced shells. Essentially most of the ore formations are oolitic or bioclastic wackestones with abundant marine faunas. Significant current action is suggested by the presence of ooliths, abraded shell debris and extensive cross-bedding indicating deposition in a high energy marine shoal environment (Hallam, 1963; Bubinicek, 1971). Taylor (1949) argued that the iron mineralization was a primary syndepositional feature. The alternative is that primary carbonate minerals were penecontemporaneously replaced by iron released from adjacent muds; the role of decaying organic matter being critical (Kimberly, 1979). This is consistent with the alkaline reducing conditions which are known to favour the formation and preservation of chamosite and siderite (Fig. 42).

V. PHOSPHATES

The element phosphorous is an essential constituent of all living matter, both plant and animal. Phosphate minerals are, therefore, extensively used as agricultural fertilizers. They are a valuable natural resource which must be located by geologists.

Table XVII summarizes the mode of occurrence of economic phosphates. This shows that, though some phosphates occur in igneous rocks as apatite, the majority of economic phosphates are in sedimentary rocks.

Table XVII
Main modes of occurrence of economic phosphates

Type		Mineral	% Total world production
Primary igneous		Apatite	24
Sedimentary	bedded / placer	Collophane and Apatite	74
Guano		Complex Phosphates and Nitrates	2
			100

Production percentages from McKelvey (1967).

The mineralogy of phosphates is complex and obscure because of their tendency to occur as microcrystalline aggregates. Typical phosphate minerals are admixtures of the phosphate radical (PO_4) with calcium, water, and traces of fluoride and uranium.

Phosphates occur in sedimentary rocks as matrix and as nodules, ooliths, pellets and phosphatized shells, bones and teeth. They also occur as a bulk replacement of limestones (Bentor, 1980).

Phosphates have been recorded from the modern sea bed, notably off the western coasts of America and Africa. Detailed studies show the wide range of occurrence of these deposits. Some phosphates have been recorded forming diagenetically within organic-rich diatomaceous oozes of the South West African shelf (Baturin, 1970). On the South African shelf diagenetic replacement by phosphate occurs in Miocene limestones which are eroded to form phosphate gravel placers (Parker and Siesser, 1972). Further north, off the West African shelf, phosphate placers occur which were derived from Eocene Moroccan and Pliocene Saharan phosphorites (Tooms *et al.*, 1971).

These three examples illustrate three characteristic features of phosphate minerals: their genesis by replacement of carbonates, their ability to be eroded and reworked, and their occurrence on continental shelves.

Sea water is generally nearly saturated with phosphate ions, ranging from about 0·3 ppm PO_4 in deep cold water to about 0·01 ppm in warm surface water. The solubility of phosphate decreases with increasing temperature and increasing pH. These changes occur, and phosphates thus tend to be precipitated, where deep cold oceanic water wells up into shallower warmer waters. These conditions occur in several situations (Fleming, 1957). The most significant locus at the present time appears to be along the western coasts of South America and Africa, where the cold currents of the Humbolt and Benguela move northwards. Rich in nutrients, including phosphates, these waters generate blooms of phytoplankton, which in turn support shoals of fish and flocks of sea birds. Phosphates removed from the sea water by organisms return again when they die and settle on the sea bed with miscellaneous organic matter. Phosphates become concentrated during early compaction of the mud. Constant agitation winnows out the lighter material to leave denser incipient phosphate mud pellets. These continue to become enriched with phosphate, as do bones, teeth and shell debris. In this manner, bedded phosphate rock—phosphorite—is formed (Fig. 70).

Most ancient phosphorite deposits can be explained in terms of this general model. The bulk of the world's bedded phosphates occur in the famous phosphate belt which stretches from Syria through the Levant, Sinai, Egypt, Morocco and into Mauritania. Phosphates occur interbedded with Upper Cretaceous and Eocene chalks and cherts. There can be little doubt that these phosphates resulted from the upwelling of oceanic currents from Tethys onto the broad continental shelves along its southern shore.

The Phosphoria Formation, which contains some of the main phosphate rocks of the USA, occurs in a similar setting. This formation, Permian in age, extends across about 260 000 km^2 of Idaho, Wyoming and Utah.

It was deposited on a marine shelf bounded by the Cordilleran geosyncline to the west. The phosphorites occur interbedded with dolomite and chert. Chert and mudstone increase westwards into the geosyncline. The dolomites grade eastward into red beds and evaporites (McKelvey, 1967; Campbell, 1962).

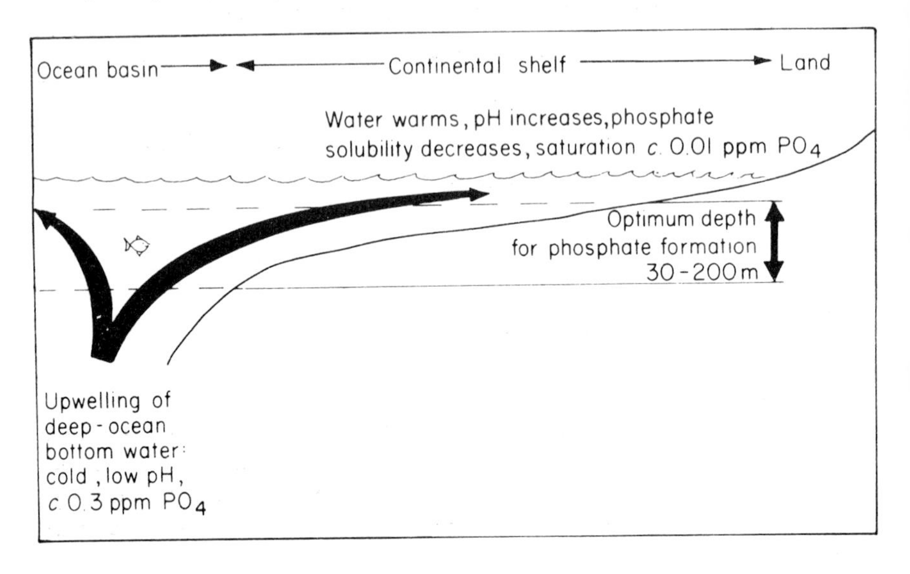

Fig. 70. Diagram to illustrate phosphate formation on a marine shelf. (Figures for optimum water depth from Buschinski, 1964.)

In addition to these bedded-shelf phosphates it is important to remember the other modes of occurrence of phosphate.

Detrital phosphate gravels occur on marine shelves, as already mentioned, and also as alluvial gravels. The "river pebble" deposits of South Carolina and Florida are a case in point. Guano is a deposit rich in phosphates and nitrates formed from the excreta of sea birds and bats. Guano has been a significant source of these minerals notably from the Chilean coasts and on Pacific islands such as Nauru, where solutions rich in phosphates have percolated down from guano to replace reefal limestones.

VI. EVAPORITES

A. Introduction

The evaporites are a group of rocks which includes the mineral salts such as anhydrite and halite. As their name implies it was once widely assumed that these rocks form by the evaporation of salt-rich fluids. Table XVIII lists some of the main evaporite minerals. This is a partial documentation, however, as the total number of evaporite minerals is vast.

Table XVIII
Some of the more common evaporite minerals

Name	Composition	
Anhydrite	$CaSO_4$	Sulphates
Gypsum	$CaSO_4.2H_2O$	
Polyhalite	$CaSO_4.MgSO_4.K_2SO_4.2H_2O$	
Epsomite	$MgSO_4.nH_2O$	
Halite	NaCl	Chlorides
Sylvite	KCl	
Carnallite	$KMg.Cl_3 6H_2O$	
Bischofite	$MgCl_2 6H_2O$	

For many years it was widely accepted that evaporites formed largely by the precipitation or crystallization of salts at the sediment:water interface (Borchert and Muir, 1964). The replacement textures shown by the microscopic fabrics of evaporites point to extensive diagenetic changes. These are to be expected in view of the chemical instability of the evaporite minerals.

A growing body of opinion holds, however, that the genesis of evaporites by diagenesis is the rule rather than the exception (Kirkland and Evans, 1973). The following account of the evaporite rocks first documents their gross geologic characteristics, then reviews their genesis and concludes with a discussion of their economic importance.

B. Evaporites: Gross Geologic Characteristics

It appears most probable that evaporites form from saline-rich fluids—brines. Brines may be generated by concentration of sea water, by evaporation or freezing, or as residual connate fluids in the subsurface. Secondary brines can form where meteoric ground water passes through and dissolves previously formed evaporites.

Normal ocean water contains 3·45% by weight of dissolved substances; 99·9% of the dissolved material comprises the nine ions shown in Table XIX.

Some of the earliest work on the genesis of evaporites was to study salts formed from the evaporation of sea water (Usiglio, 1849; Van't Hoff and Weigert, 1901). Particular attention was paid to volume and composition of the minerals which formed at particular temperatures and phases of

Table XIX
Major constituents of sea water as weight percentages of dissolved materials

Cations			Anions		
Sodium	Na^{2+}	30·61	Chloride	Cl^-	55·04
Magnesium	Mg^{2+}	3·69	Sulphate	SO_4^{2-}	7·68
Calcium	Ca^{2+}	1·16	Bicarbonate	HCO_3^-	0·41
Potassium	K^+	1·10	Bromine	Br^-	0·19
Strontium	Si^{2+}	0·03			

evaporation. These studies demonstrated two main facts; that inconceivable quantities of sea water were necessary to form observed volumes of evaporites in a closed system, and that the observed percentages of salts in an evaporite assemblage differ somewhat from those produced by the evaporation of sea water.

To amplify the first of these points: a column of sea water 1000 m high would evaporate out to form 14·85 m of salts. Many evaporite basins, however, are thousands of metres thick and thus simplistically require improbably large volumes of sea water to beget them.

Figure 71 shows the observed percentages of salts in normal sea water compared with those found in the Permian Zechstein basin of the North Sea.

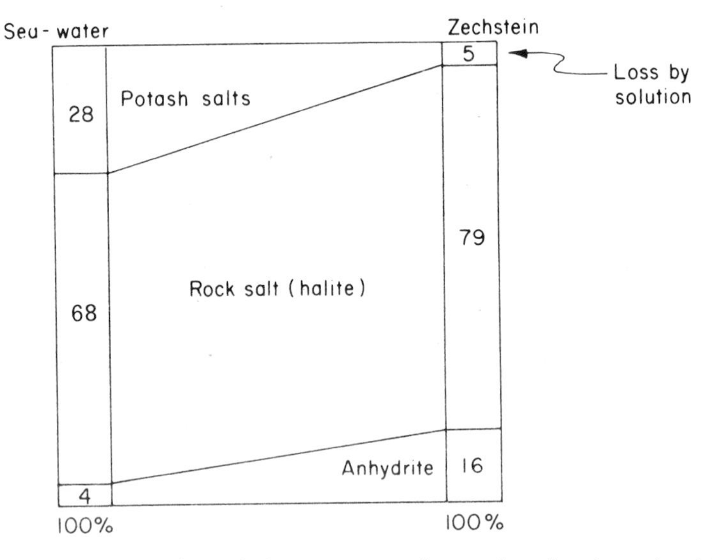

Fig. 71. Comparative sections of the percentages of evaporite minerals produced by the evaporation of average sea water, and the average observed percentages of minerals in the Zechstein evaporites of the North Sea basin.

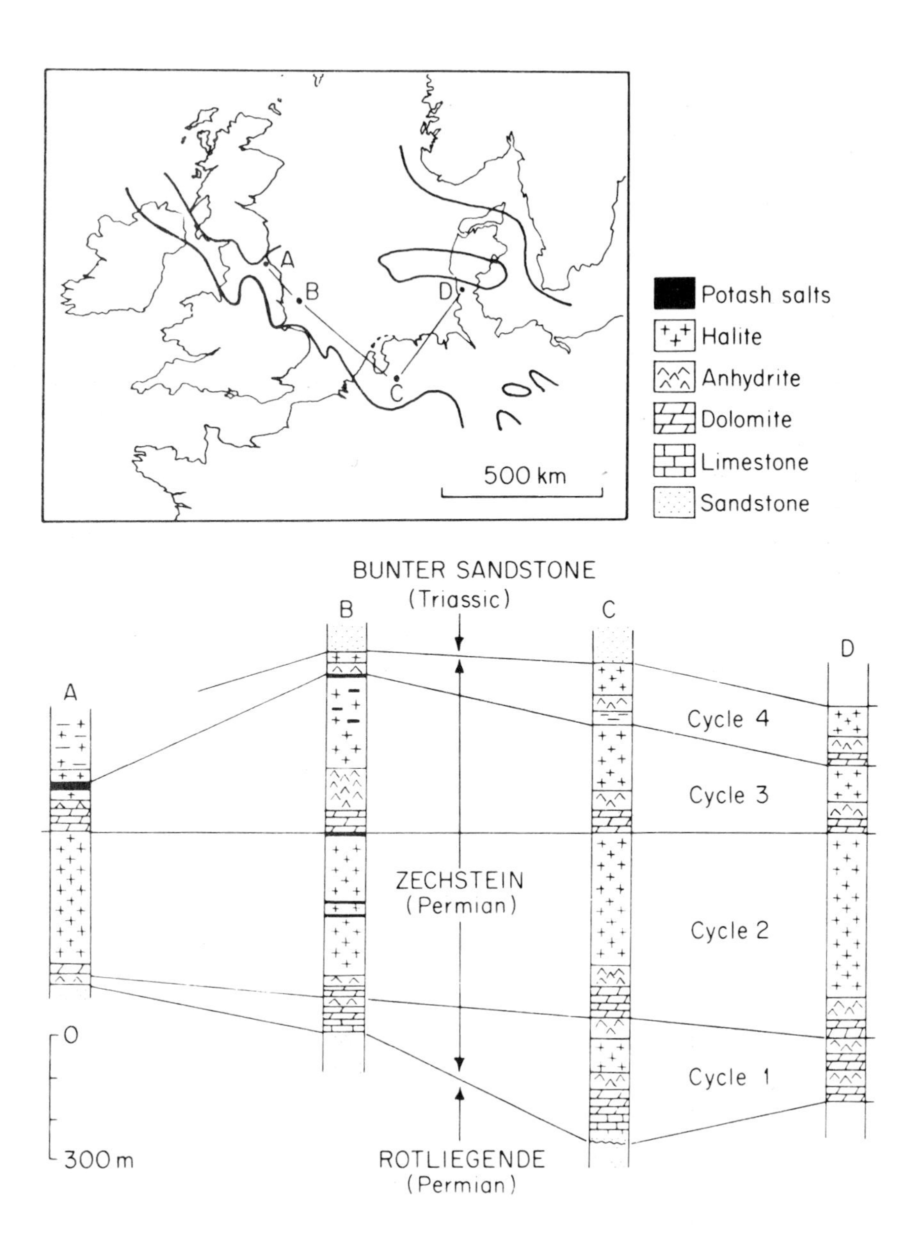

Fig. 72. Upper: map showing the approximate distribution of the Zechstein (Permian) evaporites of the North Sea basin. Lower: cross-section demonstrating the lateral continuity of four major evaporite cycles.

An attractive explanation for these two points is that evaporite formation occurs in a silled basin. It is a matter of observation that evaporite formations characteristically occur in basins which had restricted access to the sea. Examples include the Zechstein of the North Sea (Fig. 72), the Michigan basin, the Paradox salt basin and the Canadian Devonian evaporites.

In a restricted basin it is easy to see how sea water from the open ocean may flow into the basin. Here excessive evaporation causes the sea water to become concentrated. The incipient brine sinks to the basin floor because of its higher density. The sill prevents drainage of the brine out to the open sea. Continuous recycling of the brine increases concentration to the point at which evaporite minerals begin to crystallize on the basin floor (Fig. 73). This process would be aided by a fluctuating sea level which allowed

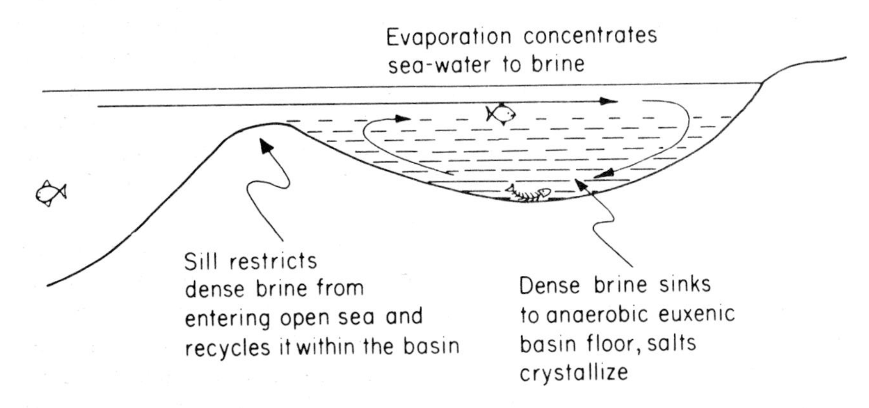

Fig. 73. Illustration of the classic barred basin model for evaporite genesis by direct crystallization from brine at the sediment/fluid interface.

repeated influxes of water over the sill, followed by a drop in water level so as to completely restrict the body of brine. This is the classic "evaporating dish" mechanism for evaporite genesis (Sloss, 1969).

Supporting evidence for this mechanism includes the fact that evaporites tend to be zonally arranged within a basin, with salts requiring higher salinity for their formation occurring towards the depocentre. Similarly evaporite minerals tend to be cyclically arranged in the same motif, i.e.

↑		
	Potassium salts (carnallite, polyhalite, etc.)	
Increasing salinity	Rocksalt (halite)	Brine
	Anhydrite	
	Dolomite	— — — — — — — —
	Limestone	Normal sea water

This cyclicity is classically demonstrated in the Zechstein evaporites of the North Sea basin (Fig. 72) but is also found in most other examples. These cycles are sometimes hundreds of metres thick when fully developed. Cyclicity is also present, however, on a much smaller scale. The monotonous repetition of interlaminated couplets of dolomite with anhydrite and of halite with potash salts will be examined more closely in the next section.

Returning again to the gross geology of evaporites, it is noticeable that they occur in two particular tectonic settings.

The first of these are the intracratonic basins (defined on p. 342), which lie within stable cratonic shields. These basins are characterized by gradual down-warping over a prolonged period of time, accompanied by infilling with diverse continental and shallow marine deposits including evaporites. Within a single basin these often range over a considerable span of time.

Thus in the Michigan basin of North America, salt formations range in age from Silurian to Lower Carboniferous (Mississippian). Similarly in the Williston basin athwart the Canadian/USA border, evaporites formed intermittently from the Devonian through to the Permian.

An additional characteristic feature of these basins is that the evaporites are closely associated with reefal limestones. Sometimes a distinct reef belt occurs around an evaporite infilled depocentre, as in the Silurian Michigan basin, and in the Permian Delaware basin of West Texas. In other instances, such as the Middle Devonian Elk Point basin, pinnacle reefs occur within the main salt depocentre (Fig. 74). Barrier reefs may also contribute to the sill on the seaward side of an intracratonic basin.

The second common tectonic setting for evaporites is in the ocean margin coastal basins and their rift valley precursors. In Chapter 9 (p. 358) it is explained how tension in the earth's crust generates rift valley systems. Where these occur in cratons, rift basins are infilled by continental facies and volcanics. As tension develops, accompanied by crustal thinning, the rift floor sinks to sea level. At this point in time conditions are favourable for extensive intermittent marine incursions and hence for evaporite formation. Ultimately the rift is split in two, and an incipient ocean forms along the axis. The fault-bounded coastal basins of the ocean margin show a vertical sequence of continental clastics, evaporites and normal marine facies. Evaporites are found in this setting on both sides of the Atlantic from the Grand Banks of Newfoundland, the Gulf of Mexico (the Louann salt) to Brazil on the west and also along the coast of West and South West Africa (p. 362).

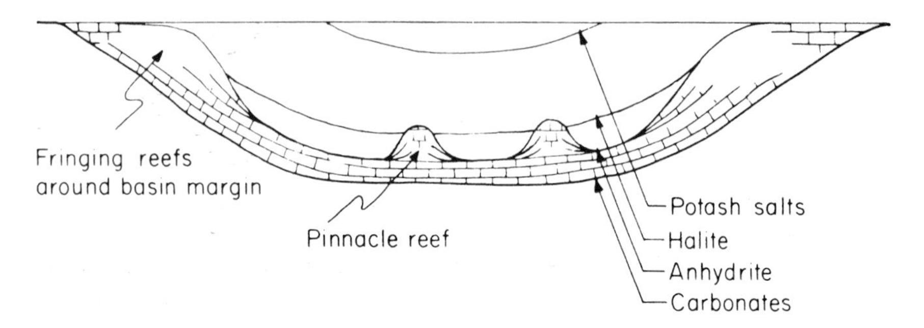

Fig. 74. Diagrammatic cross-section of an intracratonic evaporite basin and associated carbonates. This highlights the problem of whether evaporite crystallization occurred from euxenic brines on the basin floor, synchronous with carbonate reef growth in surface waters of normal salinity. Alternatively the evaporites may have formed by the replacement of sabkha supratidal carbonates when sea level dropped to the level of the basin floor.

A further particular feature of the evaporite rocks is their high degree of plasticity. When buried beneath an overburden of younger sediment salt can act as a lubricant to permit the cover to deform into eccentric structural styles. The salt acts as a lubricant and zone of decollement along which the mobile cover becomes detached from the rigid basement. Notable examples of this phenomenon occur in Iran and in the Jura Mountains of southern France.

In other instances salt forms discrete pillows, walls and domes. Salt domes or diapirs may be only a few kilometres across, but they can extend vertically up through a sedimentary cover thousands of metres thick. The crest of the dome often contains a cap rock of limestone, dolomite, anhydrite and gypsum. It is often characterized by an overhanging subvertical surface. The adjacent sediments are generally extensively faulted over the crest of the dome and a characteristic "rim syncline" may encircle the pillar (Fig. 75).

Salt domes such as these are extensively recorded from the Arabian Gulf, from north Germany and from the coastal basins previously discussed. Particularly well documented are the examples from the Gulf Coast of Louisiana and Texas (Halbouty, 1967).

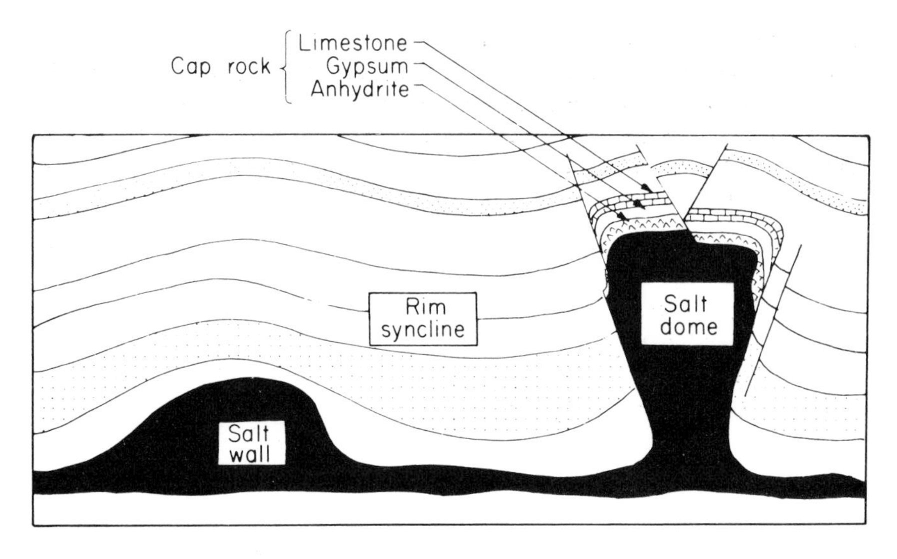

Fig. 75. Diagrammatic cross-section to show the morphology of salt-deformation structures. Salt domes (diapirs), though sometimes only a few kilometres in diameter, can penetrate over 1000 km of sedimentary cover.

C. Carbonate-Anhydrite Cycles

The general relationships of carbonate-anhydrite cycles are well exemplified in the Upper Devonian, Stetler Formation of western Canada (Fuller and Porter, 1969). Followed in the subsurface the Upper Devonian passes laterally from red beds under Saskatchewan, through anhydrite and carbonate–anhydrite rocks into open marine sediments under western Alberta (Fig. 76). In one bore-hold core, 13 rhythmic alternations of carbonate and anhydrite rocks were counted in a 15·24 m interval from the upper part of the formation. The carbonate members are in part limestone and in part dolomite. Near Calgary, a wedge of marine limestone extends eastwards into the evaporites, and this limestone, the Crossfield member, is the reservoir rock of the Olds gas field of Alberta. Anhydrite rocks are for the most part "tight" with respect to oil and gas, and in consequence make excellent seals to hydrocarbon reservoirs.

Another example of carbonate–anhydrite cycles is found in the upper part of the Madison Limestone Formation of south-east Saskatchewan (Fuller, 1956). The anhydrite units die out westwards towards the open marine facies, while some of the limestones tend to die out eastwards in the direction of the inferred shoreline. The carbonate–anhydrite cycles of the Mississippian are much thicker than those of the Middle Devonian Stetler Formation: some cycles are in excess of 30 m in thickness. Oolites and carbonate mud-pellet rocks are abundant in the limestones, suggesting that they were deposited in shallow-water environments. Some of the limestones have undergone only limited cementation, and are oil reservoirs, e.g. the Midale cycle, Fig. 77. Locally the overlying anhydrite rocks form the caps to the reservoirs, but the situation is complicated by the fact that the

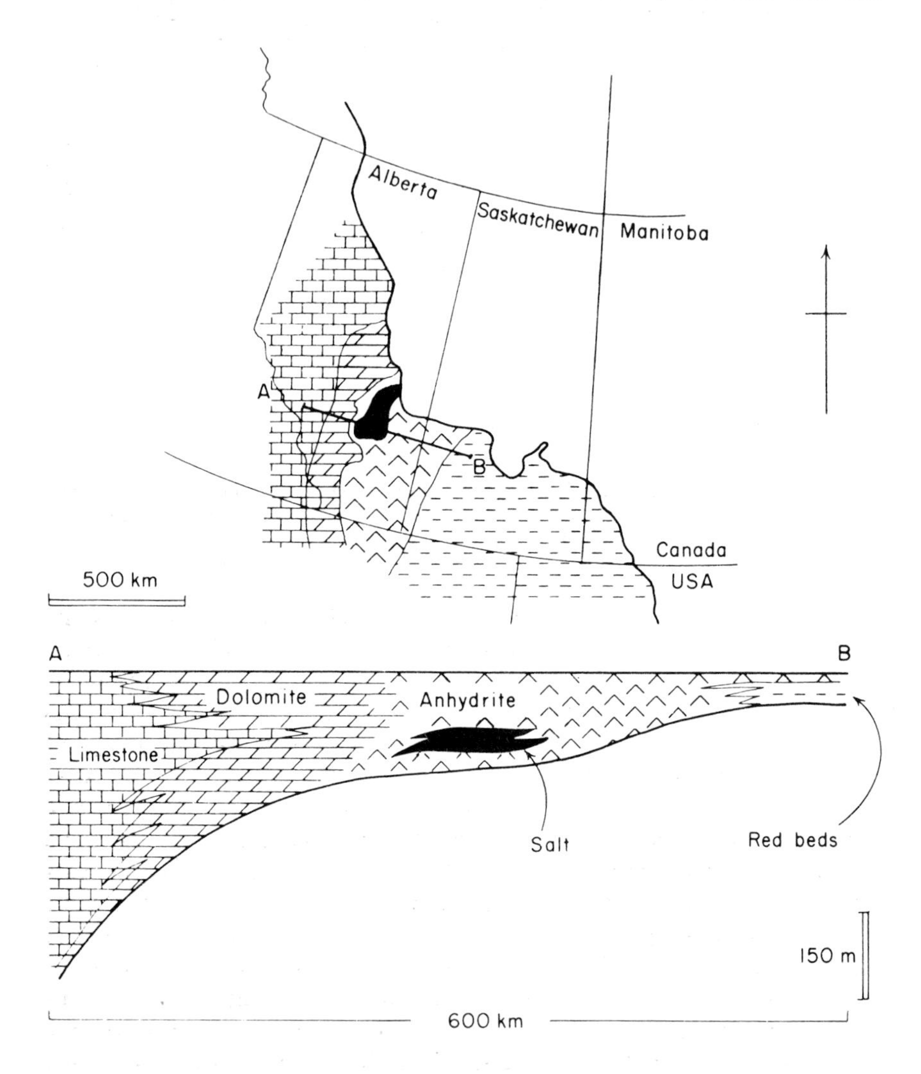

Fig. 76. Map and cross-section showing the distribution of evaporites and associated carbonates of the Stettler Formation (Upper Devonian), north-western Williston basin. (From Fuller and Porter, 1969.) See how the evaporite depocentre occurs in a restricted embayment separated by dolomite from open marine limestones to the west.

Madison Limestone Formation underlies a major unconformity. However, this is not deleterious because the rocks above the unconformity also act as oil and gas seals.

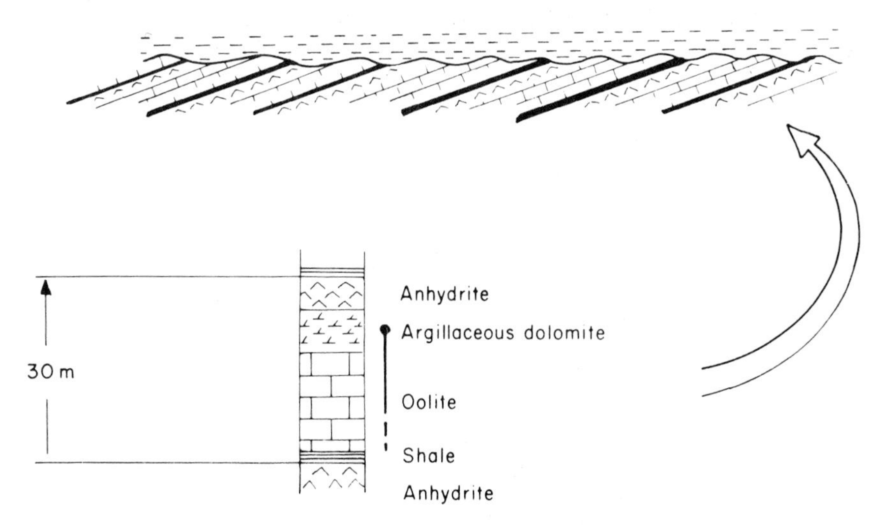

Fig. 77. Midale evaporite cycle of the Madison Formation, Mississippian (Lower Carboniferous), Saskatchewan. This shows the hydrocarbon trapping mechanism provided by the sub-Mesozoic unconformity. (From Fuller, 1956.)

Another well documented example of carbonate–anhydrite cycles, is that of the Jurassic Arab-Darb Formation of the off-shore oil field of Umm Shaif in the Persian Gulf (Wood and Wolfe, 1969). Seven carbonate anhydrite cycles were encountered in 15 m. In each cycle, the limestones shallow upwards and the top of each limestone shows an assemblage of features that characterize shallow subtidal and intertidal zone sedimentation, Fig. 78. The limestones pass up into anhydrite rocks, without a significant break, but at the top of many of the anhydrite units there is a clearly marked erosion surface. Each cycle is, therefore, a carbonate–anhydrite cycle and not an anhydrite–carbonate cycle. Since each carbonate unit is a "shallowing up" entity, with the upper part becoming intertidal, this poses the problem of the environment of formation of the anhydrite.

It is important at this stage to refer to the essential features of the anhydrite. Although the anhydrite forms units that are interbedded with the limestones, the anhydrite is characteristically not bedded but is usually nodular. At one extreme the nodules may occur scattered in a background of carbonate, while at the other the nodules are tightly packed and separated only by thin films of carbonate. In some instances the nodules

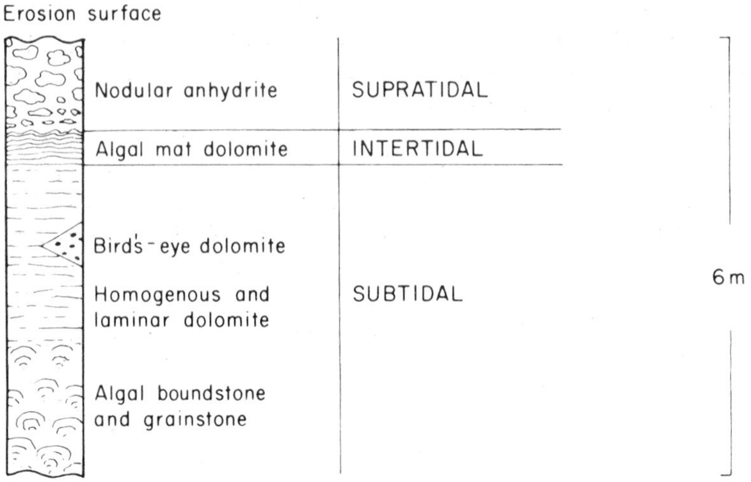

Erosion surface
Nodular anhydrite
SUPRATIDAL
Algal mat dolomite
INTERTIDAL
Bird's-eye dolomite
Homogenous and laminar dolomite
SUBTIDAL
Algal boundstone and grainstone
6 m

Fig. 78. Regressive sabkha cycle in Arab–Darb Formation, Trucial Coast. (From Wood and Wolfe, 1969.)

Nodular structure of the anhydrite has been largely obliterated as nodules coallesced to form enterolithic structure
0·5 m
Stromatolite mound disrupted by growth of nodular anhydrite within it

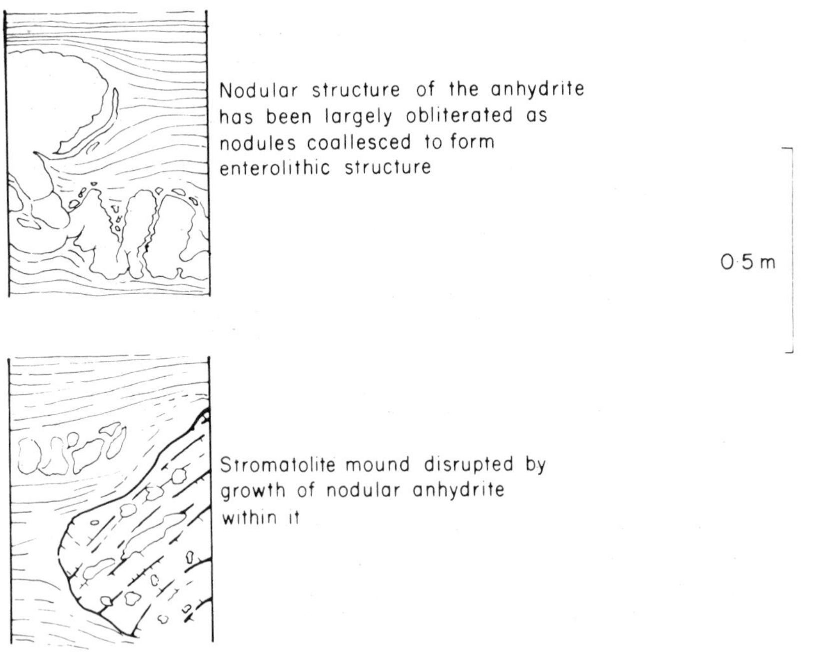

Fig. 79. Sketches of anhydrite from sabkha cycles in the Purbeckian (Upper Jurassic) of southern England. Note the displacive textures indicative of a secondary diagenetic origin. (From Shearman, 1966.)

occur in layers that simulate bedding (Fig. 79) and locally the layers show remarkable contortions, termed "enterolithic structure" because of the resemblance to the coils of an intestine. The nodules comprise masses of tiny plate-like crystals, the arrangement of which often change from place to place within any one nodule—in one part the arrangement may be subparallel, while elsewhere it may be decussate. The nodular form of the anhydrite argues against the anhydrite having been precipitated from a standing body of brine, and this led some earlier workers to argue that the nodules grew displacively within the sediment that now contains them.

Although marine evaporites are apparently not forming to any great extent at the present day, they are locally developed in the coastal areas of some desert regions. It was one of these, the Trucial Coast of the Arabian Gulf, that provides the answers to some of the problems posed by the carbonate–anhydrite cycles of the ancient rocks (Fig. 80). The Trucial Coast is a shoal-water complex of islands and lagoons

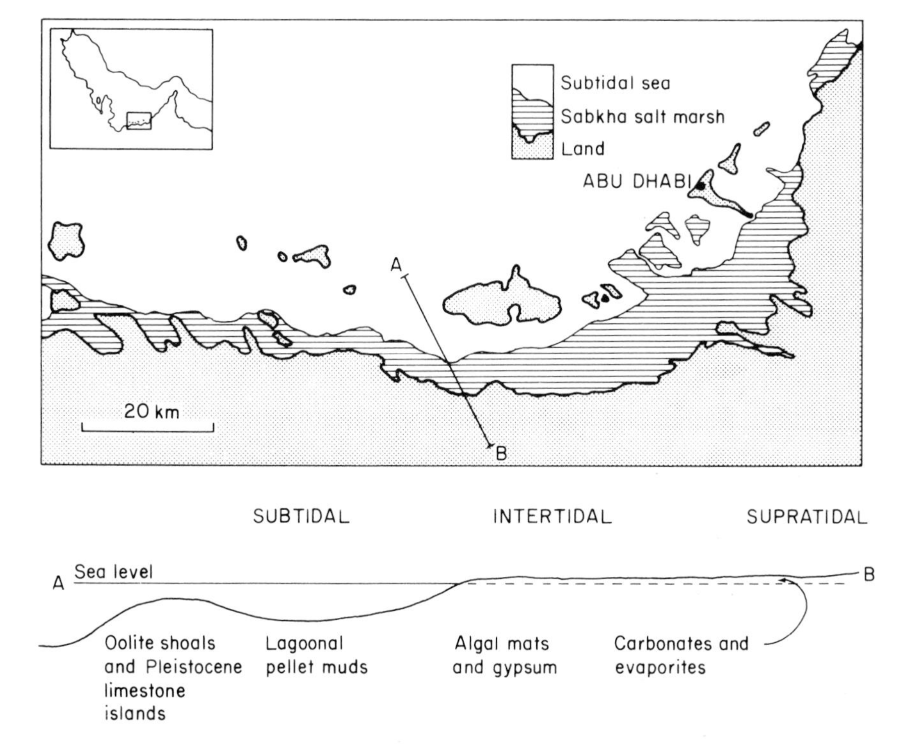

Fig. 80. Map and cross-section showing distribution of the broad sabkha salt marsh zone along the coast of Abu Dhabi, Arabian Gulf, and its associated modern deposits.

some 200 m in lateral extent in which the whole spectrum of shallow-water carbonate sediments is being formed (Evans *et al.*, 1969). The sediments include aragonite muds, pelleted muds, oolite sands, skeletal sands, and a small coral patch reef. By virtue of the shallow water and on-shore wind and waves, a wide supratidal flat has developed along much of the coast. For the greater part the flat stands only 0·5–1·0 m above normal high-tide level and in places extends inland for 20 km or so. It is a completely flat barren desert, and the arabic word "*sabkha*" can be used to describe it. However, in terms of geomorphology, this coastal sabkha is simply the desert zone analogue of the salt marshes of temperate regions. Trenches dug in the sabkha expose a simple upward succession from earlier subtidal carbonate sediments, through intertidal into supratidal sediments. This is a simple regressive cycle of shoreline sedimentation. In places the present intertidal zone is marked by a wide belt of sediment-trapping algal mats, and similar algal mats occur buried under large areas of the sabkha

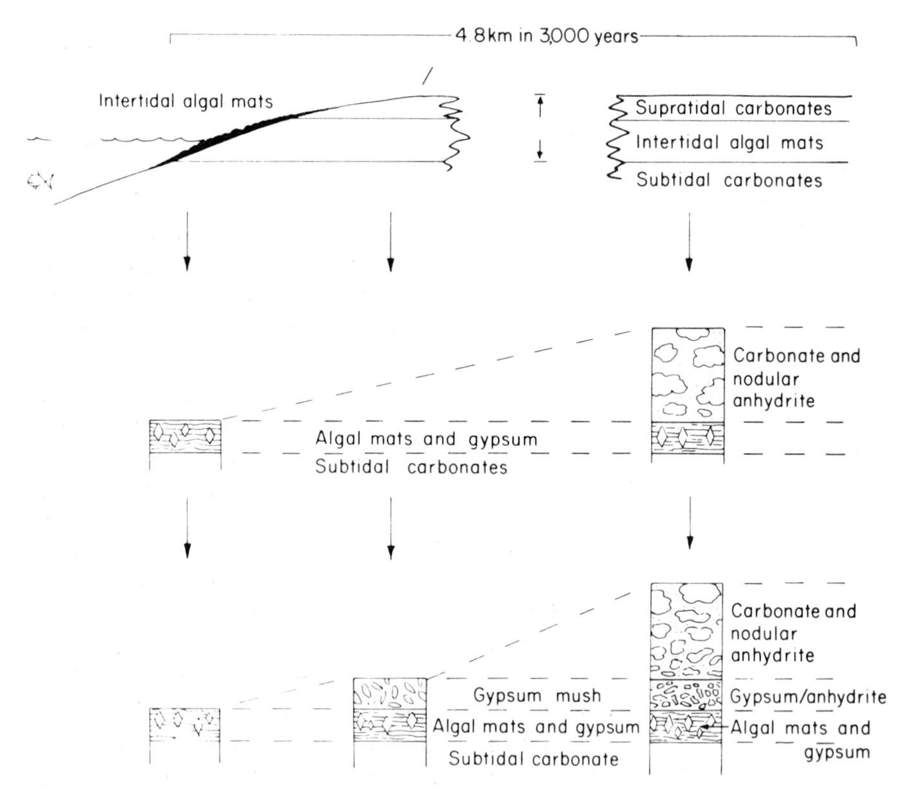

Fig. 81. Schematic cross-section of the Trucial Coast sabkha showing present-day regressive sequence with superimposed early diagenetic sabkha evaporites.

flat. It is evident that the intertidal and supratidal zones have been prograding seawards, so that the sediments of these facies change in age laterally, i.e. they are diachronous (Fig. 81).

Calcium ions are removed from the lagoonal waters by the organic precipitation of aragonite skeletal debris. Thus the water becomes progressively enriched in ions of magnesium, potassium and sulphates. Simultaneously salinity increases due to the high rate of evaporation. Thus, although the background sediment is basically carbonate, evaporite minerals occur in abundance in the sediments of the intertidal and supratidal facies. Gypsum and dolomite are present in the sediments of the present intertidal and low supratidal zones, and they are also found in earlier buried intertidal sediments, while nodules of anhydrite are locally abundant in the sediments of the supratidal facies. In all essential respects this present-day nodular anhydrite is strictly comparable with that found in anhydrite units of the ancient carbonate–anhydrite cycles. It is evident in the field that the nodules grew within the host carbonate sediment by displacement, and that features such an enterolithic folds are growth structures that results from the demand for space as more and more anhydrite crystals grew within layers of coalesced nodules.

The ground waters of the coastal sabkha are marine derived, and due to heat and aridity they have become concentrated by capillarity and evaporation to the point where they are now highly concentrated brines. It is from these interstitial brines that the evaporite minerals are formed. Prior to the discovery of this presently forming anhydrite of the Trucial Coast, it had been thought that the nodular anhydrite of the ancient carbonate–anhydrite cycles may have been formed in shallow, highly saline coastal lagoons. However, a supratidal origin now appears more likely, i.e. the carbonate–anhydrite cycles are desert zone carbonate-shoreline regressive cycles of sedimentation in which evaporites were emplaced in the intertidal and supratidal sediments by penecontemporaneous diagenesis. Vertical repetition of the cycles and the total thickness of the succession is the expression of the diastropic background of relative subsidence. It is convenient to refer to these cycles as "sabkha cycles". In a sense, sabkha cycles can be thought of as being desert zone analogues of coal measures.

Along parts of the Trucial Coast, the coastal sabkha passes into a complex of continental dune sands and inland sabkhas, that in turn pass back into alluvial fans at the front of the Oman Mountains (Fig. 82). The lateral transition from continental desert-zone sediments, through coastal sabkhas with evaporites, out into open marine sedimentation, is essentially that displayed by the facies of the Upper Devonian Stetler Formation of western Canada. In some ancient sabkha cycles, e.g. those in the Windsorian (Carboniferous) of Nova Scotia, the nodular anhydrite units are overlain by red beds. In these cases it appears that the regressions were followed by the establishment of continental conditions.

Carbonate–anhydrite sabkha cycles are present in the Purbeck evaporites of southern England (West, 1964; Shearman, 1966), and thick developments have been encountered in bore-hole cores in the Lower Carboniferous of parts of the east Midlands of England (Llewellyn *et al.*, 1969).

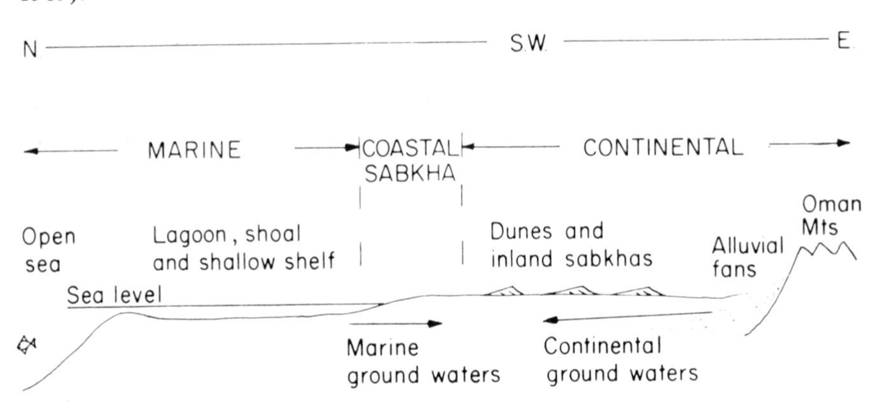

Fig. 82. Schematic cross-section across the Trucial Coast showing transition from arid continental to open marine environments.

It is appropriate to refer briefly to the chemical process that appears to operate in the formation of sabkha-type evaporites. The essential feature is that the evaporite minerals grow interstitially within earlier formed sediment. Dolomite and gypsum develop in the sediments of the present day high intertidal and low supratidal environment and they are also present in the older buried intertidal sediments under the sabkha plain, whereas anhydrite characterizes the supratidal concentration of the interstitial brines inwards through the intertidal into the supratidal zone, it appears that concentration is an important factor in determining which of the two calcium sulphate minerals will be formed. Gypsum forms during the early stages of concentration but anhydrite is not generated until high concentrations are achieved. Although much of the anhydrite appears to have formed directly as anhydrite, there are occurrences where the anhydrite pseudomorphs gypsum. In the latter cases it would appear that gypsum, formed during the early stages of concentration of the ground waters, became made-over into anhydrite as the concentration increased.

The dolomite forms by reaction between the brines and the host carbonate sediment, and the reaction releases calcium ions. Normal sea water carries more sulphate ions than are required to satisfy the calcium in sea water, so that as sea water is concentrated by evaporation, the calcium will be precipitated as calcium sulphate, but the excess sulphate ions remain in the brine. This excess sulphate is available to combine with the calcium ions released by dolomitization of the carbonate sediments, and a further crop of calcium sulphate is generated. Thus by virtue of dolomitization, the sabkha mechanism of evaporite genesis may lead to production of almost twice the amount of calcium sulphate minerals than would be formed by simple evaporation of the same volume of sea water.

Halite deposits are uncommon in carbonate-anhydrite evaporite sequences. However, they do occur occasionally, as for example in the Middle Devonian Stetler Formation of western Canada (Fig. 76). Such a large lens of halite could be accounted for if local subsidence developed at the back of an otherwise emergent sabkha plain. The ground waters, already concentrated sodium chloride brines, would break surface and form a brine pool from which halite would be precipitated.

D. Halite–Potash Evaporite Successions

The halite-potash evaporite successions differ from the cyclic carbonate-anhydrite sequences in a number of important respects. Although many of them contain significant proportions of carbonate and anhydrite rocks, they are characterized by substantial thickness of halite. Generally they comprise extensive basin-shaped accumulations that appear to have formed in large partially enclosed embayments that had only restricted access to the open sea. They are exemplified by the Permian Zechstein evaporites of north-west Europe, by the Middle Devonian evaporite complex of the Elk Point basin of western Canada, the Silurian evaporites of the Michigan basin, the Pennsylvanian evaporites of the Paradox basin Utah, and the Cambrian evaporites of Siberia. In a general way the succession of mineral salts tend to be cyclic, in the sense that they pass up from carbonate-anhydrite rocks into thick piles of halite and in some instances terminate with potassium salts. Four such cycles are developed in the Zechstein evaporites of Germany. It is the thicknesses of halite that are the remarkable feature of these deposits. The Prairies Halite of the Elk Point basin in Saskatchewan is approximately 200 m in thickness, and some 500 m of halite are present in the Zechstein of Germany. Vast quantities of sea water had to be processed to form these thick accumulations and space had to be provided to accommodate them. The latter consideration has led to long controversy as to whether the halite was formed in deep brine-filled basins or in shallow brine pools against a background of subsidence.

A common rock type in the thick halite successions is the so-called "layered halite rock". This consists of repeated alternations of layers of halite, 2–10 cm in thickness, separated by 1 mm-thick laminae of anhydrite, or anhydrite and dolomite sometimes with organic matter. Some geologists have interpreted these cyclic alternations of anhydrite and halite as being the record of annual evaporation cycles. Such cycles are termed "*Jahresringe*" by German geologists. On this basis it has been argued that the 200 m thickness of salt of the Prairie Halite of Saskatchewan was deposited in 4000 years, and that the 500 m of halite in the Zechstein accumulated in 10 000 years. The Jahresringe concept argues against a shallow-water origin, because to accommodate the observed thicknesses would require a background of subsidence of approximately 5 cm a year. Such a rate of subsidence is greatly in excess of what could reasonably be expected in any tectonic or diastropic setting, and it becomes necessary, therefore, to postulate an initially deep basin. However, it must be borne in mind that the deep brine hypothesis depends largely on the validity of the Jahresringe concept.

All too little work has been done on the petrology of salt rocks to date, but the layered halite rocks of the Prairie Halite have been well described (Wardlaw and Schwerdtner, 1966). Each halite layer is an admixture of two types of halite crystals. Some of the crystals carry abundant tiny brine inclusions; these are arrayed in planes parallel to the cube faces, and give the crystals a zoned appearance. In thin sections, under the microscope, the zoned crystals are elongate upwards and the zones appear as chevrons with their apices directed upwards. This type of fabric is indicative of competitive growth of crystals upwards from a substrate, and the abundant brine inclusions suggest intermittent episodes of rapid growth. The other type of halite crystals are clear and free of inclusions, and the mutual relationships of the two types of crystals suggest that the clear halite replaced halite with inclusions. Layered halite rocks comprising layers of halite 2–10 cm in thickness separated by thin laminae of gypsum are forming at the present day, or formed in the Recent geological past, in brine pools on coastal salt flats at the head of the Gulf of California, Mexico. The deposit is, of course, thin—rarely more than 25 cm in thickness. The halite layers are built of zone crystals that have the same arrangement as those in the layers of the Prairie Halite, but the rock is riddled with small dissolution hollows. The salt flats dry out frequently, and when they are periodically flooded by the sea, this incoming water pipes its way down into the halite rock. The way in which the dissolution pipes corrode the crystals of zoned halite is closely similar to the manner in which the clear halite appears to replace the zoned halite in the layered Prairie Halite. Indeed if the dissolution hollows in the present-day occurrence could be filled with clear halite, the two rocks would be identical. If the apparent replacement of zoned halite by clear halite in the layers of the Prairie Halite is the record of piping of the rock by dissolution and subsequent filling by a later generation of clear halite, then this would argue against a deep-water origin for the salt, because it is difficult to conceive how piping could take place beneath a deep standing body of brine. Another thing that is evident in the Recent occurrence in Mexico is that the layers are not annual but may have taken long periods of time to form. Thus interpretations of the environment of deposition of salt deposits like 2 m thickness of the Prairie Halite, range from the extremes of, on the one hand, a brine-filled basin approximately 200 m deep with rapid deposition, to slow accumulation in very shallow water against a background of gentle subsidence on the other.

Where potash salts are present in the major halite evaporite sequences, as for example in the Middle Devonian of the Elk Point basin of Canada or the Zechstein of north-west Europe, the potash salts usually occur in the upper part of the succession, or, in the case of the Zechstein, in the upper part of the major evaporite cycles. It would seem at first sight that they were precipitated residual lakes of brine left after the deposition of halite. However, the deposits pose problems, and one of these is that their overall chemical composition, as expressed by the mineralogy, is not that that would be expected by simple evaporation of sea water.

In the theoretical direct evaporation of sea water, precipitation of halite should be followed by, first, precipitation of $MgSO_4.nH_2O$ (epsomite), then by KCl (sylvite) and finally $MgCl_2.6H_2O$ (bischofite). In most occurrences, the epsomite or its mineralogical equivalent is only weakly developed or is absent, and the bischofite is also characteristically absent. Although it is reasonable to argue that bischofite is so soluble that it is unlikely to survive even if its precipitation was achieved, the depletion with respect to magnesium and sulphate demands explanation. The Middle Devonian potash deposits of Saskatchewan provide an interesting example, because they consist solely of sylvite (KCl) and carnallite ($KMgCl_3.nH_2O$). Not only is sulphate absent, but the proportion of magnesium is much lower than would be predicted. Evidently the brines had been conditioned and their chemistry modified earlier in their history. The absence of sulphate ions could be accounted for by the activities of sulphate-reducing bacteria, but such a process cannot explain the combined deficiency in both sulphate and magnesium.

It is of interest at this stage to refer back to the reactions that are taking place in the formation of the Recent evaporites of the Trucial Coast sabkhas. It will be recalled that dolomitization of the carbonate sediments releases calcium ions, and these promote precipitation of more calcium sulphate than would occur with simple evaporation of sea water. The resultant sabkha brines are, in consequence, stripped of their sulphate and depleted with respect to magnesium. Brines of this composition could generate the restricted mineral assemblage found in the potash zones of the Devonian of Saskatchewan.

The Middle Devonian evaporites of the Elk Point basin are separated from the open marine Mackenzie limestone and shale basin that lay to the north by a limestone–dolomite complex, the Presqu'ile Formation. It is a matter of opinion whether the Presqu'ile Formation was present as a physical barrier, i.e. a leaky dam with a deep-water evaporite basin behind; or, whether the carbonate complex built up as shoal banks *pari passu* with the accumulation of shallow-water evaporites behind. Whichever of the alternatives applied, the sea water that entered the Elk Point evaporite basin had to do so through the Presqu'ile carbonate barrier. The Presqu'ile Formation is largely dolomite with a complex of dolomites and anhydrite rocks behind it, and these in turn pass back mainly into halite. The possibility has to be considered that the incoming sea water was conditioned chemically by dolomitization and associated precipitation of sulphate as it passed through the barrier. In consequence the brines that passed on into the distal parts of the basin would only have been capable of generating halite and the observed potash assemblage of silvite and carnallite.

Some potash deposits evidently accumulated in highly saline lakes of residual brine that remained after the virtual drying out of the evaporite "basin". In other instances the lakes may have been formed after complete drying out, by bleeding of interstitial brines into tectonic depressions. However, in some potash deposits the crystal fabrics are not those of precipitates, but of diagenetic replacements, e.g. the Permian potash of Texas and New Mexico and the Zechstein potash of north-west England. The evidence suggests that the potash minerals were emplaced by reactions between interstitial potassium and magnesium chloride brines and earlier formed minerals.

E. Economic Significance of Evaporites

Evaporite minerals are of great economic importance for three reasons. They are an economic material in their own right, they are closely related to the genesis and entrapment of hydrocarbons, and there is a strong presumption that evaporite-associated brines play an important role in the genesis of certain metallic ores. These three points will now be examined.

Evaporites are a natural resource of great importance. They supply a large proportion of the world's requirements for the rare earth elements, notably sodium and potassium, for the halogens, principally chlorine and bromine, and for sulphur. Chemical industrial complexes thus tend to be situated adjacent to economic evaporite bodies.

Evaporites are of importance in the search for oil and natural gas for three reasons: source, structure and seal (Buzzalini *et al.*, 1969). The conditions which favour evaporite genesis are unfavourable to biological decay. In a basin with brine in its lower depths, organic matter may be preserved on the basin floor interbedded with evaporites because the conditions are hostile to bacteria. Similarly in the sabkha environment, algal laminae are preserved interbedded with evaporites and carbonates. Organic laminae are thus a common constituent of evaporites (be they basinal or sabkha in origin) and there is a strong presumption that evaporites are often potential hydrocarbon-source rocks.

Secondly, as previously pointed out, the plastic behaviour of evaporites enables them to generate structures, even in areas devoid of tectonic activity. Salt domes host a series of potential hydrocarbon traps, both domal anticlines above the cap rock, and faulted flank traps (p. 374).

Finally evaporites are significant to the petroleum geologist because they provide an ideal reservoir seal, combining a maximum of plasticity with a minimum of permeability. Thus evaporites seal reefal reservoirs, such as those of the Williston basin, and provide the seal for the gas in the Rotliegende sandstone reservoirs of the North Sea.

It has already been pointed out that a characteristic evaporite basin is rimmed by reef limestones and that those often contain oil, probably generated from and certainly sealed by, the evaporites.

In many instances the reefs also contain sour gas (H_2S) and a particular type of mineral deposit. These are the telethermal (low temperature) suite of sulphide ores of lead (galena PbS) and of zinc (sphalerite ZnS). Associated minerals include fluorite (CaF_2), baryte ($BaSO_4$), dolomite and crystalline calcite.

These minerals occur in coarse crystalline "roll front" ore bodies within the limestones. The classic example of this type of ore are in the Mississippi Valley and they are sometimes referred to collectively as of "Mississippi Valley type" ores (see p. 393).

There are many lines of evidence that suggest that the sulphur of the sour gas and of the sulphide metals were provided by the reaction between anhydrite and hydrocarbons. In its simplest terms the reaction may be written:

$$CaSO_4 + \text{hydrocarbons} \longrightarrow H_2S + \underbrace{Ca^{+} + CO_3^{2-}}_{CaCO_3} + H_2O \text{ etc.}$$

This reaction is known to be promoted by sulphate-reducing bacteria, but there is reason to believe that it can also take place in a sterile environment; in the absence of bacteria (Dunsmore, 1973). In the latter case this would be an exothermic reaction which could generate the high-temperature crystalline fabrics seen in the ores and their gangue. Hence the superficial resemblance of these bodies to hydrothermal veins.

The implications of this possibility are discussed on p. 393.

In conclusion, the evaporites are a very important group of rocks. Their origin is still a matter for debate, centring on the degree to which they form by crystal growth on the sediment/water interface of basins, and the extent to which they form by replacement of carbonate during the diagenesis of brine-soaked supratidal sabkhas.

Regardless of their precise genesis evaporites are economically important minerals, both in their own right and because of their close association with hydrocarbons and sulphide ores.

VII. REFERENCES

Anon. (1971). Massive Danian limestone key to Ekofisk success. *Wld Oil* May 1971, 51-52.

Badiozamani, K. (1973). The dorag dolomitization model—application to the Middle Ordovician of Wisconsin. *J. sedim. Petrol.* **43**, 965-984.

Bathurst, R. G. C. (1975). "Carbonate Sediments and Their Diagenesis" (2nd Edition) Elsevier, Amsterdam. 658pp.

Baturin, G. N. (1970). Recent authigenic phosphorite formation on the south-west African shelf. *In* "The Geology of the East Atlantic Continental Margin" 1: General and economic papers, 90-97. Inst. Geol. Sci. Rep. 70/13.

Bentor, Y. K. (1980). Marine phosphorites. *Spec. Publs Soci. econ. Palaeont. Miner., Tulsa.* **29**, 264pp.

Black, M. (1953). The constitution of the chalk. *Proc. geol. Soc.* **1491**, LXXXI-LXXXVI.

Borch, C. C. von der (1976). Stratigraphy and formation of Holocene dolomitic carbonate deposits of the Coorong area, South Australia. *J. sedim. Petrol.* **46**, 952-966.

Borchert, H. and Muir, R . O. (1964). "Salt Deposits—the Origin, Metamorphism and Deformation of Evaporites" Van Nostrand-Reinhold, London. 338pp.

Bromley, R. G. (1967). Some observations on burrows of Thalassinodean Crustacea in chalk hardgrounds. *Q. Jl geol. Soc. Lond.* **123**, 159-182.

Brunstrom, R. G. W. and Walmsley, P. J. (1969). Permian evaporites in North Sea basin. *Bull. Am. Ass. Petrol. Geol.* **53**, 870-883.

Bubenicek, L. (1971). Geologie due Gisement de Fer de Lorraine. *Bull. Cent. Rech. Pau.* **5**, 223-320.

Buschinski, G. I. (1964). Shallow-water origin of phosphorite evaporites. *In* "Deltaic and Shallow Marine Sediments" (L. M. J. U. Van Straaten, Ed.), 62-70. Elsevier, Amsterdam.

Butler, G. P. (1969). Modern evaporite deposition and geochemistry of coexisting brines, the Sabkha, Trucial Coast, Arabian Gulf. *J. sedim. Petrol.* **39**, 70-90.

Buzzalini, A. D., Adler, F. J. and Jodry, R. L. (Eds) (1969). Evaporites and petroleum. *Bull. Am. Ass. Petrol. Geol.* **53**, 775-1011.

Byrd, W. D. (1975). Geology of the Ekofisk Field, offshore Norway. *In* "Petroleum and the Continental Shelf of Northwest Europe". I. Geology (A. Woodland, Ed.), 439-445. Applied Science, London.

Campbell, C. V. (1962). Deposition environments of Phosphoria Formation (Permian) in southeastern Bighorn basin, Wyoming. *Bull. Am. Ass. Petrol. Geol.* **46**, 478–503.
Carozzi, A. V. and Gerber, M. S. (1978). Synsedimentary chert breccia: a Mississippian tempestite. *J. sedim. Petrol.* **48**, 705–708.
Chilingar, G. V. and Terry, R. D. (1964). Relationship between porosity and chemical composition of carbonate rocks. *Petrol. Engr* B—**54**, 341–342.
Chilingar, G. V., Bissell, H. J. and Fairbridge, R. W. (1967a). "Carbonate Rocks" (2 Vols). Elsevier, Amsterdam. 471 and 413pp.
Chilingar, G. V., Bissell, H. J. and Wolf, K. H. (1967b). Diagenesis of carbonate rocks. *In* "Diagenesis in Sediments" (G. Larsen and G. V. Chilingar, Eds), 197–322. Elsevier, Amsterdam.
Chilingar, G. V., Mannon, R. W. and Rieke, H. (1972). "Oil and Gas Production from Carbonate Rocks" Elsevier, Amsterdam. 408pp.
Choquette, P. W. and Pray, L. C. (1970). Geological nomenclature and classification of porosity in sedimentary carbonates. *Bull. Am. Ass. Petrol. Geol.* **54**, 207–250.
Dapples, E. C. and Hopkins, M. E. (Eds) (1969). Environments of coal deposition. *Spec. Pap. Geol. Soc. Am.* **114**, 204pp.
Dunham, R. J. (1962). Classification of carbonate rocks according to depositional texture. *In* "Classification of Carbonate Rocks—a Symposium" (W. E. Ham, Ed.), 108–121. Am. Ass. Petrol. Geol., Tulsa.
Dunham, R. J. (1969). Vadose pisolite in the Capitan Reef (Permian), New Mexico and Texas. Depositional Environments in Carbonate Rocks (G. M. Friedman, Ed.). *Spec. Publs Soc. econ. Paleont. Miner., Tulsa* **14**, 182–191.
Dunsmore, H. E. (1973). Diagenetic processes of lead-zinc emplacement in carbonates. *Trans. Instn Min. Metall.* Sect. B, **82**, B168–173.
Ernst, W. G. and Calvert, S. E. (1969). An experimental study of the recrystallization of porcellanite and its bearing on the origin of some bedded cherts. *Am. J. Sci.* **267**-A, 114–133.
Evamy, B. D. (1967). Dedolomitization and the development of rhombohedral pores in limestones. *J. sedim. Petrol.* **37**, 1204–1215.
Evamy, B. D. and Shearman, D. J. (1965). The developments of overgrowths on echinoderm fragments in limestones. *Sedimentology* **5**, 211–233.
Evamy, B. D. and Shearman, D. J. (1969). Early stages in development of overgrowths on echinoderm fragments in limestones. *Sedimentology* **12**, 317–322.
Evans, G. E., Schmidt, V., Bush, P. and Nelson, H. (1969). Stratigraphy and geologic history of the sabkha, Abu Dhabi, Persian Gulf. *Sedimentology* **12**, 145–159.
Fairbridge, R. W. (1967). Phases of diagenesis and authigenesis. *In* "Diagenesis in Sediments" (G. Larsen, and G. V. Chilingar, Eds), 19–89. Elsevier, Amsterdam.
Falls, D. L. and Textoris, D. A. (1972). Size, grain type and mineralogical relationships in Recent marine calcareous beach sands. *Sedimentary Geol.* **7**, 89–102.
Fleming, R. H. (1957). General features of the oceans. *In* "Treatise on Marine Ecology and Paleoecology". I. Ecology (J. W. Hedgpeth, Ed.), Vol.1, 87–108. Mem. geol. Soc. Am. No. 67.
Folk, R. D. (1962). Spectral subdivision of limestone types. *In* "Classification of Carbonate Rocks—a Symposium" (W. E. Ham, Ed.) 62–84. Am. Ass. Petrol. Geol., Tulsa.
Folk, R. L. (1965). Some aspects of recrystallization in ancient limestones.

Dolomitization and Limestone Diagenesis (L. C. Pray and R. C. Murray, Eds). *Spec. Publs Soc. econ. Palaeont. Miner., Tulsa*, **13**, 14–48.

Folk, R. L. (1973). Evidence for peritidal deposition of the Devonian Caballos Novaculite, Marathon, Texas. *Bull. Am. Ass. Petrol. Geol.* **57**, 702–725.

Folk, R. L. and Land, L. S. (1974). Mg/Ca ratio and salinity: two controls over crystallization of dolomite. *Bull. Am. Ass. Petrol. Geol.* **59**, 60–68.

Folk, R. L., and McBride, E. F. (1976). The Caballos novaculite revisited. Part I: Origin of novaculite members. *J. sedim. Petrol.* **46**, 659–669.

Forsman, J. P. and Hunt, J. M. (1958). Kerogen in sedimentary rocks. *In* "The Habitat of Oil" (L. G. Weeks, Ed.), 747–778. Am. Ass. Petrol. Geol., Tulsa.

Friedman, G. M. (1964). Early diagenesis and lithifaction in carbonate sediments. *J. sedim. Petrol.* **34**, 777–813.

Friedman, G. M. (1979). Dolomite is Evaporite Mineral—Evidence from Rock Record and from Sea-Marginal Pools of Red Sea. *Bull. Am. Ass. Petrol. Geol.* **63**, 453.

Fuller, J. G. C. (1956). Mississippian rocks and oil fields in southeastern Saskatchewan. *Sask. Dept Min. Res.* **19**, 72pp.

Fuller, J. G. C. M. and Porter, J. W. (1969). Evaporite formations with petroleum reservoirs in Devonian and Mississippian of Alberta, Saskatchewan and North Dakota. *Bull. Am. Ass. Petrol. Geol.* **53**, 909–926.

Goodwin, A. M. (1973). Plate tectonics and evolution of Pre-Cambrian crust. *In* "Implications of Continental Drift to the Earth Sciences" (D. H. Tarling and S. K. Runcorn, Eds), Vol. 2, 1047–1069. Academic Press, London and New York.

Goudarzi, G. H. (1971). Geology of the Shatti Valley area iron deposit. *In* "The Geology of Libya" (C. Grey, Ed.), 491–500. University of Libya, Tripoli.

Gross, G. A. (1972). Primary features in cherty iron-formations. *Sedimentary Geol.* **7**, 241–262.

Grunau, H. R. (1965). Radiolarian cherts and associated rocks in space and time. *Eclog. geol. Helv.* **58**, 157–208.

Halbouty, M. T. (1967). "Saltdomes—Gulf region, United States and Mexico" Gulf Publishing. 425pp.

Hallam, A. (1963). Observations on the palaeoecology and ammonite sequences of the Frodingham Ironstone (Lower Jurassic). *Palaeontology* **6**, 554–574.

Ham, W. E. (Ed.) (1962). "Classification of Carbonate Rocks—a Symposium" Am. Ass. Petrol. Geol., Tulsa. 279pp.

Ham, W. E. and Pray, L. C. (1962). Modern concepts and classifications of carbonate rocks. *In* "Classification of Carbonate Rocks" (W. E. Ham, Ed.). Mem. Am. Ass. Petrol. Geol. No. 1, 279pp.

Heur, M. D. (1980). Chalk reservoir of the West Ekofisk field. *In* "Sedimentation of North Sea Reservoir Rocks" Norwegian Petroleum Society, Oslo. 19pp.

Hilt, C. (1873). Die Beziehung zwischen der Zusammensetzung und der technischen Eigenschaften de Steinkohlen. *Z. Ver. dt. Ing.* **17**, 194–202.

Holland, H. D. (1972). The geologic history of seawater—an attempt to solve the problem. *Geochim. cosmochim. Acta* **36**, 637–652.

Hoskin, C. M. (1966). Coral pinnacle cementation, Alacran Reef lagoon, Mexico. *J. sedim. Petrol.* **36**, 1058–1074.

Hoskin, C. M. (1971). Size modes in biogenic carbonate sediment, southeastern Alaska. *J. sedim. Petrol.* **41**, 1026–1037.

Jefferies, R. (1963). The stratigraphy of the *Actinocamax plenus* subzone (Turonian) in the Anglo-Paris Basin. *Proc. geol. Ass.* **74**, 1–34.

Khvorova, I. V. (1968). Geosynclinal siliceous rocks and some problems of their origin. Rep. 23rd Int. Geol. Cong., Prague. Sect. 8, 105–112.

Kimberley, M. M. (1979). Origin of oolitic iron formations. *J. sedim. Petrol.* **49**, 111–132.

Kirkland, D. W. and Evans, R. (1973). "Marine evaporites: Origins, Diagenesis and Geochemistry" Benchmark papers in geology. Dowden, Hutchinson and Ross, and Strondsburg, Pennsylvania. 426pp.

Langres, G. L., Robertson, J. O. and Chilingar, G. V. (1972). "Secondary Recovery and Carbonate Reservoirs" Elsevier, Amsterdam. 250pp.

Larsen, G. and Chilingar, G. V. (1967). "Diagenesis in Sediments" Elsevier, Amsterdam. 551pp.

Leighton, M. W. and Pendexter, G. (1962). Carbonate rock type. *In* "Classification of Carbonate Rocks" (W. E. Ham, Ed.), 33–61. Mem. Am. Ass. Petrol. Geol. No. 1.

Lindstrom, M. (1979). Diagenesis of Lower Ordovician hardgrounds in Sweden. *Geol. Palaeontologica* **13**, 9–30.

Llewellyn, P. G., Backhouse, J. and Hoskin, I. R. (1969). Lower-Middle Tournaisian miospores from the Hathern Anhydrite Series, Carboniferous Limestone, Leicestershire. *Proc. geol. Soc.* **1655**, 85–92.

Longman, M. W. (1980). Carbonate Diagenetic Textures from Nearshore Diagenetic Environments. *Bull. Am. Ass. Petrol. Geol.* **64**, 461–487.

Lowenstam, H. A. (1963). Biologic problems relating to the composition and diagenesis of sediments. *In* "The Earth Sciences—Problems and Progress in Current Research" (T. W. Donelly, Ed.), 137–195. University of Chicago Press, Chicago.

McKee, E. D. and Gutschick, R. C. (1969). History of Redwall Limestone of Northern Arizona. No. 114. Mem. Geol. Soc. Am. 726pp.

McKelvey, V. E. (1967). Phosphate deposits. *Bull. U.S. geol. Surv.* **1252**-D. 21pp.

Murray, R. C. (1960). Origin of porosity in carbonate rocks. *J. sedim. Petrol.* **30**, 59–84.

Milliman, J. D. (1974). "Marine Carbonates" Springer-Verlag. Berlin. 375pp.

Park, W., and Schot, D. H. (1968). Stylolitization in carbonate rocks. *In* "Carbonate Sedimentology in Central Europe" (G. M. Friedman and G. Muller, Eds), 66–74. Springer-Verlag, Berlin.

Parker, R. J. and Siesser, W. G. (1972). Petrology and origin of some phosphorites from the Southern African continental margin. *J. sedim. Petrol.* **42**, 434–440.

Patijn, R. J. H. (1964). Die Entstehung von Erdgas—Erdol und Kohle. *Erdgas und Petrochemie* **17**, 2–9.

Peterson, M. N. A. and Von der Borch, C. C. (1965). Chert: modern inorganic deposition in a carbonate-precipitating locality. *Science, N.Y.* **149**, 1501–1503.

Purser, B. (1978). Early Diagenesis and the Preservation of Porosity in Jurassic Limestones. *J. Petrol. Geol.* **1**, 83–94.

Scholle, P. A. (1977). Chalk Diagenesis and Its Relation to Petroleum Exploration: Oil from Chalks, a Modern Miracle. *Bull. Am. Ass. Petrol. Geol.* **61**, 982–1009.

Shearman, D. J. (1966). Origin of marine evaporites by diagenesis. *Trans. Instn Min. Metall.* Ser. B, **75**, 208–215.

Shearman, D. J., Khouri, J. and Taha, S. (1961). On the replacement of dolomite by calcite in some Mesozoic limestones from the French Jura. *Proc. geol. Ass.* **72**, 1–12.

Sloss, L. L. (1969). Evaporite deposition from layered solutions. *Bull. Am. Ass. Petrol. Geol.* **53**, 776–789.

Stopes, M. C. (1919). On the four visible ingredients in banded bituminous coal. *Proc. R. Soc.* B, **90**, 69–87.

Taylor, J. H. (1959). *In* "Petrology of the Northampton Sand Ironstone Formation." Mem. Geol. Surv. Great Britain. 111pp.

Terry, C. E. and Williams, J. J. (1969). The Idris "A" Bioherm and oilfield, Sirte basin, Libya—its commercial development, regional Paleocene geologic setting and stratigraphy. *In* "The Exploration for Petroleum in Europe and North Africa" (p. Hepple, Ed.), 31–48. Inst. Petrol. London.

Tooms, J. S., Summerhayes, C. P. and McMaster, R. L. (1971). Marine geological studies on the north-west African margin: Rabat-Dakar. *In* "The Geology of the East Atlantic Continental Margin" Vol. 4: Africa, 11–25. Inst. Geol. Sci. Rept. 70/16.

Trendall, A. F. (1968). Three great basins of PreCambrian banded iron formation deposition: a systematic comparison. *Bull. geol. Soc. Am.* **79**, 1527–1544.

Unesco (1973). *In* "Genesis of PreCambrian Iron and Manganese Deposits." Proc. Kiev. Symposium 1970. 382pp.

Usiglio, J. (1849). Analyse de l'eau de la Mediterranée sur les Cotes de France. *Annls Chim. Phys.* **27**, 92–107.

Van't Hoff, J. H. and Weigert, F. (1901). Untersuchungen uber die Bildungsverhaltnisse der oceanischen Salzablagerungen, unsbesondere des Stass furter Salzlagers. *Sber. preuss Akad. Wiss.* **23**, 1140–1148.

Wardlaw, N. C. and Schwerdtner, W. M. (1966). Halite-anhydrite seasonal layers in the Middle Devonian Prairie Evaporite Formation, Saskatchewan, Canada. *Bull. geol. Soc. Am.* **77**, 331–342.

Wells, A. J. and Illing, L. V. (1964). Present day precipitation of calcium carbonate in the Persian Gulf. *In* "Deltaic and Shallow Marine Deposits" (L. M. J. U. Van Straaten, Ed.), 429–435. Elsevier, Amsterdam.

Wenk, E. (1949). Die Assoziation von Radiolarienhornsteinen mit ophiolithischen Erstarrungsgesteinen als petrogenetisches. *Problem. Experimentia* **6**, 226–232.

West, I. (1964). Evaporite diagenesis in the Lower Purbeck beds of Dorset. *Proc. Yorks. geol. Soc.* **34**, 315–330.

Weyl, P. K. (1958). The solution kinetics of calcite. *J. Geol.* **66**, 163–176.

Wood, G. V. and Wolfe, M. J. (1969). Sabkha cycles in Arab/Darb formation of the Trucial Coast of Arabia. *Sedimentology* **12**, 165–191.

Zenger, D. H. (Ed.) (1980). Concepts and models of dolomitization. *Spec. Publs Soc. econ. Palaeont. Miner., Tulsa* **28**, 300pp.

6 Transportation and Sedimentation

I. INTRODUCTION

The transportation and deposition of sediments are governed by the laws of physics. The behaviour of granular solids in fluids have been extensively studied by physicists and by engineers of various types. Much of this work has been documented and will be found in texts on hydraulics and fluid dynamics (Dailey and Harleman, 1966; Henderson, 1966; Leliavsky, 1959). Accounts of the physical processes of sedimentation seen from a geological standpoint have been given by Bagnold (1966, 1979) and Allen (1970).

This chapter introduces some of the fundamental concepts of sedimentation as a means to understanding the fabric and structures of the deposits which they generate.

Sedimentation is, literally, the settling out of solid matter in a liquid. To the geologist, however, sedimentary processes are generally understood as those which both transport and deposit sediment. They include the work of water, wind, ice and gravity.

The physics of granular solids in fluids will now be described. This is followed by accounts of sediment transport and deposition by these four processes.

Matter occurs in three phases: solid, liquid and gaseous. The physicist considers gases and liquids together as fluids, on the grounds that, unlike solids, they both lack shear strength. The behaviour of granular solids in liquids and gases are comparable. The similarity of the bed forms and structures of wind-blown and water-laid deposits are the root problem of differentiating them in sedimentary rocks.

To start with, let us consider the behaviour of a sediment particle moving through a fluid. The physics of this situation is expressed by the Reynolds equation, from which is derived a dimensionless coefficient, the Reynolds number, thus:

$$R = \frac{Udp}{\mu}$$

where R is the Reynolds number, U is the velocity of the particle, d is the diameter of the particle, p is the density of the particle and μ is the viscosity of the fluid.

For a given situation, the Reynolds number can be used to differentiate two different types of fluid behaviour at the solid boundary; be it a sphere or a confining surface such as a tube or channel wall.

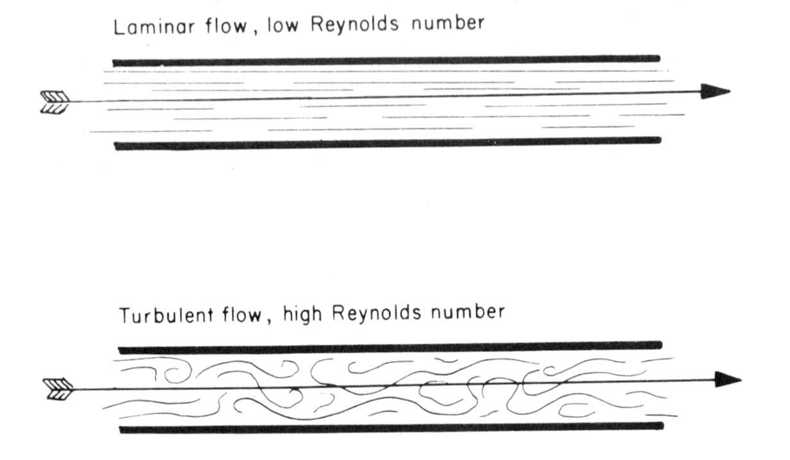

Fig. 83. Illustrative of the difference between laminar and turbulent flow in tubes.

For low Reynolds numbers the fluid flow is laminar, flow lines running parallel to the boundary surface; for high Reynolds numbers the flow is turbulent, generating eddies and vortices (Fig. 83). For flow in tubes the critical Reynolds number separating laminar and turbulent flow is about 2000. For a particle in a fluid the critical number is about one. Stokes law of settling, discussed on p. 13, is derived from the Reynolds equation. Note that turbulence is proportional to velocity, but inversely proportional to viscosity.

A second important coefficient of fluid dynamics is the Froude number. This is essentially the ratio between the force required to stop a moving particle and the force of gravity; that is the ratio of the force of inertia and the acceleration due to gravity. Hence:

$$F = \frac{U}{\sqrt{gL}}$$

where U is the velocity of the particle, L is the force of inertia, i.e. the length travelled by the particle before it comes to rest and g is the acceleration due to gravity.

For flow in open channels the Froude number is expressed thus:

$$F = \frac{U}{\sqrt{gD}}$$

where D is the depth of the channel and U is the average current velocity.

A Froude number of one separates two distinct types of fluid flow in open channels. Each flow regime generates specific bed forms and sediment structures. These will be described in detail in the section on aqueous traction sedimentation.

Let us now consider the mechanics of particle movement. Essentially a grain can move through a fluid (liquid or gaseous) in three different ways: by rolling, bouncing or in suspension (Fig. 84).

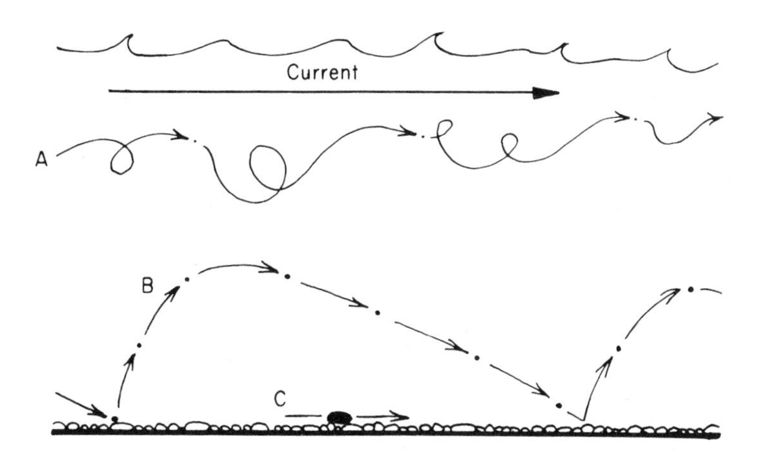

Fig. 84. The mechanics of particle movement. A: suspension; B: bouncing (saltation); C: rolling.

In a given situation, the heaviest particles are never lifted from the ground. They remain in contact with their colleagues, but are rolled along by the current. At the same velocity, lighter particles move down-current with steep upward trajectories and gentler downward glide paths. This process is known as saltation. At the same velocity the lightest particles are borne along by the current in suspension. They are carried within the fluid in erratic but essentially down-flow paths never touching the bottom or ground.

In a situation such as a river channel, therefore, gravel will be rolling along the bottom, sand will sedately saltate, and silt and clay will be carried in suspension. Sand and gravel are generally referred to as the traction carpet or the channel bed load. The silt and clay, loosely termed "fines", are referred to as the suspended load.

Considerable importance is attached to the critical flow-velocity needed to start a particle into motion. The critical flow-velocity for a particle is a function of the variables contained in the Froude and Reynolds equations. A number of empirical experimental and theoretical studies have been made to determine the critical flow-velocity for varying sediment grades notably by Shields (1936) and Hjulstrom (in Sundborg, 1956). Of these the latter is the best known (Fig. 85). As one might expect, the critical fluid-flow

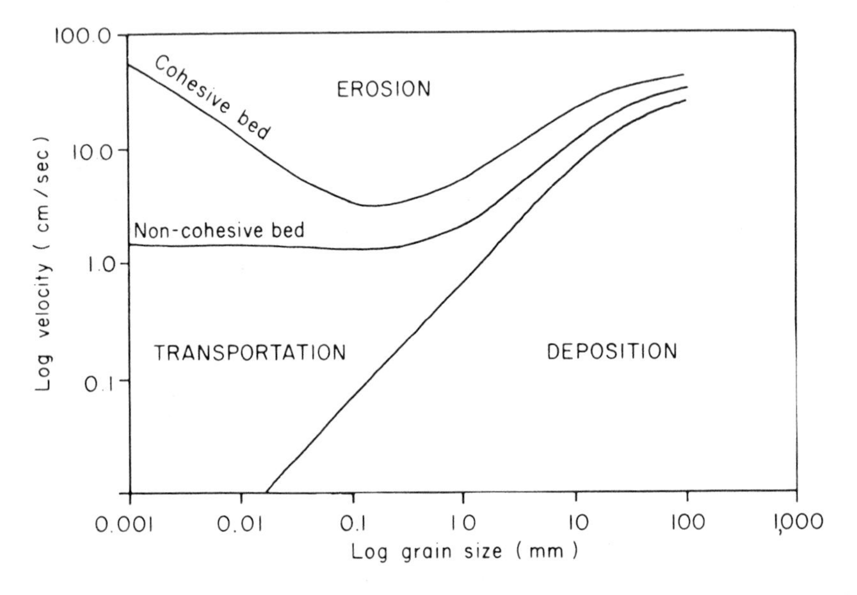

Fig. 85. Hjulstroms graph showing the critical velocities required to erode, transport and deposit sediments of varying grades. (After Sundborg, 1956.)

increases with grain size. An exception to this rule is noted for cohesive clay bottoms. Because of their resistance to friction, they need rather higher velocities to erode them and commence movement than for silt and very fine sand. This anomaly, termed the Hjulstrom effect, is responsible for the preservation of delicate clay laminae in tidal deposits.

Let us now consider the sediment types that result from different sorts of fluid flow in both liquid and gaseous states, i.e. in water and air, for the purpose of the geologist. Three types can be recognized: traction deposits, density current deposits and suspension deposits. Transportation in a traction current is mainly by rolling and saltating bed load. The fabric and structure of sediment deposited from a traction carpet reflect this manner of transport. They are generally cross-bedded sands. Traction currents may be generated by gravity (as for example in a river), or by wind or tidal forces in the sea. Desert sand dunes are also traction deposits.

The deposits of density currents, by contrast, originate from a combination of traction and suspension. Their fabric and structures are correspondingly different from those of traction deposits. They are characterized by mixtures of sand, silt and clay, which lack cross-bedding and typically show graded bedding. Density currents are caused by differences in density in fluids both liquid or gaseous. This may arise from thermal layering or from differences in salinity or turbidity in liquids. The result is for the denser fluid to flow by gravity beneath the less dense fluid and to traverse the sediment substrate. Geologically the most important density flow is the turbidity current, a predominantly subaqueous phenomenon. Eolian turbid-flows include nuées ardentes and certain types of high-velocity avalanches and mud flows. They are rare and not volumetrically significant depositional mechanisms. The fundamental differentiation of many sediments into cross-bedded traction-current deposits and graded turbidities was recognized by Bailey in 1930.

Table XX
A tabulation of sedimentary processes and the sediment types which they generate

Subaereal	Traction deposits	Predominantly cross-bedded sands
	Density deposits	Nuées ardentes, etc.
	Suspension deposits	Loess
Subaqueous	Traction deposits	Predominantly cross-bedded sands
	Density (turbidity) deposits	Graded sands, silts and clays
	Suspension deposits	Nepheloid clays
Mass gravity transport	Subaerial	Generally unstratified poorly sorted deposits of boulder to clay grade (diamictites)
	Subaqueous	
Glacial transport		

The third group of sedimentary deposits are those which settle out from suspension. These are fine-grained silts and clays and include wind-blown loess and the pelagic detrital muds or nepheloids of ocean basins.

A fourth major group of sediment types is the diamictites (Flint *et al.*, 1960). These are extremely poorly sorted rocks which show a complete range of grain size from boulders down to clay. Diamictites are formed both from glacial processes and also by mud flows, both subaerial and subaqueous.

Table XX summarizes the main sedimentary processes and shows how these generate different depositional textures. The various processes will now be described in more detail.

II. AQUEOUS PROCESSES

A. Sedimentation From Traction Currents

Let us consider now one of the most important processes for transporting and depositing sediment. Traction currents are those which, as already defined, move sediment along by rolling and saltation as bed load in a traction carpet. Continuous reworking winnows out the silt and clay particles which are carried off in suspension. Finer, lighter sand grains are transported faster than larger heavier ones. The sediments of a unidirectional traction current thus tend to show a down-stream decrease in grain size, termed "size grading". Traction currents may be unidirectional, as in river channels. In estuaries, and in the open sea, sediment may be subjected to the to-and-fro action of tidal traction currents, or to even more complex systems.

The basic approach to understanding traction-current sedimentation has been through experimental studies of unidirectional flow in confined channels. These can be made in artificial channels, termed flumes, which are now standard equipment in many geological laboratories. The phenomenon now to be described has been documented by Simons *et al.* (1965) and is demonstrated as an integral part of many university courses on sedimentology.

The experiment commences with the flume at rest. Sand on the flume floor is flat, current velocity is zero. Water is allowed to move down the flume at a gradually increasing velocity. The sand grains begin to roll and saltate as the critical threshold-velocity is crossed. The sand is sculpted into a rippled bed form. Steep slopes face down-current, gentle back slopes face up-current. Sand grains are eroded from the back slope and are deposited on the down-stream slope. Thus the ripples slowly move down-stream depositing cross-laminated sand.

With increasing current-velocity the bed form changes from ripples to dunes. These are similar to ripples in their shape, mode of migration and internal structure. They differ in scale, however, being measurable in decimetres rather than in centimetres.

Through these phases of ripple and dune formation, the Froude number, introduced in the previous section, is less than one. With increasing velocity the Froude number approaches one. This value separates two flow regimes. The lower flow regime, with a Froude number of less than one, generates cross-laminated and cross-bedded sand from ripples and dunes. As the velocity increases to a Froude number of one the dunes are smoothed out and the bed form assumes a planar surface. Sand is still being transported and deposited, however. Now it is laid down in horizontal beds with the grains aligned parallel to the current. This stage is termed "shooting flow". As the current velocity increases further, the Froude number exceeds unity. The plane bed form changes into rounded mounds, termed "antidunes".

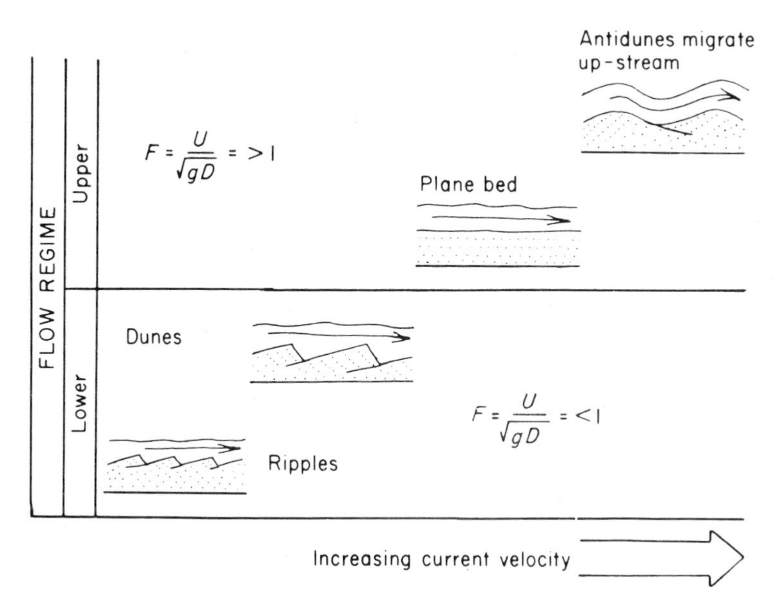

Fig. 86. Bed forms and sedimentary structures for different flow regimes. (After Harms and Fahnestock, 1965 and Simons *et al.*, 1965.)

In contrast to dunes, these tend to be symmetric in cross-section. They may be stable or they may migrate up-current to deposit up-current dipping cross-beds. Plane bed and antidunes are attributed to the upper flow-regime.

With decreasing current velocity, the sequence of bed forms is reversed. Antidunes wash out to a plane bed, the dunes form, then ripples and so back to plane bed and quiescence at zero current velocity (Fig. 86).

This experiment elegantly demonstrates the relationship between stream power, bed form and sedimentary structures. It is, however, difficult to quantify this relationship and to enumerate flow parameters from the study of grain size and sedimentary structures in ancient deposits.

Experiments have shown that if the fluid parameters are held constant (i.e. velocity and viscosity) then the thresholds at which one bed form changes to another varies with grain size. Most important of all, it is an experimental observation that the ripple phase is absent in sediments with a fall diameter of more than about 0·65 mm. (The fall diameter is a function of the particle diameter and the viscosity of the fluid. The fall diameter of a particle decreases with increasing viscosity. For example, one sand grain may have the same fall diameter as a larger particle in a more viscous fluid.) It is a matter of field observation that ripple lamination is absent in sediments with a diameter of over about 0·5 mm.

A second important point to note is the way in which temperature affects sedimentary structures. Harms and Fahnestock (1965), in their study of reaches of the Rio Grande, showed how, for similar discharges, either plane bed or dunes could be present. The main controlling variable seemed to be temperature. This controlled the fluid viscosity and hence the fall diameter of the sediment.

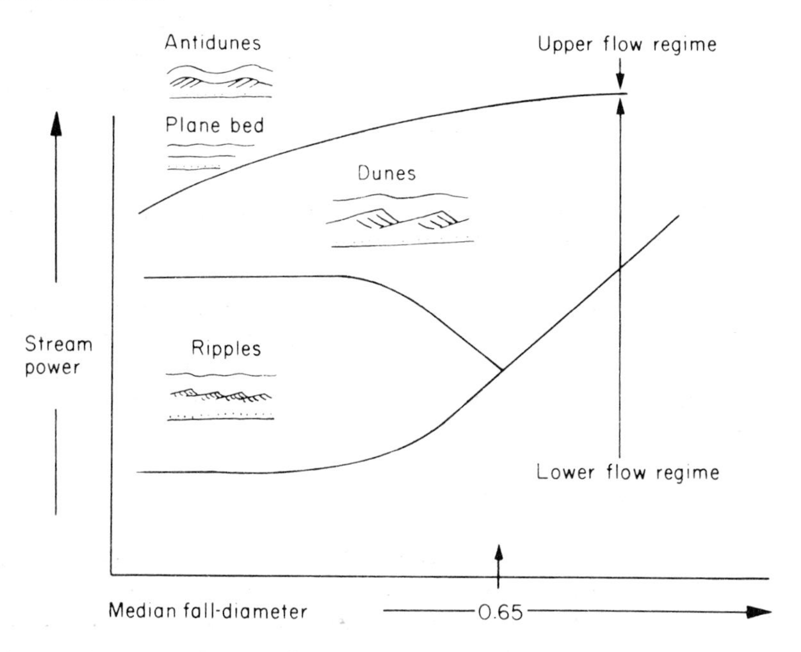

Fig. 87. The relationship between stream power, fall diameter, bed form and sedimentary structure in a unidirectional traction current system. (After Simons *et al.*, 1965.)

Figure 87 shows the relationship between stream power (current velocity, more or less), fall diameter (grain size, more or less), bed form and sedimentary structures. Figure 88 shows the relationship between grain size and sedimentary structures in a typical ancient fluvial deposit laid down by unidirectional traction currents.

The previous account was concerned solely with unidirectional currents such as occur in fluvial channel systems. Bidirectional current systems, their bed forms and internal structures are far more complex. These occur in the

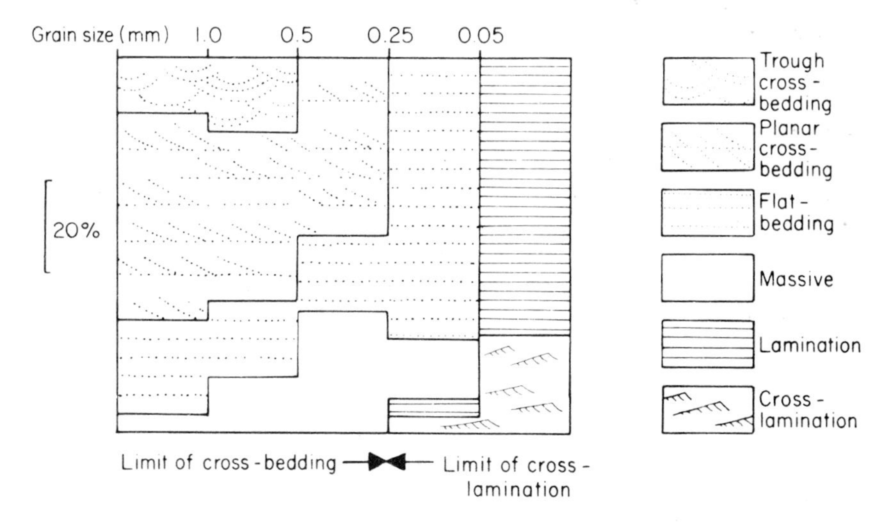

Fig. 88. Relationship between grain size and sedimentary structure in alluvial sandstones of the Torridon Group, Scotland. (Data from Selley, 1969.)

tidal regimes of estuarine channels and continental shelves. Obviously such environments are less susceptible to observation, and to controlled laboratory experiments. Nonetheless a series of models for marine sandwaves and their internal structures have been put together by Allen (1980).

B. Sedimentation From Turbidity Currents

The concept of the density flow has already been introduced. Where two fluid bodies of different density are mixed together, the less dense fluid will tend to move above the denser one. Conversely, the denser fluid will tend to flow downwards. Aqueous density flows may be caused by differences of temperature, salinity and suspended sediment. Glacial melt streams and certain polar currents tend to flow under gravity beneath warmer, less-dense water bodies. Water discharged by rivers in temperate latitudes often flows out for considerable distances from the shore above the denser, more saline sea water.

Turbid bodies of water with large loads of suspended sediment frequently move as density flows beneath clear water. This particular variety of density current, termed a turbidity current, is of great interest to geologists.

Turbidity currents are widely believed to be a major process for the transportation and deposition of a significant percentage of the world's sedimentary cover.

The concept of the turbidity flow was introduced to geology by Bell (1942). The process was originally invoked as an erosional agent capable of scouring the submarine canyons on delta and continental slope margins. Later it was claimed as the generator of flysch deposits (Kuenen and Migliorini, 1950). This facies, typical of geosynclinal troughs, is characterized by thick sequences of interbedded sand and shale. The sands have abrupt bases, transitional tops and tend to fine upwards. Sands of this type are often genetically termed "turbidites" (see pp. 89 and 311).

This section begins by briefly discussing the hydrodynamics of turbidity flows. The evidence for turbidity currents in the real world is then described, and this is followed by a discussion of the parameters of ancient "turbidites".

The hydrodynamics of density flows have been studied for a number of years. Attempts to equate the various physical parameters which govern density flows have been based on both theoretical and experimental grounds.

The conditions under which a turbidity current will flow down a slope may be expressed thus:

$$S_1 + S_2 = (d_2 - d_1)g \,.\, h \,.\, \alpha$$

where S_1 and S_2 are the shear stress between the turbid flow and the floor beneath and the fluid above, d_2 and d_1 are the density of the turbidity flows and of the ambient fluid respectively ($d_2 > d_1$), g. is the acceleration due to gravity, h. is the height of the flow and α is the bottom slope.

The relationship between the speed of a density flow and other parameters may be expressed as:

$$V = \frac{2(d_2 - d_1)}{d_1} g.h.$$

where V is the velocity.

Derivations and discussions of these equations will be found in Middleton (1966a, b) and Allen (1970).

It can be seen, therefore, that the behaviour of a turbid flow is governed by the difference in density between it and the ambient fluid, by the shear stresses of its upper and lower boundary, by its height, and the angle of slope down which it flows. Additional important factors are whether it is a steady flow, as a turbid river entering a lake, or whether it is a unique event of limited duration as in the case of a liquified slump.

Let us now see to what extent this mechanism can be recognized in the real world. Modern turbidity currents have been studied experimentally in the laboratory and in present-day lakes and seas.

The early experimental studies of Bell (1942) and Kuenen (1937, 1948) showed that when muddy suspensions of sand were suddenly introduced to a flume they rushed down-slope in a turbulent cloud to cover the bottom. Sand settled out first, followed by silt and then clay. Thus beds were deposited with sharp basal contacts which showed an upward-fining grain size profile from sand to clay in the space of a few centimetres. Additional experiments by Kuenen (1965) produced laminated and rippled graded sands from turbid flows in a circular flume. Experiments by Dzulinski and Walton (1963) showed how small turbidity flows could, in the laboratory, generate many of the erosional features scoured beneath ancient turbidite sands.

Experiments such as these have been criticized because particle fall-diameter and fluid viscosity were not proportional to each other and to the flow size. More recently, however, Middleton (1966a, b, 1967) has carried out carefully scaled experiments with plastic beads with a density of $1\cdot52$ g/cm^3 and diameters of about $0\cdot18$ mm. Suspensions of beads released into a standing body of water generated graded beds similar to those of less scientific experiments.

Turning from the laboratory to the outside world, there are numerous cases of modern turbidity flows. They have been described from lakes, such as Lake Mead, by Gould (1951) and from Norwegian fjords by Holtedahl (1965). In these instances it was possible to demonstrate direct relationships between inflows of muddy river water and extensive layered deposits on the floors of the lakes and fjords.

The evidence for the existence of modern marine turbidity flows is equivocal, but still impressive. A common feature of continental shelves and delta fronts is that they are incised by submarine valleys. Where these terminate at the base of the slope it is common to find a radiating fan-shaped body of sediment. Submarine telegraph cables which cross these regions tend to be broken rather frequently. Daly (1936) postulated that these submarine canyons were eroded by turbidity currents, which snapped the cables. One famous and oft-quoted instance of this was the celebrated Grand Banks earthquake of 1929. On the 18th of November there was an earthquake with an epicentre at the edge of the Grand Banks off Nova Scotia. Within the next few hours, 13 submarine telegraph cables were broken on the slopes and ocean floor at the foot of the Banks. No cables were broken on the Banks themselves. Subsequently, Heezen and Ewing (1952) attributed these breakages to a turbidity current. They postulated that the earthquake triggered slumps on the continental slope of the Grand

Banks. These liquified as they fell and mixed with the sea water until they acquired the physical properties of a turbidity flow. According to the sequence and timing of cable breaks, this flow moved out onto the ocean floor at speeds of up to 100 km/h, ultimately covering an area of some 280 000 km^2. Subsequent coring has revealed an extensive, sharp based, clean-graded silt bed over this region.

Many other studies have been published attributing deep-sea sands to turbidity current deposition. These have been described from the Californian Coast (e.g. Hand and Emery, 1964), the Gulf of Mexico (Conolly and Ewing, 1967), the Mediterranean (Van Straaten, 1964), and the Antarctic (Payne *et al.*, 1972). Modern bioclastic deep-sea sands have been attributed to turbidite transportation from adjacent carbonate shelves by Rusnak and Nesteroff (1964) and Bornhold and Pilkey (1971).

These modern "turbidite" sands show a variety of features. They have abrupt, often erosional bases, but none of the characteristic bottom structures found under ancient turbidites. This may be due to the problems of collecting small cores of unconsolidated sediment. Upward size-grading is sometimes but not always present. The sands are frequently clean; an interstitial clay matrix is generally absent. Internally the sands are massive, laminated or cross-laminated. A shallow-water fauna is sometimes present, especially in the bioclastic sands; this contrasts markedly with the pelagic fauna of the intervening muds.

By analogy with modern experimental and lacustrine turbidites, it can be convincingly argued that deep-sea sands such as these were transported from the continental shelves by turbidity currents. These moved down submarine canyons cut in the continental margins and out onto the ocean floors. The decrease in gradient would cause the flow to lose velocity and deposit its load in a graded bed, the coarsest particles settling out first.

This attractive mechanism has been criticized for a number of reasons. First, it has been pointed out that many deep-sea sands do not conform to the ideal turbidite model. Some are clean and well sorted, and are internally cross-laminated. These features could indicate that the sand was deposited by a traction current.

Studies of modern continental rises show the existence of currents which flow parallel to the slope. These are termed geostrophic or contour currents. Photographs and cores show that these currents deposit cross-laminated clean sand. Seismic data show the existence of stacked megaripples, tens of metres high, whose axes parallel the slope (Heezen and Hollister, 1971; Hollister and Heezen, 1972; Bouma, 1972).

An additional argument against the turbidity current mechanism for deep-sea sand transport is to be found in the submarine canyons down which they are believed to flow. Attempts to trigger turbidites by explosions in canyon heads have been unsuccessful (Dill, 1964). The sediments of the canyons themselves often suggest transportation by normal traction currents aided by some slumping and grain flow (Shepard and Dill, 1966; Shepard *et al.*, 1969).

Other critiques of the turbidity current mechanism have been made by Hubert (1964) and Van der Lingen (1969).

Let us turn now from the problems of deciding the role played by turbidity currents in modern deep-sea sands to their ancient analogues.

There is a particular sedimentary facies termed "flysch", which is described in Chapter 8. Many geologists use this term interchangeably with "turbidite", implying that these rocks were deposited from turbidity currents. The sediments termed turbidites show the following features: they are generally thick sequences of regularly interbedded sandstones and shales. These typically occur in orogenic belts or in fault-bounded marine basins. The sands have abrupt basal contacts and show a variety of erosional and deformational structures which are more fully described in Chapter 7. These include pear-shaped hollows, termed "flutes", which taper down-current, and various erosional grooves and tool marks caused by debris marking the soft mud beneath the turbidite sand. The mud often

Fig. 89. Graded bedding in Khreim Group (Silurian) sandstones. Southern Desert of Jordan.

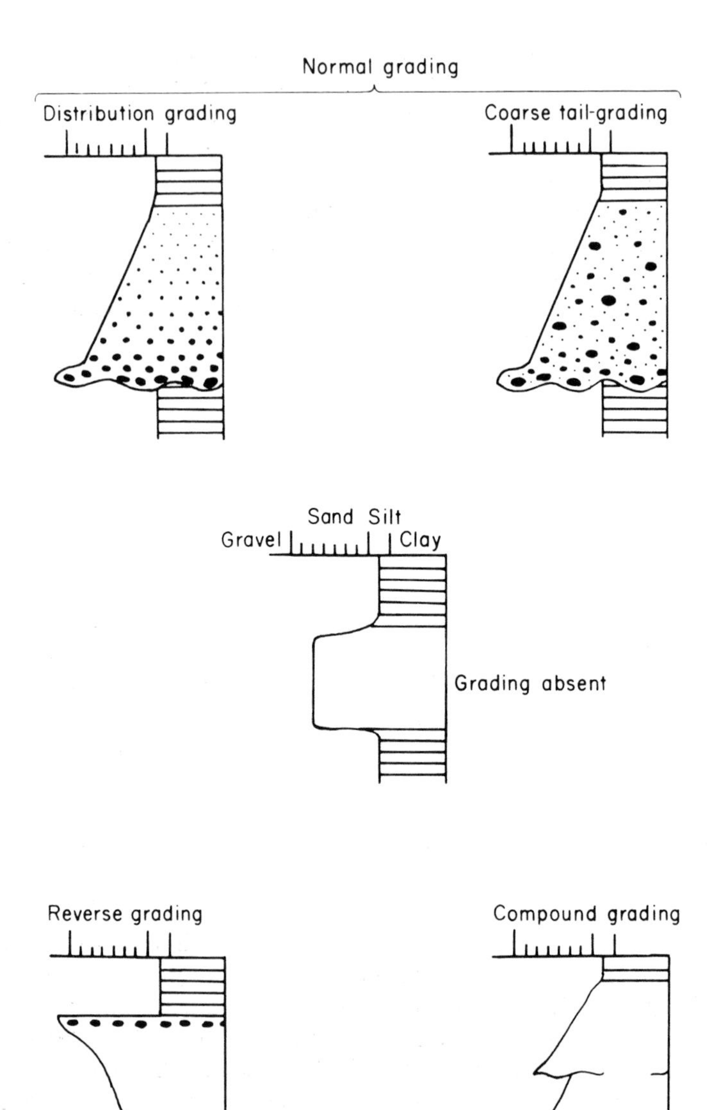

Fig. 90. Various types of graded bedding.

shows deformation caused by differential movement of the overlying sand. This generates load structures, pseudonodules, slides and slumps.

Internally, the sands tend to show an upward-fining of grain size, termed graded bedding (Fig. 89). There are a variety of graded bedding types. Distribution grading shows a gradual vertical decrease of grain size, while maintaining the same distribution, i.e. sorting of the sediment. Coarse tail-grading shows a gradual vertical decrease in the maximum grain size. Hence there is a vertical improvement in sorting. These textural differences may relate to different types of density flow (Allen, 1970, p. 194). Compound grading may be present within a single sandstone bed. Reverse grading has been observed. An absence of grading in a turbidite may indicate a source of uniform sediment grade (Fig. 90).

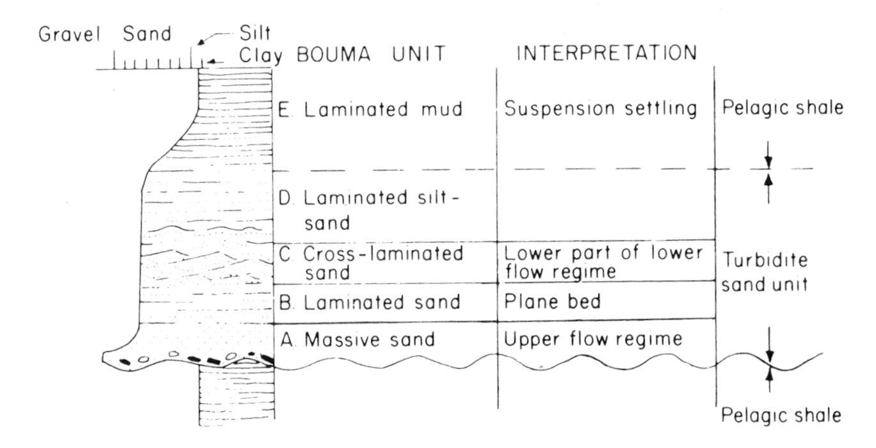

Fig. 91. Turbidite unit showing the complete Bouma sequence and its interpretation in terms of flow regimes.

The internal structures of turbidite beds are few in number, and tend to be arranged in a regular motif, termed a "Bouma sequence" (Bouma, 1962) (Fig. 91). In the ideal model, five zones can be recognized from A to E, and these have been interpreted in flow-regime terms by Walker (1965), Harms and Fahnestock (1965), and Hubert (1967). The scoured surface at the base of the bed is frequently succeeded by a conglomerate of extraneous pebbles and locally derived mud clasts. This indicates the initial high-power erosive phase of the current. Ideally this is overlain by a massive unit attributed to sedimentation from antidunes in the upper flow regime. Antidune up-current dipping foresets in this unit have been discovered by Walker (1967b) and Skipper (1971). The massive A unit is overlain by the laminated B unit, attributable to a shooting flow regime with deposition from a planar bed form. This is succeeded by a cross-laminated C unit, which often shows convolute deformational structures due to penecontemporaneous dewatering (see p. 234). The cross-laminated zone reflects sedimentation from a lower flow regime. This C unit is overlain by a second laminated zone, the D unit, which grades up into the pelagic muds of the E unit, which settle from suspension.

Detailed statistical studies of whole turbidite formations show that there are a number of predictable directional changes (e.g. Walker, 1970; Pett and Walker, 1971; Parkash, 1970). Moving down-current or down-section, the following trends are commonly found: grain size and sand bed thickness diminish. Bottom structures tend to change from channelling to flute marks, groove marks and finally tool marks. The internal structures of the turbidite appear to undergo progressive truncation from the base upward. First, the massive A unit is over-stepped by the laminated B unit. This is in turn over-stepped by the cross-laminated C unit, and so on (Fig. 92).

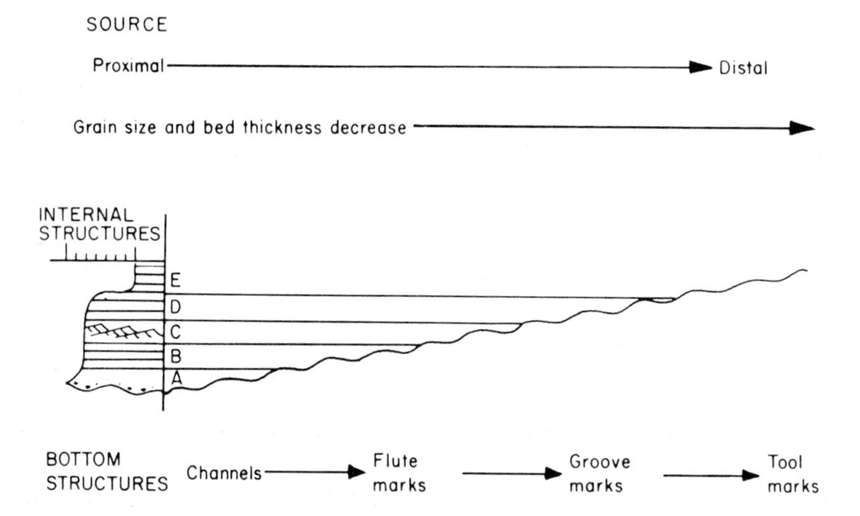

Fig. 92. Down-current variation in the sedimentary structures of turbidites. (Based on data due to Walker, 1967a, b.)

Walker (1967a) has proposed a statistical coefficient, the P index, as a measure of the proximal (i.e. near source) or distal position of beds within a turbidite formation. The concepts that turbidite formations change their sedimentologic parameters upwards and sourcewards throws considerable light on their genesis.

Field geologists (e.g. Bailey, 1930) are impressed by the great difference between traction deposits and turbidites. The former are cross-bedded, clean and laterally restricted. Turbidites, on the other hand, are flat-bedded, graded, argillaceous and often laterally extensive. One tends to contrast the lateral sedimentation of the traction current with the essentially vertical sedimentation of the turbidite. This is an over-simplification and, as further sections in this chapter will show, turbidites form part of a continuous spectrum of sedimentary processes which range from a land-slide to a cloud of suspended clay.

C. Sedimentation From Low Density Turbidity Currents

Fine-grained clay and silt are seldom if ever deposited from traction currents because they tend to be transported in suspension rather than as bed load. A certain amount of sand and silt is deposited at the distal end and waning phases of turbidites.

The bulk of subaqueous silt and clay is transported by a third mechanism: suspension. Suspension-deposited mudrocks can occur interbedded or interlaminated with turbidites or with traction deposits. Three types of suspension can be defined, though the divisions between them are arbitrary.

First we may consider the fine sediments of distal turbidites as suspension deposits. These are thinly interlaminated, laterally extensive laminae of silt and clay. Examples of this sediment type occur in deep marine basins, but are more characteristic of lacustrine environments. Such "varved" deposits, as these are called, are a feature of many Pleistocene glacial lakes (Smith, 1959). Each varve consists of a silt–clay couplet and is, by definition, considered as the product of one year's sedimentation. The silt lamina represents suspended load settled out from the summer melt water. The clay lamina, often rich in lime and organic matter, settled from suspension in winter when the lake and its environs were frozen and there was no terrigenous transportation into the lake.

Varve-like interlaminated silt and clay also occur in older lake deposits where fossil evidence shows them to have originated in diverse and certainly non-glacial climates. The Tertiary Green River Shale of Wyoming and Utah is a famous example (Bradley, 1931).

A second type of suspension deposits is that which originates from what are termed "nepheloid" layers. These are bodies of turbid water whose density differential with the ambient fluid is not sufficiently great for them to sink to the bottom as a conventional turbidity flow, yet they are sufficiently dense to form a cohesive turbid layer suspended within the ambient fluid.

Nepheloid layers have been discovered off the Atlantic coast of North America (Ewing and Thorndike, 1965). Bodies of this type may transport clay and organic matter far out into the oceans where the fine sediment settles out of suspension on to the sea bed in the pelagic environment (see p. 316).

The third main type of suspension deposit occurs where turbid flows enter bodies of water with no significant density difference. This situation, termed hypopycnal flow by Bates (1953), allows a complete mixing of the two water masses. Fine material then settles out of suspension from the admixture of water bodies. This sedimentation is accelerated where muddy fresh water mixes with sea water. The salts cause the clay particles to flocculate and settle more rapidly than if they were still dispersed.

It is not hard to establish that fine-grained deposits settled out of subaqueous suspensions. It is virtually impossible to determine whether the transporting mechanism was the last gasp of turbidity current, a nepheloid layer, or a hypopycnal flow situation. The end product of all three is a claystone with varying amounts of silt. These sediments may be massive or laminated due to vertical variations in grain size or chemical composition. Silt and very fine sand ripples may sometimes be present testifying to occasional traction current activity. Other sedimentary structures include slumps, slides and syneresis cracks. The slides and slumps occur because suspension deposits can form on slopes which are inherently unstable. Sedimentation continues until a critical point is reached at which the mud slides and slumps down-slope to be resedimented from suspension in a more stable environment. This may completely disturb the lamination of superficial sediments. Deeper, more cohesive muds may retain their lamination though this may have been disturbed into slump and slide structures due to mass movement down-slope.

On level bottoms spontaneous dewatering of clays leads to the formation of syneresis cracks (see p. 239), which are themselves infilled by more mud.

III. EOLIAN PROCESSES

At the beginning of this chapter it was pointed out that eolian and aqueous transportation and sedimentation showed many points in common. This is because both processes are essentially concerned with the behaviour of granular solids in a fluid medium. Gases and liquids both lack shear strength and share many other physical properties.

Eolian processes involve both traction carpets and suspensions (dust clouds). Turbidity flows are essentially unknown except in volcanic gas clouds termed "*nuée ardentes*". These are masses of hot volcanic gases with suspended ash and glass shards. These masses move down the sides of volcanoes at great speed. The resultant deposit, termed an ignimbrite, may have been formed at sufficiently high temperature for the ash particles to be welded together (Smith, 1960).

Modern eolian sediment transport and deposition occurs in three situations. They are found in the arid desert areas of the world, such as the Sahara. They are found erratically developed around ice caps, where the climate may have considerable precipitation, but this is often seasonal and ice-bound. Dunes also occur on the crests of barrier islands and beaches in diverse climates.

The bulk of eolian sediments consist either of traction deposited sands or of suspension deposited silt.

These two types will now be described.

A. Eolian Sedimentation From Traction Carpets

The foundation for the study of sand dunes was laid down in a classic book by Bagnold (1954) and updated in 1979. Additional significant work on the physics of eolian sand transport has been described by Williams (1964), Owen (1964), Wilson (1972) and McKee (1979).

These studies have shown how sediment, blown by the wind, moves by sliding and saltation just like particles in water. Silt and clay is winnowed from the traction carpet and carried off in dust clouds. Studies of the threshold velocity needed to commence grain movement show that, as with aqueous transport, the threshold velocity increases with increasing grain size. Quartz particles of about 0·10 mm (very fine sand) are the first to move in a rising wind. Silt and clay need velocities as strong as those for fine–medium sand to initiate movement (Horikowa and Shen in Allen, 1970). This is analogous to the Hjulstrom effect for the threshold of particle movement in aqueous flows.

A relationship between bed form and wind velocity has not been worked out in the same way that the flow regime concept unifies these variables for aqueous flow.

Ripples, dunes and plane beds are all common eolian sand bed forms. It is a matter of observation that ripples are blown out on both dunes and plane beds during sandstorms, to be rebuilt as the wind wanes.

The factors which control the areal distribution of sand plains and sand dune fields, or sand seas, are little understood. Attempts have been made, however, to define a model which integrates wind velocity and direction with net sand flow-paths (Wilson, 1971, 1972).

Particular attention has been paid to the geometry and genesis of sand dunes. Four main morphological types can be defined.

The most beautiful and dramatic sand dune is the barchan or lunate dune. This is arcuate in plan, convex to the prevailing wind direction, with the two horns pointing down-wind. They have a steep slip face in their concave down-wind side (Fig. 93a). Barchans occur in isolation or as outriders around the edges of sand seas. They generally overlie playa mud or granule deflation surfaces. This suggests that lunate dunes form where sand is in short supply. They are bed forms of transportation, not of net deposition. It is unlikely that they are often preserved in the geological record.

The second type of dune to consider is the stellate, pyramidal or Matterhorn type (Fig. 93b). These consist of a series of sinuous, sharp, rising sand ridges, which merge together in a high peak from which wind often blows a plume of sand, making the dune look as if it were smoking.

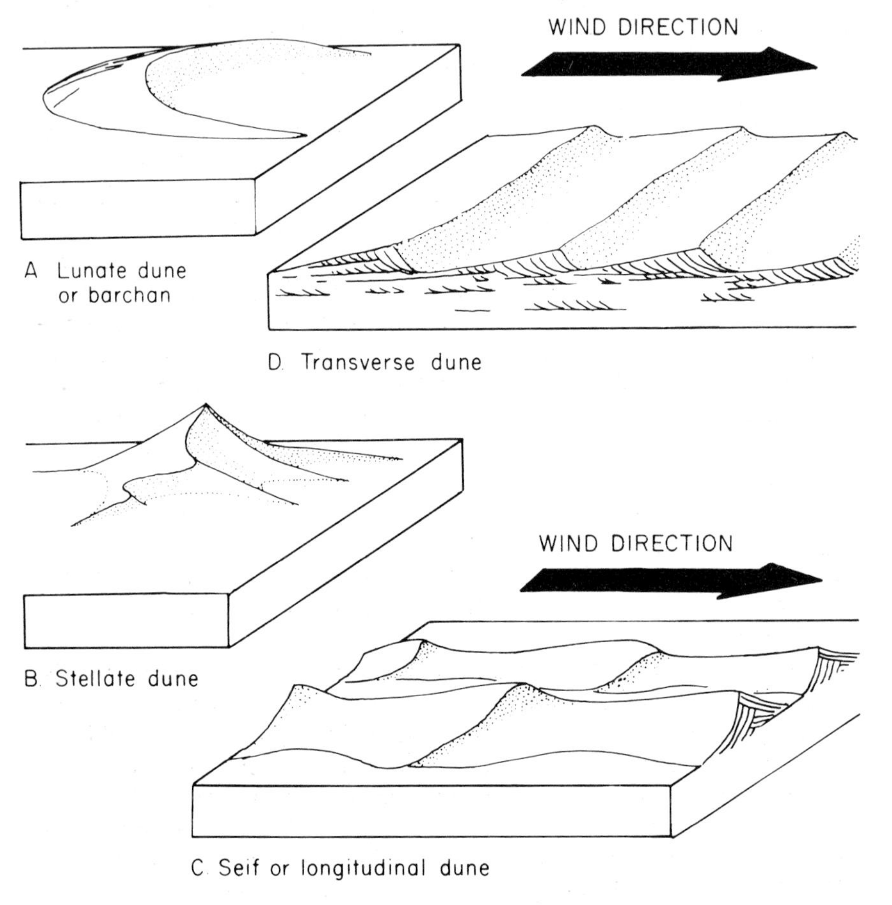

Fig. 93. The four main morphological types of sand dune. Only the transverse dunes are probably responsible for net sand deposition. The other three types occur in environments of equilibrium where sand is constantly transported and reworked, but where there is little net sedimentation.

These stellate dunes are sometimes hundreds of metres high. They often form at the boundary of sand seas and jebels, suggesting an origin due to interference of the wind-stream by resistant topography. This is not the whole story, however, as they also occur, apparently randomly, mixed in with dunes of other types.

The third type of dune is the longitudinal or seif dunes (Fig. 93c). These are long, thin, sand dunes with sharp median ridges. Individual dunes may be traced for up to 200 km, with occasional convergence of adjacent seifs in a downwind direction. Individual dunes are up to 50 m high, and adjacent dunes are spread up to 1 or 2 km apart.

The simple gross morphology of seif dunes is often modified by smaller parasitic bed forms. These are sometimes regularly spaced stellate peaks, separated by long gentle saddles, or they may be inclined transverse dunes with steep down-wind slip faces.

Seif dunes occur both on gravel deflation surfaces and on sand sheets.

The origin of seif dunes has been discussed by Bagnold (1953), Hanna (1969) and Folk (1971). The consensus of opinion is that they are formed from helicoidal flow cells set up by unidirectional wind systems.

The fourth type of eolian dune is the transverse dune. These dunes are straight, or slightly sinuous, crested dunes which strike perpendicular to the mean wind direction (Fig. 93d). Their steep slip-off faces are directed down-wind. Transverse dunes seldom occur on deflation surfaces. They are typically gregarious, climbing up the back-side of the next dune down-wind. This strongly suggests that transverse dunes are the eolian bed form which actually deposits sand. The other three dune types seem to be largely transportational bed forms.

These conclusions are based on several years spent travelling around, and sometimes through, Arabian and Saharan sand seas.

Considerable attention has been directed towards the study of the internal structure of dunes (McKee and Tibbitts, 1964; McKee, 1966, 1979; Glennie, 1970; Bigarella, 1972).

In ripples and low subaqueous dunes the height of foresets ultimately deposited is not much less than the dune height. This situation does not always apply to eolian dunes. The observational data presently available do not suggest that 50 m-high dunes are always composed of a single set of the same height. Many dunes which have been bulldozed and trenched show that they are composed of numerous sets of very varying heights. Individual sets are separated by erosion surfaces or units of flat or incline-bedded sand.

This heterogenous bedding reflects erratic variations in gross dune morphology. The cross-beds were deposited on down-wind migrating slip faces. The flat and subhorizontal bedded units were deposited on the gentle back-slope of the dune.

These remarks apply primarily to lunate and transverse dunes. Dome-shaped dunes in White Sands National Monument do show set heights corresponding to dune heights of up to 10 m.

Seif dunes are internally composed of mutually opposed foresets which dip perpendicular to the prevailing wind direction.

Penecontemporaneous deformation of bedding is quite common in dunes. It takes the form of both fracturing and slipping, as well as various fold structures (McKee, 1971).

In conclusion, one can see that eolian traction deposits are broadly similar to aqueous eolian ones. They have similar modes of grain transport and similar bed forms. Much remains to be learnt of the factors which control the morphology of eolian dunes, and of their elusive internal structure.

B. Eolian Sedimentation From Suspension

Most eolian sediment originates from wind blowing over desiccated alluvium. The gravel remains behind. The sand saltates into dunes. The silt and clay are blown away in suspension. This much is known, and there is no question of the competence of wind to transport large amounts of silt and clay in suspension. It has been calculated that between 25 and 37 million tonnes of dust are transported from the Sahara throughout the longitude of Barbados each year. This quantity of dust is sufficient to maintain the present rate of pelagic sedimentation in the entire North Atlantic (Prospero and Carlson, 1972).

Dust is transported in the winds of deserts, but little is actually deposited out of suspension in the way that mud settles on the sea floor. Most silt and clay actually get deposited in playas, following rains and flash floods. Desiccation and cohesion tend to prevent this material from being recycled.

The dust suspensions of periglacial deserts are different from those of tropic deserts. Clay is largely absent. They are high in silica particles produced by glacial action. This material is termed "loess" (Berg, 1964). Extensive deposits of Pleistocene loess occur in a belt right across the Northern Hemisphere to the south of the maximum extent of the ice sheets. Loess occurs in laterally extensive layers, often of great thickness. It is slightly calcareous, massive and weathers into characteristic polygonal shrinkage cracks. While most authorities agree that loess was transported by eolian suspensions (i.e. as dust clouds), there is some debate as to whether it settled out of free air, or whether it was actually deposited as a result of fluvial action (e.g. Smalley, 1972). Some authorities argue for a polygenetic origin of loess; suggesting that it may form by wind action or by *in situ* cryogenesis of diverse rock types (Popov, 1972).

IV. GLACIAL PROCESSES

Several types of sedimentary deposit are associated with glaciation. These include loess, just described, varved glacial-like clays and the sands and gravels of fluvioglacial outwash plains. These deposits though associated with glaciation, are actually eolian, aqueous suspension and traction-current deposits respectively.

Ice itself transports and deposits one rock type only, termed diamictite (Flint *et al.*, 1960). This is a poorly sorted sediment ranging from boulders down to clay grade. Much of the clay material is composed of diverse minerals, but largely silica, formed by glacial pulverization. Clay minerals are a minor constituent. The boulders show a wide size range, are often angular, and are rarely grooved due to the ice causing the sharp corner of one boulder to scratch across a neighbour. Statistical analysis of the orientation of the larger particles shows that their long axes parallel the direction of ice movement (e.g. Andrews and Smith, 1970). Glacial diamictites tend to occur as laterally extensive sheets, seldom more than a few metres thick. They overlie glacially striated surfaces and have hummocky upper surfaces. They also occur interbedded with the periglacial deposits listed above.

Let us now consider the mechanics of glacial transport and deposition. Ice, formed from compacted snow, moves both in response to gravity, in valley glaciers, and in response to horizontal pressures in continental ice sheets. Ice movement is very slow, compared with aqueous or eolian flow velocities. On the other hand it is highly erosive, breaking off boulders from the rocks over which it moves. The detritus which is caught up in the base of a glacier is transported in the direction of ice flow. It is unaffected by the sorting action found in eolian and aqueous action.

When the climate ameliorates ice movement ceases and the ice begins to melt *in situ*. Its load of sediment is dumped where it is as a heterogenous structureless diamictite.

Modern and Pleistocene diamictites are widely known by a variety of terms such as till, boulder clay, drift and moraine (see Harland *et al.*, 1966 for a review of terminology).

There are many ancient diamictites. Many of these have been interpreted as ancient glacial deposits (tillites). As the next section of the book shows, however, not all diamictites are of glacial origin. The genetic name tillite should not be applied unless a glacial origin may be clearly demonstrated (Crowell, 1957).

Ancient diamictites for which a glacial origin can be convincingly demonstrated occur at geographically widespread localities at certain geologic times. Late Pre-Cambrian tillites associated with periglacial sediments occur in Canada, Greenland, Norway, Northern Ireland and Scotland (Spencer, 1971; Reading and Walker, 1966).

There is also extensive evidence for a glaciation of the Southern Hemisphere in the Permo-Carboniferous. This deposited tillites in South America, Australia, South Africa (the Dwyka tillite) and India (the Talchir boulder beds). Notable descriptions of diamictites from the Falkland Islands and New South Wales have been given by Frakes and Crowell (1967) and Whetten (1965), respectively.

Ancient glacial deposits are rare, but they provide a fruitful source for academic controversy out of all proportion to their volume.

V. GRAVITATIONAL PROCESSES

The force of gravity is an integral part of all sedimentary processes; aqueous, eolian or glacial. Gravity can, on its own, act as an agent of sediment transport, but for this to acquire a horizontal component it requires some additional mechanism. There is a continuous spectrum of depositional processes which grade from pure gravity deposition to the turbidity flows which have already been described.

This spectrum may be arbitrarily classified into four main groups: rock falls, slides and slumps, mass flows and turbidites (Dott, 1963).

Rock falls, or avalanches, are examples of essentially vertical gravitational sedimentation with virtually no horizontal transport component (Varnes, 1958). The resultant sediment is a scree composed of poorly sorted angular boulders with a high primary porosity. Subsequently weathering may round the boulders *in situ*, while wind or water transport finer sediment to infill the voids.

Rock falls may occur both on land and under the sea. They may be triggered off by earthquakes and, on land, by heavy rain and freeze–thaw action in cold climates. The actual site of a rock fall necessitates a steep cliff from which the sediment may collapse. This may be an erosional feature or a fault scarp.

The second type of gravitational process to consider is the slide or slump. These occur on gentler slopes than rock falls, down to even less than a degree. Slides and slumps can be both subaerial or subaqueous. The sedimentary processes of sliding involve the lateral transportation of sediment along subhorizontal shear planes. Water is generally needed as a lubricant to reduce friction and to permit movement along the slide surfaces. The fabric of the sediment is essentially undisturbed. Slumping involves the sideways downslope movement of sediment in such a way that the original bedding is disturbed, contorted and sometimes completely destroyed. The detailed morphology of the sedimentary structures which sliding and slumping produce are described in the next chapter.

Sliding and slumping become progressively more effective transporting agents with increasing water content. Many sediments which are slump susceptible are deposited on a slope with a loose packing. Once movement is initiated the sediment packing is disturbed and the packing tends to tighten. Porosity decreases, therefore, and the pore pressure increases. This has the effect of decreasing intergranular friction and allowing the sediment to flow more freely. Thus with increasing water content and, hence, decreasing shear strength, slumping grades into the third mechanism of the spectrum: mass flow or grain flow (Fisher, 1971). This process embraces a wide range of phenomena known by such names as sand flows, grain flows, fluidized flows, mud flows, debris flows and their resultant deposits, fluxoturbidites, diamictites and pebbly mudstones.

A. Debris Flow

A debris flow is defined as a "highly-concentrated non-Newtonian sediment dispersion of low yield strength" (Rupke, in Reading, 1978, p. 378). This includes mud flows though not all debris flows are muddy. Debris flows occur in a wide range of environments ranging from deserts to continental slopes. In the former situation torrential rainfall is generally required to initiate movement; in the latter, earthquakes tide or storm surges are required. In all situations a slope is a necessary pre-requisite.

In desert environments the mud flow was first described by Blackwelder (1928). This is a mass of rock, sand and mud which, liquified by heavy rain, moves down mountain sides. Movement may initially be slow, but with increasing water content can accelerate to a fast-flowing flood of debris which carries all before it. Large mud flows are a truly catastrophic process which destroy animals, trees and houses (e.g. Scott, 1971). Blackwelder lists the four prerequisites of a mud flow as an abundance of unconsolidated detritus, steep gradients, a lack of vegetation and heavy rainfall.

The deposits of mud flows range from boulders to gravel, sand, silt and clay. Where sediment of only one grade is available at source, then the resultant deposit will be of that grade and well sorted. Characteristically, however, mud-flow deposits are poorly sorted. They are variously named pebbly mudstones (Crowell, 1957), diamictites (Flint *et al.*, 1960) or fluxoturbidites (Kuenen, 1958). The last of these, though widely used, is a naughty genetic word for a little-known process, as Walker (1967a) has enthusiastically pointed out.

Mud flows may occur both on land and under water. On land, mud flows grade with increasing water content into sheet floods which are midway between grain flows and traction currents. Sheet floods, or flash floods, deposit subhorizontally bedded coarse sands and gravels with intermittent channelling (Ives, 1936; Hooke, 1967). These characteristically occur on alluvial fans and desert pediments.

Subaqueous mud flows are typically found in submarine canyons on delta fronts and continental margins (Stanley and Unrug, 1972). Modern and ancient diamictites from these settings are described in Chapter 8 (p. 311).

B. Grainflow

The concept of the grain flow was expounded by Bagnold (1954, 1966). Grainflows are liquified cohesionless particle flows in which the intergranular friction between sand grains is reduced by their continuous agitation. This is believed by some to be non-turbulent and to involve considerable horizontal shearing.

Grainflows have been observed in modern submarine channels. They appear to require a high gradient to initiate them, and a confined space (i.e. a channel) to retain the high pore pressure required for their maintenance. Unlike most debris flows, grain flows are typically well sorted and clay free, though they may contain scattered clasts. Internally, grain flows are generally massive with no vertical size grading, though grain orientation may be parallel to flow. Bases are abrupt, often loaded, but seldom erosional. Tops are also sharp (Fig. 314). Individual units may be a metre or so in thickness, but multistorey sequences of grain flows may attain tens of metres.

Hendry (1972) has described graded massive marine breccias from the Alpine Jurassic and has interpreted them as grain-flow deposits. Stauffer (1967) has defined the typical features of sand-flow deposits and pointed out the differences between them and true turbidity-current deposits, viz. sand-flow deposits are more typical of the submarine channel, whereas turbidites occur more generally on the fan or basin floor. Individual sand-flow beds have erosional bases but lack the suite of sole marks characteristic of turbidites. Sand-flow beds are massive or faintly bedded with clasts up to cobble size, scattered throughout them. They are not graded. Turbidites, by contrast, are graded, laminated and/or cross-laminated and their coarsest clasts are restricted to the base of the bed (Fig. 94).

C. Fluidized Flows

Fluidization of a sand bed occurs when the upward drag exerted by moving pore fluids exceeds the effective weight of the grains. When this upward movement exceeds the minimum fluidization velocity, the bed expands rapidly, porosity increases and the bed becomes liquified and fluid-supported rather than grain supported. The sediments produced by fluidization are similar in many ways to grain flows and may in practice be indistinguishable. They occur in thick non-graded clean sands, with abrupt tops and bottoms. Because of their high porosities, however, fluidized beds

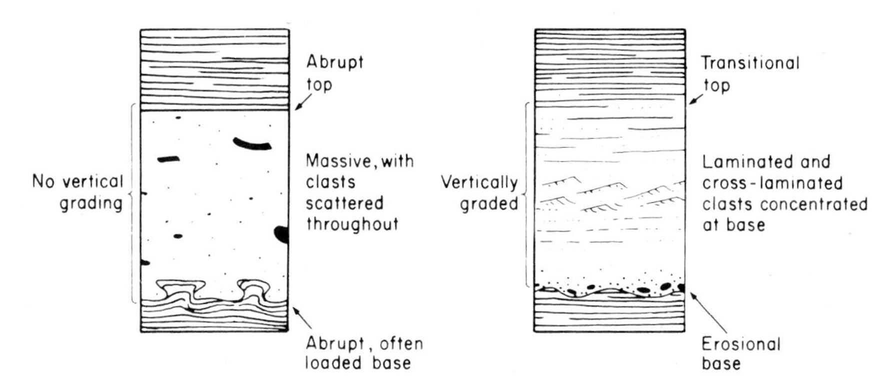

Fig. 94. Comparison of the internal characteristics of grain flows (A), and turbidites (B). (From Stauffer, 1967.)

frequently contain sand pipes and "dish structures" due to post-depositional de-watering (Lowe, 1976). Like grainflows, fluidized flows appear to require a slope and trigger to initiate them, and a channel to retain pore pressure. Observations of ancient deep-sea sands show that, once initiated, both grainflows and fluidized flows may move down channels far across basin floors with minimal gradient.

In conclusion it must be stated that gravity-related sedimentary processes are not well understood, though they may be observed at work today, and their deposits may be studied. They range from avalanches, via debris flows, grain flows and/or fluidized flows to turbidites. This spectrum is marked by decreasing gradient and sand concentration and increasing velocity and water content (Hampton, 1972).

VI. REFERENCES

Allen, J. R. L. (1970). "Physical Processes of Sedimentation" Allen and Unwin, London. 248pp.

Allen, J. R. L. (1979). A model for the interpretation of wave ripple-marks using their wavelength, textural composition, and shape. *J. geol. Soc. Lond.* **136**, 673–682.

Andrews, J. T. and Smith, D. I. (1970). Statistical analysis of till fabric methodology, local and regional variability. *Q. Jl geol. Soc. Lond.* **125**, 503–542.

Bagnold, R. A. (1953). The surface movement of blown sand in relation to meteorology. *In* "Desert Research", 89–96. Unesco, Jerusalem.

Bagnold, R. A. (1954). "The Physics of Blown Sand and Desert Dunes" Methuen, London. 265pp.

Bagnold, R. A. (1966). An approach to the sediment transport problem from general physics. *Prof. Pap. U.S. geol. Surv.* **422**-I. 37pp.

Bagnold, R. A. (1979). Sediment transport by wind and water. *Nordic Hydrology* **10**, 309–322.

Bailey, Sir E. B. (1930). New light on sedimentation and tectonics. *Geol. Mag.* **67**, 77–92.

Bates, C. C. (1953). Rational theory of delta formation. *Bull. Am. Ass. Petrol. Geol.* **37**, 2119–2162.

Bell, H. S. (1942). Density currents as agents for transporting sediments. *J. Geol.* **L**, 512–547.

Berg, L. S. (1964). "Loess as a Product of Weathering and Soil Formation" Israel Program for Scientific Translations, Jerusalem. 205pp.

Bigarella, J. J. (1972). Eolian environments, their characteristics, recognition and importance. Recognition of Ancient Sedimentary Environments (J. K. Rigby and W. K. Hamblin, Eds). *Spec. Publs Soc. econ. Palaeont. Miner., Tulsa* **16**, 12–62.

Blackwelder, E. (1928). Mudflow as a geologic agent in semi-arid mountains. *Bull. geol. Soc. Am.* **39**, 465–483.

Bornhold, B. D. and Pilkey, O. H. (1971). Bioclastic turbidite sedimentation in Columbus Basin, Bahama. *Bull. geol. Soc. Am.* **82**, 1254–1341.

Bouma, A. H. (1962). "Sedimentology of some Flysch Deposits" Elsevier, Amsterdam. 168pp.

Bouma, A. H. (1972). Recent and Ancient turbidites and contourites. *Trans. Gulf-Cst Ass. geol. Socs* **22**, 205–221.

Bradley, W. H. (1931). Non-glacial marine varves. *Am. J. Sci.* **22**, 318–330.

Conolly, J. R. and Ewing, M. (1967). Sedimentation in the Puerto Rico Trench. *J. sedim. Petrol.* **37**, 44–59.

Crowell, J. C. (1957). Origin of pebbly mudstones. *Bull. geol. Soc. Am.* **68**, 993–1010.

Dailey, J. W. and Harleman, D. R. F. (1966). "Fluid Dynamics" Addison-Wesley, Reading, Mass. 454pp.

Daly, R. A. (1936). Origin of submarine canyons. *Am. J. Sci.* **31**, 401–420.

Dill, R. F. (1964). Sedimentation and erosion in Scripps submarine canyon head. *In* "Marine Geology" (R. L. Miller, Ed.), 23–41. Macmillan, London.

Dott, R. H. (1963). Dynamics of subaqueous gravity depositional processes. *Bull. Am. Ass. Petrol. Geol.* **47**, 104–128.

Dzulinski, S. and Walton, E. K. (1963). Experimental production of Sole markings. *Trans. Edinb. geol. Soc.* **19**, 279–305.

Emery, K. O. and Ross, D. A. (1968). Topography and sediments of a small area of the continental slope south of Marthas Vineyard. *Deep Sea Res.* **15**, 415–422.

Ewing, M. and Thorndike, E. M. (1965). Suspended matter in deep-ocean water. *Science, N.Y.* **147**, 1291–1294.

Fisher, R. V. (1971). Features of coarse-grained, high-concentration fluids and their deposits. *J. sedim. Petrol.* **41**, 916–927.

Flint, R. F., Sanders, J. E. and Rodgers, J. (1960). Diamictite: a substitute term for symmictite. *Bull. geol. Soc. Am.* **71**, 1809–1810.

Folk, R. L. (1971). Longitudinal dunes of the north-western edge of the Simpson desert, Northern Territory, Australia. 1. Geomorphology and grainsize relationships. *Sedimentology* **16**, 5–54.

Frakes, L. A. and J. C. Crowell (1967). Facies and palaeogeography of Late Palaeozoic diamictite, Falkland islands. *Bull. geol. Soc. Am.* **78**, No. 1, 37–58.

Friend, P. F. (1965). Fluviatile sedimentary structures in the Wood Bay Series (Devonian) of Spitzbergen. *Sedimentology* **5**, 39–68.

Glennie, K. W. (1970). "Desert Sedimentary Environments" Elsevier, Amsterdam. 222pp.

Gould, H. R. (1951). Some quantitative aspects of Lake Mead turbidity currents. Turbidity Currents and the Transportation of Coarse Sediments to Deep Water. *Spec. Publs Soc. econ. Palaeont. Miner., Tulsa* **2**, 34–52.

Hampton, M. A. (1972). The role of subaqueous debris flow in generating turbidity currents. *J. sedim. Petrol.* **42**, 775–793.

Hand, B. M. and Emery, K. O. (1964). Turbidites and topography of north end of San Diego Trough, California. *J. Geol.* **72**, 526–552.

Hanna, S. R. (1969). The formation of longitudinal sand dunes by large helical eddies in the atmosphere. *J. appl. Met.* **8**, 874–883.

Harland, W. B., Herod, K. N. and Krinsley, D. H. (1966). The definition and identification of tills and tillites. *Earth Sci. Rev.* **2**, 225–256.

Harms, J. C. and Fahnestock, R. K. (1965). Stratification, bed forms and flow phenomena (with an example from the Rio Grande). Primary Sedimentary Structures and their Hydrodynamic Interpretation (G. V. Middleton, Ed.), *Spec. Publs Soc. econ. Palaeont. Miner., Tulsa* **12**, 84–155.

Heezen, B. C. and Ewing, M. (1952). Turbidity currents and submarine slumps and the 1929 Grand Banks earthquake. *Am. J. Sci.* **250**, 849–873.

Heezen, B. C. and Hollister, C. D. (1971). "The Face of the Deep" Oxford University Press. 659pp.

Henderson, F. M. (1966). "Open Channel Flow" Macmillan, London. 522pp.

Hendry, H. E. (1972). Breccias deposited by mass flow in the Breccia Nappe of the French Pre-Alpa. *Sedimentology* **18**, 277–292.

Hollister, C. D. and Heezen, B. C. (1972). Geologic effects of ocean bottom currents: western North Atlantic. *In* "Studies in Physical Oceanography" (A. L. Gordon, Ed.), Gordon and Breach, New York.

Holtedahl, H. (1965). Recent turbidities in the Hardangerfjord, Norway. *In* "Submarine Geology and Geophysics" (W. F. Whittard and R. Bradshaw, Eds), 107–142. Butterworth, London.

Hooke, R. LeB. (1967). Processes on arid-region alluvial fans. *J. Geol.* **75**, 438–460.

Hubert, J. F. (1964). Textural evidence for deposition of many western N. Atlantic deep sea sands by ocean bottom currents rather than turbidity currents. *J. Geol.* **72**, 757–785.

Hubert, J. F. (1967). Sedimentology of pre-Alpine Flysch sequences, Switzerland. *J. sedim. Petrol.* **37**, 885–907.

Ives, J. C. (1936). Desert floods in the Sonoyta Valley, Northern Mexico. *Am. J. Sci.* **32**, 102–135.

Kuenen, P. H. (1937). Experiments in convection with Daly's hypothesis on the formation of submarine canyons. *Leid. geol. Meded.* **8**, 327–335.

Kuenen, P. H. (1948). Turbidity currents of high density. Rep. 18th Int. geol. Cong. U.K. Pt. 8, 44–52.

Kuenen, P. H. (1958). Problems concerning course and transportation of flysch sediments. *Geologie Mijnb.* **20**, 329–339.

Kuenen, P. H. (1965). Experiments in connection with turbidity currents and clay suspensions. *In* "Submarine Geology and Geophysics" (W. F. Whittard and R. Bradshaw, Eds), 47–74. Butterworth, London.

Kuenen, P. H. and Migliorini, C. I. (1950). Turbidity currents as a cause of graded bedding. *J. Geol.* **58**, 91–128.

Leliavsky, S. (1959). "An Introduction to Fluvial Hydraulics" Dover, New York. 257pp.

Lowe, D. R. (1976). Grain flow and grain flow deposits *J. sedim. Petrol.* **46**, 188–199.

McKee, E. D. (1966). Dune structures. *Sedimentology* **7**, 3–69.

McKee, E. D. (1971). Primary structures in dune sand and their significance. *In* "The Geology of Libya" (C. Grey, Ed.), 401–408. University of Libya.

McKee, E. D. and Tibbitts, G. C. (1964). Primary structures of a seif dune and associated deposits in Libya. *J. sedim. Petrol.* **34**, No. 1, 5–17.

Middleton, G. V. (1966a). Experiments on density and turbidity currents, I. *Can. J. Earth Sci.* **3**, 523–546.

Middleton, G. V. (1966b). Experiments on density and turbidity currents, II. *Can. J. Earth Sci.* **3**, 627–637.

Middleton, G. V. (1967). Experiments on density and turbidity currents, III. *Can. J. Earth Sci.* **4**, 475–505.

Owen, P. R. (1964). Saltation of uniform grains in air. *J. Fluid Mech.* **20**, 225–242.

Parkash, B. (1970). Downcurrent changes in sedimentary structures in Ordovician turbidite greywackes. *J. sedim. Petrol.* **40**, 572–590.

Payne, R. R., Conolly, J. R. and Abbott, W. H. (1972). Turbidite muds within diatom ooze off Antarctica: Pleistocene sediment variation defined by closely spaced piston cores. *Bull. geol. Soc. Am.* **83**, 481–486.

Pett, J. W. and Walker, R. G. (1971). Relationship of flute cast morphology to internal sedimentary structures in turbidites. *J. sedim. Petrol.* **41**, 114–128.

Popov, A. I. (1972). Les loess et dépots loessoides, produit des processes cryolithogènes. *Biul. Peryglac.* **21**, 193–200.

Prospero, J. M. and Carlson, T. N. (1972). Vertical and areal distribution of Saharan dust over the Western Equatorial North Atlantic Ocean. *J. geophys. Res.* **77**, 5255–5265.

Reading, H. G. (Ed.) (1978). "Sedimentary Environments and Facies" Blackwell Scientific, Oxford. 557pp.

Reading, H. G. and Walker, R. G. (1966). Sedimentation of Eocambrian tillites and associated sediments in Finmark, Northern Norway. *Paleogeog. Paleoclimatol. Paleoecol.* **2**, 177–212.

Rusnak, G. A. and Nesteroff, W. D. (1964). Modern turbidites: terrigenous abyssal plain versus bioclastic basin. *In* "Marine Geology" (L. R. Miller, Ed.), 488–503. Macmillan, New York.

Scott, K. M. (1971). Origin and sedimentology of 1969 debris flows near Glendora, California. *Prof. Pap. U.S. geol. Surv.* **750**-C, C242–247.

Selley, R. C. (1969). Torridonian alluvium and quicksands. *Scott. J. Geol.* **5**, 328–346.

Shepard, F. P. and Dill, R. F. (1966). "Submarine Canyons and Other Sea Valleys" Rand McNally, Chicago. 381pp.

Shepard, F. P., Dill, R. F. and Von Rad, U. (1969). Physiography and sedimentary processes of La Jolla submarine fan and Fan Valley, California. *Bull. Am. Ass. Petrol. Geol.* **53**, 390–420.

Shields, A. (1936). Anwendung der Ahnlich keitsmechanik und der Turbulenz forschung auf die Geschiebebewegung. *Mitt. preuss. Vers Anst. Wasserb. Schiffb.* Berlin. Heft **26**, 26pp. (See also Vanoni, V. A., 1964.) Measurements of critical shear stress for entraining fine sediments in a boundary layer: Caltech. W. M. Kech Lab. Rept K. H.—R—7, 47pp.

Simons, D. B., Richardson, E. V. and Nordin, C. F. (1965). Sedimentary structures generated by flow in alluvial channels. Primary Sedimentary Structures and their Hydrodynamic Interpretation (G. Middleton, Ed.). *Spec. Publs Soc. econ. Palaeont. Miner., Tulsa*, **12**, 34–52.

Skipper, K. (1971). Antidune cross-stratification in a turbidite sequence, Cloridorme Formation, Gaspé, Quebec. *Sedimentology* **17**, 51–68.

Smalley, I. J. (1972). The interaction of great rivers and large deposits of primary loess. *Trans. N.Y. Acad. Sci.* **34**, 534–542.

Smith, A. J. (1959). Structures in the stratified late-glacial clays of Windermere, England. *J. sedim. Petrol.* **29**, 447–453.

Smith, R. L. (1960). Ash flows. *Bull. geol. Soc. Am.* **71**, 795–842.

Spencer, A. M. (1971). Late PreCambrian glaciation in Scotland. Geol. Soc. Lond. Mem. No. 6, 100pp.

Stanley, D. J. and Unrug, R. (1972). Submarine channel deposits, Fluxoturbidites and other indicators of slope and base of slope environments in modern and ancient marine basins. Recognition of Ancient Sedimentary Environments (J. K. Rigby and W. K. Hamblin, Eds). *Spec. Publs Soc. econ. Palaeont. Miner., Tulsa* **16**, 287–340.

Stauffer, P. H. (1967). Grainflow deposits and their implications, Santa Ynez Mountains, California. *J. sedim. Petrol.* **37**, 487–508.

Sundborg, A. (1956). The River Klaralven, a study in fluvial processes. *Geogr. Annlr* **38**, 125–316.

Van der Lingen, G. J. (1969). The turbidite problem. *N.Z. Jl Geol. Geophys.* **12**, 7–50.

Van Straaten, L. M. J. U. (1964). Turbidite sediments in the southeastern Adriatic Sea. *In* "Turbidites" (A. H. Bouma and A. Brouwer, Eds), 142–147. Elsevier, Amsterdam.

Varnes, D. J. (1958). Landslide types and processes. *In* "Landslides and Engineering Practice", 20–47. Highw. Res. Board Spec. Rept. 29, U.S. Natl Res. Council Pub. No. 544.

Walker, R. G. (1965). The origin and significance of the internal sedimentary structures of turbidites. *Proc. Yorks. geol. Soc.* **35**, 1–32.

Walker, R. G. (1967a). Turbidite sedimentary structures and their relationship to proximal and distal depositional environments. *J. sedim. Petrol.* **37**, 25–43.

Walker, R. G. (1967b). Upper flow regime bedforms in turbidites of the Hatch Formation, Devonian of New York State. *J. sedim. Petrol.* **37**, 1052–1058.

Walker, R. G. (1970). Review of the geometry and facies organization of turbidites and turbidite-bearing basins. *Spec. Pap. geol. Ass. Can.* **7**, 219–251.

Whetten, J. T. (1965). Carboniferous glacial rocks from the Werrie Basin, New South Wales, Australia. *Bull. geol. Soc. Am.* **76**, 43–56.

Williams, G. (1964). Some aspects of the eolian saltation load. *Sedimentology* **3**, 257–287.

Wilson, I. G. (1971). Desert sandflow basins and a model for the development of Ergs. *Geogrl J.* **137**, 180–199.

Wilson, I. G. (1972). Aeolian bedforms—their development and origins. *Sedimentology* **19**, 173–210.

7 Sedimentary Structures

I. INTRODUCTION

This chapter describes the internal megascopic features of a sediment. These are termed sedimentary structures, and are distinguished from the microscopic structural features of a sediment, termed the fabric, described in the second chapter.

Sedimentary structures are arbitrarily divided into primary and secondary classes. Primary structures are those generated in a sediment during or shortly after deposition. They result mainly from the physical processes described in the previous chapter. Examples of primary structures include ripples, cross-bedding and slumps. Secondary sedimentary structures are those which formed sometime after sedimentation. They result from essentially chemical processes, such as those which lead to the diagenetic formation of concretions. Primary sedimentary structures are divisible into inorganic structures, including those already mentioned, and organic structures, such as burrows, trails and borings.

Table XXI shows the relationships between the various structures just defined. In common with most classifications of geological data, this one will not stand careful scrutiny. The divisions between the various groups are ill-defined and debatable. Nevertheless, it provides a useful framework on which to build the analysis of sedimentary structures contained in this chapter.

First let us consider the definition of a sedimentary structure. Colloquially, a sedimentary structure is deemed to be a primary depositional feature of a sediment which is large enough to be seen by the naked eye, yet small enough to be carried by a group of healthy students, or at least to be contained in one quarry. A channel is thus generally considered to be a sedimentary structure—just. A sand dune is generally considered a sedimentary structure; but what about an off-shore bar or a barrier island, which is just a large and very complex dune? Surely a coral reef is a sedimentary structure? But is it organic or inorganic? These questions highlight the problems of defining exactly what is a sedimentary structure.

Table XXI
A classification of sedimentary structures

I. Primary (physical)	Inorganic	Fabric	Microscopic
		Cross-bedding, ripples, etc.	Megascopic
	Organic	Burrows and trails	Megascopic
II. Secondary (chemical)	Diagenetic	Concretions, etc.	Megascopic

Conventionally a sedimentary structure is considered as a smaller scale feature which is best illustrated by the examples of ripples, cross-beds and slumps already mentioned. It is about such phenomena that this chapter is concerned, with an additional section on trace fossils.

First let us consider the observation, interpretation and classification of inorganic sedimentary structures.

Sedimentary structures can be studied at outcrop, in cliffs, quarries and stream sections. They can also be studied in cores taken from wells. Sedimentary structures in cores are the easiest to describe because of the small size of the sample to be observed. Where large expanses of rock are available for analysis, the problem of what is a sedimentary structure is at once apparent. Cross-beds are seen to be grouped in large units; ripples form an integral part of a bed; a slump structure is composed of contorted beds with diverse types of sedimentary structure. It is at once clear that sedimentary structures do not occur in isolation. Hence the problems of observing, defining, and classifying sedimentary structures. There are two basic approaches to observing sedimentary structures.

The first approach is to pretend the outcrop is a bore hole and to measure a detailed sedimentological log. This records a vertical section of limited lateral extent. The methodology of this approach is described in Bouma (1962) and Selley (1968a).

The second method is a two-dimensional survey of all, or a major part of the outcrop. This may be recorded on graph paper, using a tape measure and an Abney level for accurately locating inaccessible reference points on cliff faces. This method is aided by photography, especially by on-the-spot polaroid photos, on which significant features and sample points may be located.

Let us consider now the interpretation of sedimentary structures. They are the most useful of sedimentary features to use in environmental interpretation because, unlike texture, lithology and fossils, they cannot be recycled. They unequivocally reflect the depositional process which laid down the sediment. The interpretation of the origin of sedimentary structures is based on studies of their modern counterparts, on laboratory experiments and on theoretical physics. Examples of these different methods will be shown at appropriate points in the chapter.

II. BIOGENIC SEDIMENTARY STRUCTURES

A great variety of structures in sedimentary rocks can be attributed to the work of organisms. These structures are referred to as biogenic, in contrast to the inorganic sedimentary structures. Biogenic structures include plant rootlets, vertebrate footprints (tracks), trails (due to invertebrates), soft sediment burrows and hard rock borings. These phenomena are collectively known as trace fossils and their study is referred to as ichnology. Important texts on ichnology include those by Hantzschel (1962), Crimes and Harper (1970, 1977) and Frey (1975).

An individual morphological type of trace fossil is termed an ichnogenus. One of the basic principles of trace fossil analysis is that similar ichnogenera may be produced by a wide variety of organisms. The shape of a trace fossil reflects environment rather than creator. This means that trace fossils can be very important indicators of the origin of the sediments in which they are found because of their close environmental control. Furthermore, trace fossils always occur *in situ* and cannot be reworked like most other fossils.

Sedimentologists therefore need to know something about trace fossils. Some basic principles of occurrence and nomenclature will be described before analysing the relationship between trace fossils and environments. The various types of ichnogenera cannot be grouped phyllogenetically because, as already pointed out, various organisms produce similar traces. Ichnofossils have been grouped according to the activity which made them (Seilacher, 1964) or according to their topology (Martinsson, 1965). The topological scheme essentially describes the relationship of the trace to the adjacent beds (Fig. 95). Table XXII equates the two systems side by side. Martinsson's descriptive scheme is easy to apply, whereas Seilacher's necessitates some interpretation. The difference between a feeding burrow and a dwelling burrow, for example, is often subtle (e.g. Bromley, 1975).

The most useful aspect of trace fossils is the broad correlation between depositional environment and characterisic trace fossil assemblages, termed ichnofacies. Schemes relating ichnofacies to environments have been drawn up by Seilacher (1964, 1967), Rodriguez and Gutschick (1970) and Heckel (1972). Figure 96 is a blend of these various schemes.

The most landward ichnofacies to be defined consists largely of vertebrate tracks. These include the footprints of birds and terrestrial animals.

Dinosaur tracks are particularly well-studied examples. The preservation potential of such tracks is low. They are most commonly found on dried-up lake beds, river bottoms and tidal flats.

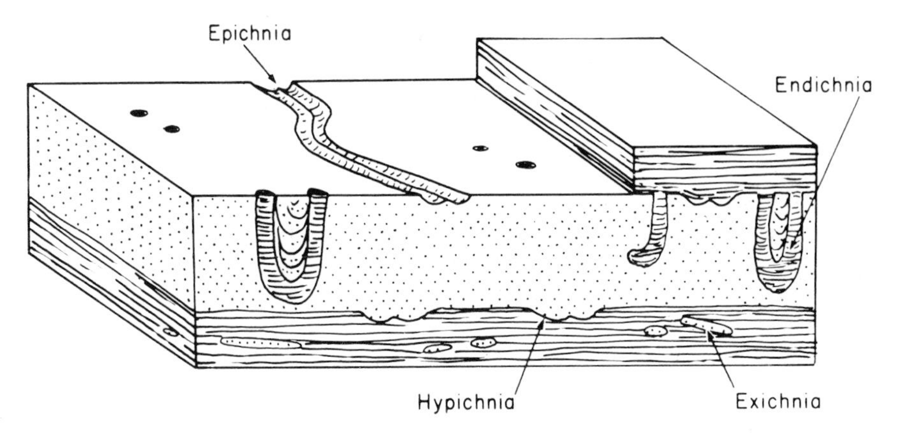

Fig. 95. Topological nomenclature for trace fossils. (After Martinsson, 1965.)

Table XXII
Nomenclature for trace fossil types

Activity nomenclature (Seilacher, 1964)		Topological nomenclature (Martinsson, 1965)
Repichnia :	crawling burrows	*Endichnia* and *Exichnia*
Domichnia :	dwelling burrows	
Fodichnia :	feeding burrows	
Pascichnia :	feeding trails	*Epichnia* and *Hypichnia*
Cubichnia :	resting trails	

Moving towards the sea a well-defined ichnofacies occurs in the tidal zone. This is often named the Scolithos assemblage because it is dominated by deep vertical burrows of the ichnogenus *Scolithos* (syn. *Monocraterion*, *Tigillites* and *Sabellarifex*). In this environment the sediment substrate is commonly subjected to scouring current action which often erodes and reworks sediment. Because of this the various invertebrates of the tidal zone —be they worms, bivalves or crabs, etc.—tend to live in crawling, dwelling and feeding burrows. These exit at the sediment: water interface, but go down deep to provide shelter for the little beasts during erosive phases. The burrows may be simple vertical tubes, like *Scolithus*, vertical U-tubes like *Diplocraterion yoyo* (so named because of its tendency to move up and down), or complex networks of passageways such as *Ophiomorpha*.

In subtidal and shallow marine environments, *Cruziana* and *Zoophycus*

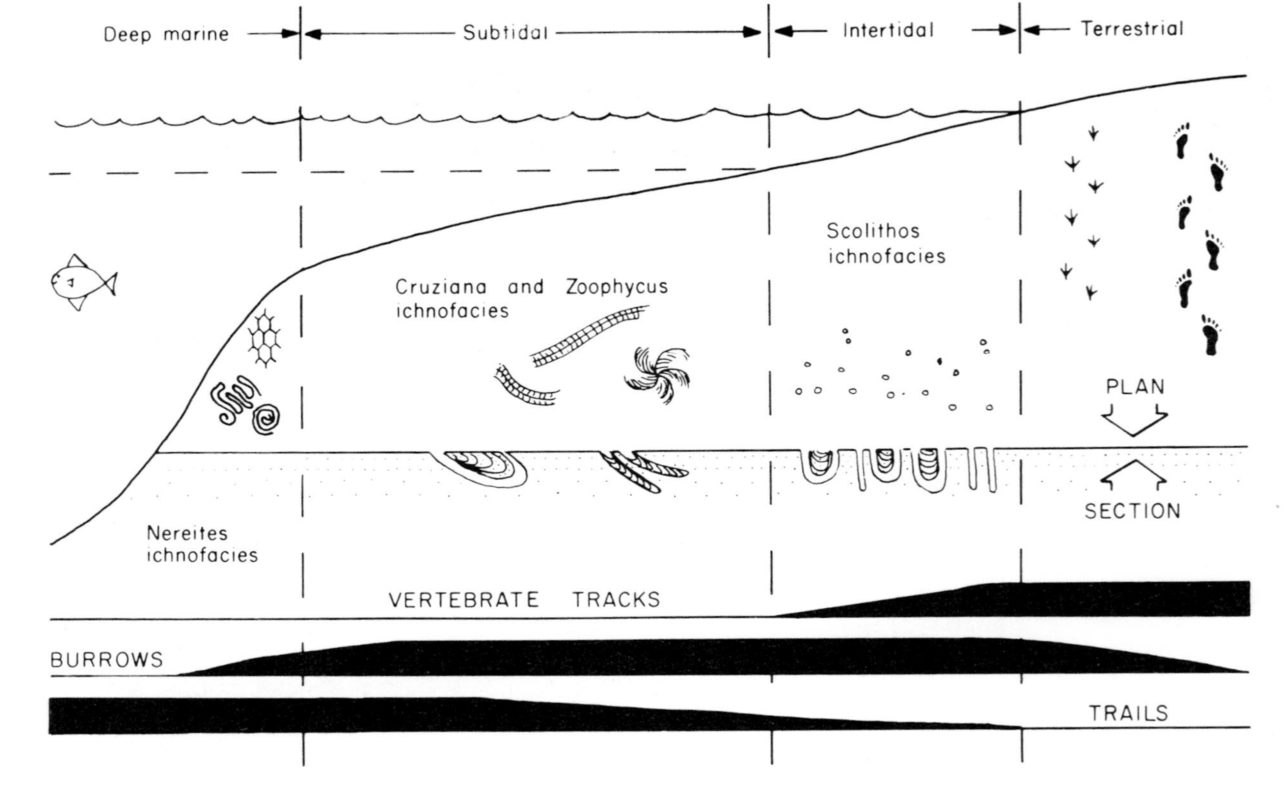

Fig. 96. Relationship between ichnofacies and environments, based on schemes proposed by Seilacher (1964, 1967), Rodriguez and Gutschick (1970) and Heckel (1972).

ichnofacies have been defined. In this zone, where marine action is less destructive, invertebrates crawl over the sea bed to feed in shallow grooves. They also make burrows, but these tend to be shallower and oriented obliquely or subhorizontally.

The Cruziana ichnofacies is characterized by the bilobate trail of that name (Fig. 97). This is generally referred to the action of trilobites. *Cruziana* has, however, been found in post-Palaeozoic strata and has been recorded from fluvial formations (e.g. Selley, 1970; Bromley and Asgaard, 1972). The environmental significance of this particular ichnogenus must be interpreted carefully. Camel flies make excellent Cruziana trails on modern sand dunes.

Zoophycus is a trace fossil with a characteristic helical spiral form in plan view. It is generally present at sand/shale interfaces. The detailed morphology of *Zoophycus* and the identity of its creator are a matter for

Fig. 97. Bilobate trails of *Cruziana*, attributable to trilobites, viewed from beneath. Found in shallow marine shales of the Um Sahm Formation (Ordovician), in the Southern Desert of Jordan. Length of specimen is 12 cm.

debate (see Crimes and Harper, 1970). Nevertheless, there is general agreement that it occurs in subtidal, shallow marine deposits.

Other trace fossils which characterize the Zoophycus and Cruziana ichnofacies include the subhorizontal burrows *Rhizocorallium* and *Harlania* (syn. *Arthrophycus*; syn. *Phycodes*).

Moving into deep quiet water, a further characteristic ichnofacies is named after *Nereites*. In this environment invertebrates live on, rather than in, the sediment substrate. Burrows are largely absent and surface trails predominate. Characteristic meandriform traces include *Nereites*, *Helminthoida* and *Cosmorhaphe*. Polygonal reticulate trails such as *Paleodictyon* are also characteristic of this ichnofacies.

This brief review of ichnofacies shows that the concept of environment restricted assemblages is of great use to sedimentologists. The forms are found *in situ* and are small enough to be studied from subsurface cores as well as at outcrop. When interpreted in their sedimentological context they are a useful tool in facies analysis.

One final point to note about biogenic sedimentary structures is the way in which they disrupt primary inorganic sedimentary structures. Intense burrowing, termed bioturbation, leads to the progressive disruption of bedding until a uniformly mottled sand is left. This is particularly characteristic of intertidal and subtidal sand bodies. Vertical burrows in interlaminated sands and shales may increase the vertical permeability of such beds; a point of some significance if they are aquifers or hydrocarbon reservoir formations.

III. PRIMARY INORGANIC SEDIMENTARY STRUCTURES

A. Introduction

Before proceeding to the actual descriptions, let us consider the classification of primary inorganic sedimentary structures. The problems of classification will be apparent from the preceding discussions. Attempts at classification and atlases of sedimentary structures have been made by Pettijohn and Potter (1964), Gubler (1966) and Conybeare and Crook (1968).

Three main groups can be defined by their morphology and time of formation (Table XXIII).

The first group of structures are pre-depositional with respect to the beds which immediately overlie them. These structures occur on surfaces between beds. Pedants may prefer to term them interbed structures, though they were formed before the deposition of the overlying bed. This group of

Table XXIII
A classification of inorganic megascopic primary sedimentary structures

Group	Examples	Origin
I. Pre-depositional (interbed)	Channels Scour-and-fill Flute marks Groove marks Tool marks	Predominantly erosional
II. Syndepositional (intrabed)	Massive Flat-bedding (including parting lineation) Cross-bedding Lamination Cross-lamination	Predominantly depositional
III. Post-depositional (deform interbed and intrabed structures)	Slump Slide Convolute lamination Convolute bedding Recumbent foresets Load structures	Predominantly deformation
IV. Miscellaneous	Rain prints Shrinkage cracks	

structures largely consists of erosional features such as scour-and-fill, flutes and grooves. These are sometimes collectively called sole marks or bottom structures.

The second group of structures are syndepositional in time of origin. These are depositional bed forms like cross-lamination, cross-bedding and flat-bedding. To avoid a genetic connotation, this group may be collectively termed intrabed structures, to distinguish them from pre-depositional interbed phenomena.

The third group of structures are post-depositional in time of origin. These are deformational structures which disturb and disrupt pre- and syndepositional inter- and intrabed structures. This third group of structures includes slumps and slides.

To these three moderately well-defined groups of sedimentary structures must be added a fourth. This last category, named "miscellaneous", is for those diverse structures which cannot be logically fitted into the scheme defined above.

The morphology and genesis of the various types of sedimentary structure will now be described in each of these four groupings.

B. Pre-Depositional (Interbed) Structures

Pre-depositional sedimentary structures occur on surfaces between beds. They were formed before the deposition of the overlying bed. The majority of this group of structures are erosional in origin.

Before describing these structures, there are two points to notice. The first is one of terminology. When the interface between two beds is split open, the convex structures which depend from the upper bed are termed casts. The concave hollows in the underlying bed into which these fit are termed moulds (Fig. 98). The second point to note is that the ease and frequency with which these bottom structures are seen is related to the degree of lithification.

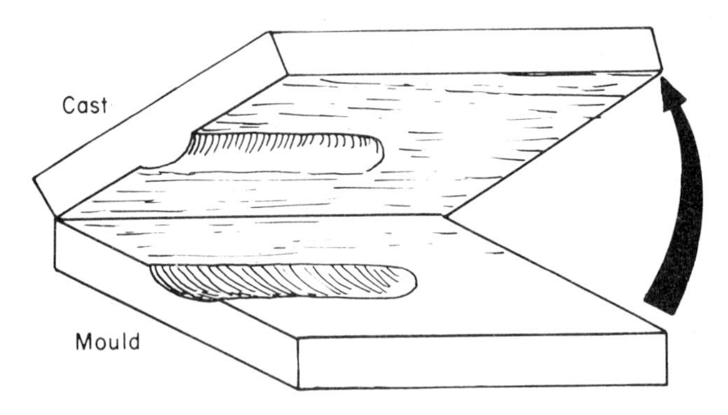

Fig. 98. Nomenclature for the occurrence of bed interface sedimentary structures (sole markings).

Unconsolidated sediments seldom split along interbed boundaries, because of their friable nature. This may explain the apparent absence of bottom structures in modern deep-sea turbidite sands. On the other extreme, well-lithified metasediments in fold belts tend to split more readily along subvertical tectonic fractures rather than along bedding planes. From this it can be seen that bottom structures are best observed in moderately well-lithified sediments which split along bedding surfaces.

The largest of this group of structures are channels. These may be kilometres wide and hundreds of metres deep. They occur in diverse environments ranging from subaerial alluvial plains to submarine continental margins. Channelling is initiated by localized linear erosion by fluid flow aided by corrosive bed load. Once a channel is established, however, a horizontal component of erosion develops due to undercutting of the channel bank followed by collapse of the overhanging sector.

The best studied channels are those of fluvial systems (see p. 268) and particular attention has been paid to the genesis of channel meandering and the mathematical relationships between sinuosity, channel width, depth, gradient and discharge (e.g. Schumm, 1969; Rust, 1978). In ancient channels, depth, width, sediment grade and flow direction can easily be established. Sinuosity can also be measured where there is adequate exposure or well control. Using these data, attempts have been made to calculate the gradient and stream power of ancient channel system (e.g. Friend and Moody-Stuart, 1972).

Channels are of great economic importance for several reasons. They can be hydrocarbon reservoirs and aquifers, they can contain placer and replacement mineral ore bodies, and they can cut out coal seams. Instances of these economic aspects of channels are documented in Chapter 10.

Smaller and less dramatic are the interbed structures termed "scour-and-fill". These are small scale channels whose dimensions are measured in decimetres rather than metres. They too occur in diverse environments.

There is a vast nomenclature for the numerous small interbed erosional structures. Reference should be made to the atlases of structures previously cited for exhaustive details of these. The following account describes the three most common varieties: flutes, grooves and tool marks. Flutes and grooves are scoured by the current alone, aided by granular bed load; while tool marks are made by single particles generally of pebble grade.

1. Flute marks

Flutes are heel-shaped hollows, scoured into mud bottoms. Each hollow is generally infilled by sand, contiguous with the overlying bed (Fig. 99). The rounded part of the flute is at the up-current end. The flared end points down-current. Flutes are about 1–5 cm wide and 5–20 cm long. They are

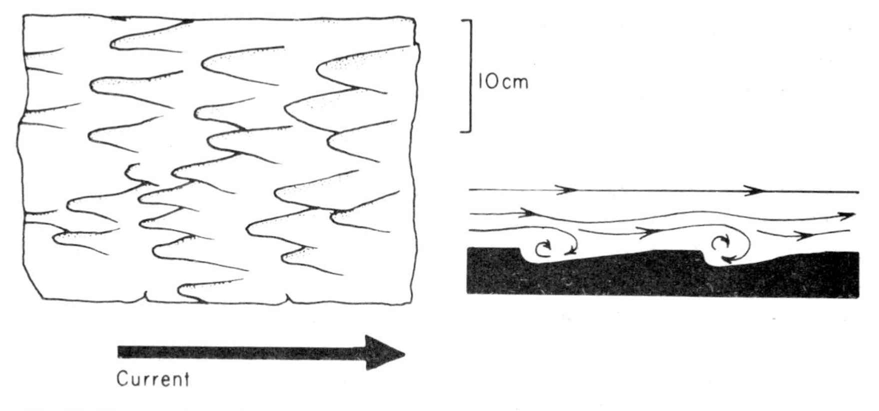

Fig. 99. Flute marks. Left: plan view. Right: section showing scouring of fluted hollows in soft mud by current vortices.

typically gregarious. Fluting has long been attributed to the localized scouring action of a current moving over an unconsolidated mud bottom. As the current velocity declines, flute erosion ceases and the hollows are buried beneath a bed of sand.

Allen (1968a, 1969, 1970, p. 82, 1971) has described experiments which explain the hydraulic conditions which generate flutes. The technique used in this and other experiments described in this chapter is as follows. A bed form is carved onto a plaster of Paris (gypsum) surface. Small, regularly spaced pits are marked on this. The slab is then placed in a flume and water passed over it. The pits become elongated in a down-current direction. The pit trends show how the current flow, immediately at the bed form/water interface, diverges from the mean flow-direction. Allen's experiments show that the flow pattern for flute erosion consists of two horizontal corkscrew vortices which lie beneath a zone of fluid separation at the top of the flute.

2. *Groove marks*

The second important type of erosional interbed structure is groove marks. These, like flutes, tend to be cut into mud and overlain by sand. They are long, thin, straight erosional marks. They are seldom more than a few millimetres deep or wide, but they may continue uninterrupted for a metre of more. In cross-section, the grooves are angular or rounded. Grooves occur where sands overlie muds in diverse environmental settings. Like flutes, they are especially characteristic of turbidite sands and trend parallel to the current direction determined from flutes and sedimentary structures within the sands. Grooves are seldom associated with flutes, however, and as discussed on p. 192, they are best developed in a more distal (down-current) situation than flutes.

It is clear that grooves are erosional features cut parallel to the current. Their straightness suggests laminar rather than turbulent flow-conditions. It has been argued that the grooves are carved by objects borne along in the current. This is not easy to prove, however, if the tools are not found. Furthermore, the linearity of the grooves proves that the tools were not saltating or rotating, but that they were transported down-current at a constant orientation and at an almost constant elevation with respect to the sediment substrate.

3. *Tool marks*

A third variety of bottom structure can, however, be attributed to such an origin and are termed "tool marks". These are erosional features cut in soft mud bottoms like flutes and grooves. They are, however, extremely irregular in shape, both in plan and cross-section, though they are roughly oriented parallel with the palaeocurrent. In ideal circumstances it has been possible to find the tool which cut these markings at their down-current end. Tools which have been found include pebbles (especially mud pellets ripped up from the bottom), wood and plant fragments, shells and fish vertebrae (Dzulinski and Slaczka, 1959).

Flutes, grooves and tool marks are three of the commonest sole markings found as interbed sedimentary structures. All are erosional, all are best seen, but not exclusive to, turbidite facies. A variety of other sole markings have been described and picturesquely named. Reviews of these will be found in Potter and Pettijohn (1963, pp. 114–128), in Dzulinski and Sanders (1962), and in Dzulinski and Walton (1965). It was pointed out (p. 192) that several detailed studies of turbidite formations show a correlation between current direction and sedimentary structures. Proximal turbidites (fluxoturbidites) tend to occur within channels. Moving down-current channels give way to tabular basin floor turbidites under which the sequence of erosional bottom structures grades from scour-and-fill to flute marks, groove marks and finally tool marks at their distal end. These changes are related to the down-current decrease in flow velocity of any individual turbidite.

C. Syndepositional (Intrabed) Structures

Syndepositional structures are those formed actually during sedimentation. They are, therefore, essentially constructional structures which are present within sedimentary beds.

At this point it is necessary to define and discuss just what is meant by a bed or bedding. Bedding, stratification or layering is probably the most fundamental and diagnostic feature of sedimentary rocks. Layering is not exclusive to sediments. It occurs in lavas, plutonic and metamorphic rocks. Conversely, bedding is sometimes absent in thick diamictites, reefs and some very well-sorted sand formations. Nevertheless some kind of parallelism is present in most sediments. Bedding is due to vertical differences in lithology, grain size or, more rarely, grain shape, packing or orientation. Though bedding is so obvious to see it is hard to define what is meant by the terms bed and bedding and few geologists have analysed this fundamental property (Payne, 1942; McKee and Weir, 1953; Campbell, 1967).

One of the most useful approaches to this problem is the concept of the "sedimentation unit". This was defined by Otto (1935) as "that thickness of sediment which appears to have been deposited under essentially constant physical conditions". Examples of sedimentation units are a single cross-bedded stratum, a varve or a mud flow diamictite.

A useful rule of thumb definition is that beds are distinguished from one another by lithological changes. Shale beds thus typically occur as thick uninterrupted sequences. Sandstones and carbonates, though they may occur in thick sections, are generally divisible into beds by shale laminae. Two more arbitrary but useful definitions are:

(a) bedding is layering within beds on a scale of about 1 or 2 cm;

(b) lamination is layering within beds on a scale of 1 or 2 mm.

Using these dogmatic definitions, the synsedimentary intrabed structures are of five categories: massive, flat-bedded, cross-bedded, laminated and cross-laminated.

The morphology and origin of these will now be described.

1. Massive bedding

An apparent absence of any form of sedimentary structure is found in various types of sedimentation unit. It is due to a variety of causes. First, a bed may be massive due to diagenesis. This is particularly characteristic of certain limestones and dolomites which have been extensively recrystallized. Secondly, primary sedimentary structures may be completely destroyed in a bed by intensive organic burrowing.

Genuine depositional massive bedding is often seen in fine-grained, low-energy environment deposits, such as some claystones, marls, chalks and calcilutites. *In situ* reef rock (biolithite) also commonly lacks bedding.

In sandstones massive bedding is rare. It is most frequently seen in very well-sorted sands, where sedimentary structures cannot be delineated by textural variations. It has been demonstrated, however, that some sands which appear structureless to the naked eye are in fact bedded or cross-bedded when X-rayed (Hamblin, 1962).

Genuine structureless sand beds may be restricted to the deposits of mud flows, grain flows and the lower (A unit) part of turbidites, though these may be size graded.

2. Flat-bedding

One of the simplest intrabed structures is flat-bedding or horizontal bedding. This, as its name implies, is bedding which parallels the major bedding surface. It is generally deposited horizontally. Flat-bedding grades, however, via subhorizontal bedding, into cross-bedding. The critical angles of dips which separate these categories are undefined. Flat-bedding occurs in diverse sedimentary environments ranging from fluvial channels to beaches and delta fronts. It occurs in sand-grade sediment both terrigenous and carbonate. Flat-bedding is attributed to sedimentation from a planar bed form. This occurs under shooting flow or a transitional flow regime with a Froude number of approximately one. Sand deposited under these conditions is arranged with the long-axes of the grains parallel to the flow direction. Moderately well-indurated sandstones easily split along flat-bedding surfaces to reveal a preferred lineation or graining of the exposed layer (Fig. 100). This feature is termed parting lineation, or primary current lineation (Allen, 1964). It is important to remember that, like many of the bed sole markings previously described, parting lineation will not be seen in friable unconsolidated sands, nor in low-grade metamorphic sediments.

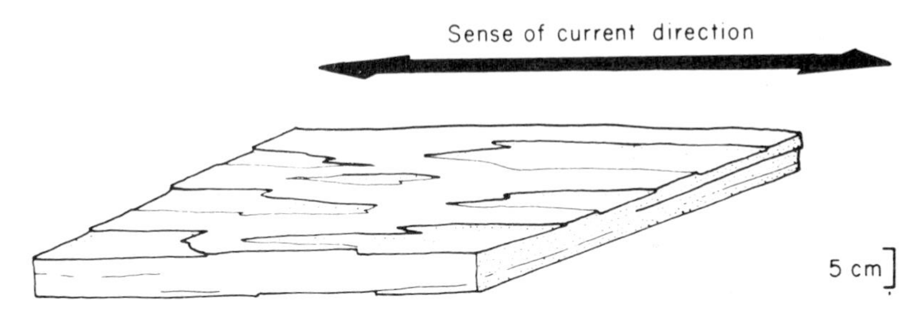

Fig. 100. Sketch of current or parting lineation. This appears as a graining on bedding planes of moderately-cemented fissile sandstones. It indicates the sense, but not the direction, of the depositing current.

3. Cross-bedding

Cross-bedding is one of the commonest and most important of all sedimentary structures. It is ubiquitous in traction current deposits in diverse environments. Cross-bedding, as its name implies, consists of inclined dipping bedding, bounded by subhorizontal surfaces. Each of these units is termed a set. Vertically contiguous sets are termed cosets (Fig. 101). The inclined bedding is referred to as a foreset. Foresets may grade with decreasing dip angle into a bottomset or toeset. At its top a foreset may grade with decreasing dip angle into a topset. In nature toesets are rare and topsets are virtually non-existent.

Foresets may be termed heterogenous if the layering is due to variations in grain size, or homogenous if it is not. Two other descriptive terms applied to foresets are avalanche and accretion (Bagnold, 1954, pp. 127, 238). Avalanche foresets are planar in vertical section and are graded towards the base of the set. Accretion foresets are ungraded, homogenous and have asymptotically curved toesets.

Many workers have recorded the angle of dip of foresets (see Potter and Pettijohn, 1963, p. 79). A wide range of values have been obtained with a mode of between 20–25° for ancient sediments. The foreset dip reflects the critical angle of rest of the sand when it was deposited. This will be a function of the grade, sorting and shape of the sediment as well as the viscosity of the ambient fluid. Legend has it that eolian sands have higher angles of rest than subaqueous sands. There are some data to support this.

Dips of 30–35° have been recorded from modern eolian dunes (e.g. Bigarella, 1972; McBride and Hayes, 1962). Dips in modern subaqueous cross-bedded sands seldom appear to exceed 30° (e.g. Harms and Fahnestock, 1965; Imbrie and Buchanan, 1965).

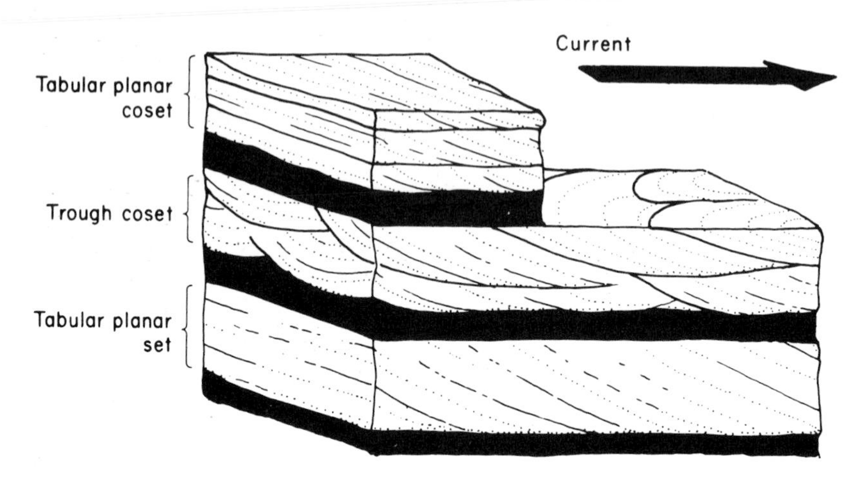

Fig. 101. Basic nomenclature of cross-bedding. Tabular planar cross-beds have subplanar foresets. Trough cross-beds have spoon-shaped foresets. Isolated cross-beds are referred to as sets. Vertically grouped foresets constitute a coset.

The rather lower angles recorded from ancient sediments may be due to a variety of factors. These include the fact that set-bounding surfaces are generally not valid palaeohorizon data. They frequently dip up-current and thus tend to diminish the maximum measured angle of foreset dip. A second factor is that unless the amount of dip is recorded from a face which is exactly perpendicular to the dip direction, then an apparent dip will be recorded which is less than the true dip. A third factor which may decrease the depositional dip of a foreset is compaction (Rittenhouse, 1972).

The diverse geometric relationships which may exist between foresets and their bounding surfaces has lead to a rich blossoming of geological nomenclature and classificatory schemes (e.g. Allen, 1963). Basically, two main types of cross-bedding can be defined by the geometry of the foresets and their bounding surfaces: tabular planar cross-bedding and trough cross-bedding (McKee and Weir, 1953). In tabular planar cross-bedding, planar foresets are bounded above and below by subparallel subhorizontal set boundaries (Fig. 102). In trough cross-bedding, upward concave foresets lie within erosional scours which are elongated parallel to current flow, closed up-current and truncated down-current by further troughs (Fig. 103). Additional details on cross-bedding morphology and nomenclature are

Fig. 102. Tabular planar cross-bedding in fluvial Cambro-Ordovician sandstones, Jebel Gehennah, southern Libya.

Fig. 103. Trough cross-bedding in fluvial Cambro-Ordovician sandstones, Jebel Dohone, southern Libya.

given in Potter and Pettijohn (1963, pp. 62–80). Full descriptions of fluvial cross-bedding have been given by Frazier and Osanik (1961), Harms *et al.* (1963) and Harms and Fahnestock (1965). Cross-bedding in tidal sand bodies has been described by Hulsemann (1955), Reineck (1961) and Imbrie and Buchanan (1965). The last of these studies shows that the internal sedimentary structures of carbonate sands are no different from those of terrigenous deposits.

The genesis of cross-bedding has been studied empirically from ancient deposits, experimentally in laboratories and in modern sediments. It appears that much cross-bedding is formed from the migration of sand dunes or megaripples. Flume experiments (described on p. 182) showed how these bed forms migrate down-current depositing foresets of sand in their down-current hollows. If sedimentation is sufficiently great then the erosional scour surface in front of a dune will be higher than that of its predecessor and a cross-bedded set of sand will be preserved (Fig. 104).

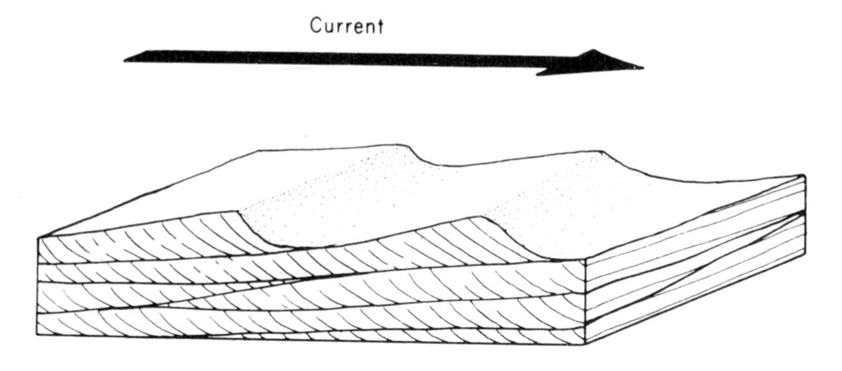

Fig. 104. Formation of tabular planar cross-bedding occurs where dunes migrate down-current, and where stoss side erosion is less than sedimentation of the foreset. Note that the preserved thickness of each set is less than the height of the dune from which it was deposited.

Tabular planar cross-bedding will thus form from straight crested dunes. Trough cross-bedding will form in the rounded hollows of more complex dune systems.

There are, however, several other ways in which cross-bedding may form and three of these should be noted in particular.

In river channels, especially those of braided type (see p. 272), the course consists of an alternation of shoals and pools through which the axial part or parts of the channel (termed the thalweg) makes its path. Where the thalweg suddenly enters a pool there is a drop in stream power and a subaqueous sand delta is built out. Given time, sediment and the right flow conditions, this delta may completely infill the pool with a single set of cross-strata (Jopling, 1965), e.g. Fig. 105.

A second important way in which cross-bedding forms is seen in channels. Infilling of a channel may deposit cross-bedding paralleling the channel margin. Alternatively cross-bed deposition occurs on the inner curves of meandering channels synchronous with erosion on the outer curve

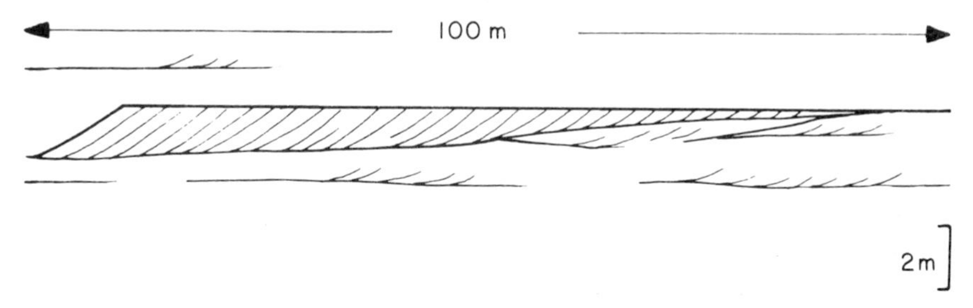

Fig. 105. Single large foreset deposited in braided channel chute-pool. Cambro-Ordovician, Wadi Rum, Jordan.

(see p. 274). By this means, a tabular set of cross-strata may be deposited in which the foresets strike parallel to the flow direction (Lyell, 1865, p. 17).

This type of lateral cross-bedding is rather larger than most types (sets may be several metres high) and their base is marked by a conglomerate. Close examination of the foresets often shows that they are composed of second order cross-beds or cross-laminae which do in fact reflect the true current direction (Fig. 106).

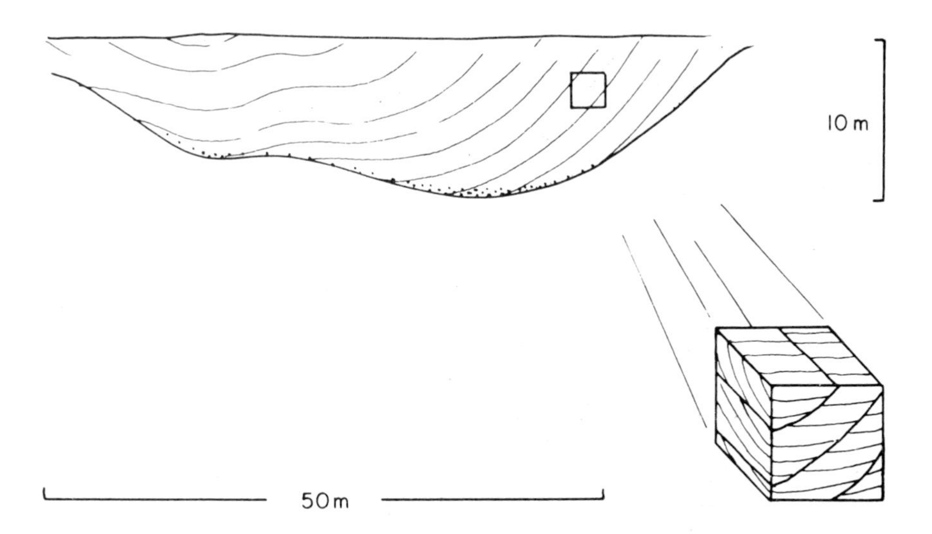

Fig. 106. Channel showing complex cross-bedded fill. Major trough cross-sets are themselves composed of smaller tabular planar sets. Miocene, Jebel Zelten, Libya.

A third important variety of cross-bedding is that formed by antidunes in upper flow-regime conditions. It has been pointed out that at very high current velocities sand dunes develop which migrate up-current (p. 183). These deposit up-current dipping foresets. The foresets of these antidunes, as they are called, are seldom preserved. As the current wanes prior to net sedimentation, antidunes tend to be obliterated as the bed form changes to a plane bed or dunes.

Exceptions have been noted, however, notably from turbidite sands (Skipper, 1971; Hand *et al.*, 1972). Figure 107 illustrates an example from an alluvial environment.

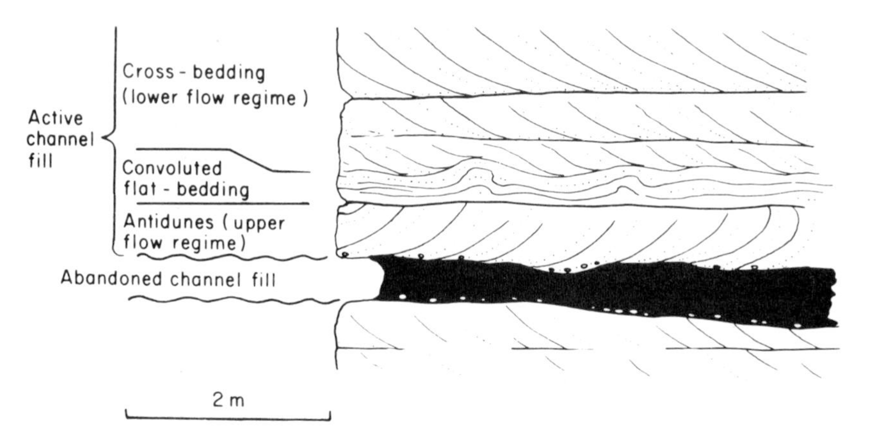

Fig. 107. Antidune cross-bedding at base of braided channel sequence. Cambro-Ordovician, Jordan.

These descriptions show that cross-bedding is a very complex sedimentary structure. More properly it is a group of structures of diverse morphology and genesis.

Particular attention has been paid by geologists to determine depositional environment from the type of cross-bedding. This has not been notably successful because, though the structural morphology is closely related to hydrodynamic conditions, the same set of hydrodynamic parameters can occur in various environments.

In shallow marine environments it is not uncommon to find "herring-bone" cross-bedding in which bimodally dipping foresets reflect the to and fro movement of ebb and flood currents (Fig. 108). Allen (1980) has illustrated the spectrum of cross-bedding and associated structures which reflect the degree of symmetry of tidal currents.

Another line of approach has been to try to determine water depth from set height. This has not been very successful, for a number of reasons. Not the least of these is that the preserved set height is controlled by the degree of erosion which occurred after a set was laid down. Nevertheless, in underwater cross-bedding water depth cannot have been less than the preserved set height.

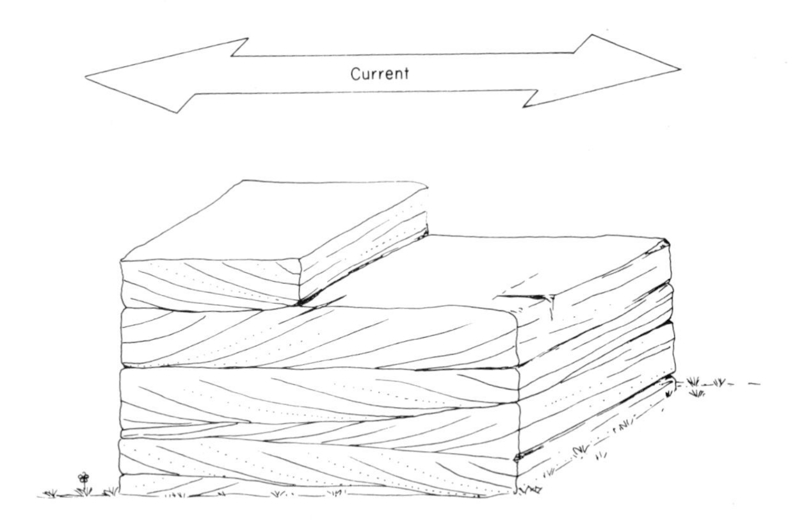

Fig. 108. Sketch of herring-bone cross-bedding due to tidal currents.

Set height has also been used to try to distinguish eolian from subaqueous dunes. The folk-lore is that eolian dunes deposit higher set heights than subaqueous ones. This is not universally true as the studies cited on p. 197 show. No satisfactory height limit for subaqueous cross-bedding can be fixed as the internal morphology of submarine dunes is so little known (see p. 198).

One of the most important things that can be learnt from cross-bedding is the flow directions of the currents which deposited them. This can give important clues to the environment, palaeogeography and structural setting of the beds in which they occur. This important topic of palaeocurrent analysis is discussed later in the chapter.

4. Ripples and cross-lamination

Ripples are a wave-like bed form which occur in fine sands subjected to gentle traction currents (Fig. 109). Migrating ripples deposit cross-laminated sediment.

Ripple marking in modern and ancient sediments has attracted the interest of many geologists. Studies of historical significance include those of Sorby (1859), Darwin (1883), Kindle (1917), Bucher (1919) and Allen (1968). The last of these is a definitive work of fundamental importance.

The following account describes the association of ripple bed form and internal cross-lamination and then discusses their crigin.

Fig. 109. Ripples in lacustrine Torridon Group (Pre-Cambrian) sandstones. Raasay, north-west Scotland.

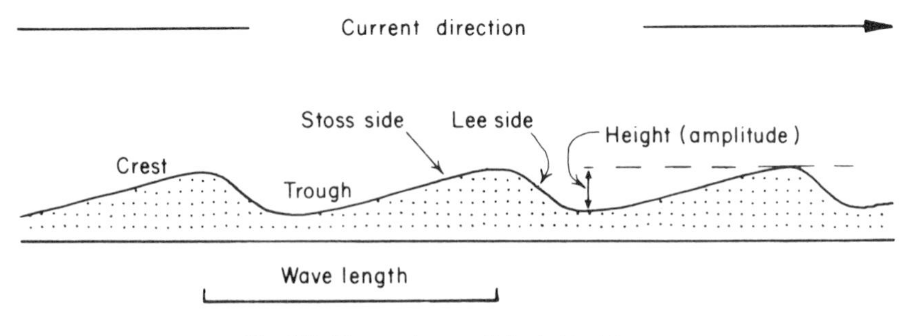

Fig. 110. Nomenclature of rippled bed forms.

Figure 110 illustrates the nomenclature of ripples. This particular case shows asymmetric ripples formed by a traction current. In cross-section a ripple consists of a gentle up-current stoss side and a steep down-current facing lee side. The highest points of the ripples are the crests. The lowest points are the troughs. The height of the ripple is the vertical distance from trough to crest. The wavelength of a train of ripples (their collective term) is the horizontal distance between two crests or troughs. A statistical parameter termed the ripple index (abbreviated to RI) is calculated by dividing the wavelength by the ripple height. Other numerical indices derived from ripple form are described by Tanner (1967) and Allen (1968b).

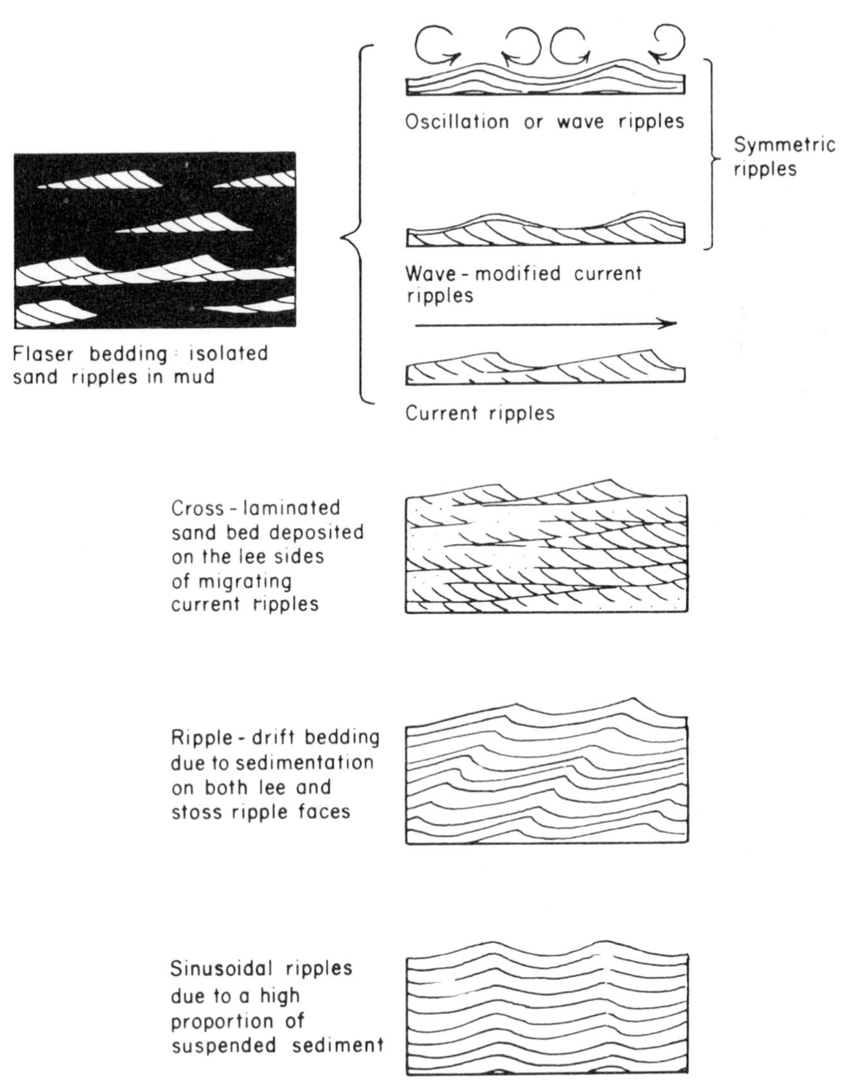

Fig. 111. Various types of sedimentary structures produced from ripple bed forms.

Various types of ripple bedding are defined by a combination of their external shape (bed form) and internal structure (Fig. 111).

In cross-section ripples are divisible into those with symmetric and asymmetric profiles. Symmetrical ripples, sometimes termed oscillation or wave ripples, show two types of internal lamination. One variety has laminae conformable with the surface form. The second shows cross-laminae unrelated to the surface form. The first of these types suggests that these symmetric ripples formed by oscillation of the fluid with no net horizontal component of sediment of fluid transport. The second variety of symmetric ripple clearly results from the modification of asymmetric current ripples by oscillatory movement.

Asymmetric ripples, by contrast to symmetric ones, show a clearly differentiated low-angle stoss side and steep-angle lee side. Internally they are cross-laminated, with the cross-laminae concordant with the lee face.

Asymmetric ripples may be produced by unidirectional traction currents as, for example, in a river channel. They may also form in lakes, lagoons, and shallow marine environments by waves. Waves generate currents on sand beds which cause asymmetric ripples to migrate in the direction of wave movement.

Symmetric ripples are not produced by simple symmetric waves, as has been popularly believed. This process simply moves sand to and fro. Acting on a previously rippled surface however, these waves will redistribute sand from ripple troughs to crests. Thus asymmetric ripples may be moulded in to symmetric ones, with a veneer of symmetric laminae over unidirectional cross-laminae (Allen, 1979).

Assymmetric ripples may be due to traction currents, and to wave-induced currents. Symmetric ripples are produced by oscillation waves reworking pre-existing ripples.

Both asymmetric and symmetric ripples can occur with isolated lenses of mudstone. This is termed flaser bedding (Reineck and Wunderlich, 1968; Terwindt and Breusers, 1972).

With gradually increasing sand content, flaser bedding can grade into beds composed entirely of cross-laminated sand in which ripple profiles are absent, though they are sometimes preserved on the top of the bed. A varied terminology has been proposed for these sedimentary structures, including cross-lamination, climbing ripples and ripple-drift bedding. Jopling and Walker (1968) have defined a spectrum of ripple types which are related to the ratio of suspended to traction load material which is deposited (Fig. 111). Normal traction currents deposit sand on the lee side of the ripple only. With increasing suspended load, sedimentation also occurs on the stoss side. This generates a series of ripple profiles whose crests migrate obliquely upward down-current. With excessive suspended load, sinusoidal ripple lamination develops from the vertical accretion of symmetric ripple profiles. Jopling and Walker point out that these symmetric ripples which deposit continuous laminae of sediment are distinct from the isolated symmetric ripples formed by wave oscillation.

Having described ripple morphology in cross-section, let us now consider their plan view. Ripples seen in modern sediments or exposed on ancient bedding surfaces show a diversity of shapes. Certain dominant types do tend to occur and these have been named (Fig. 112).

Simplest of all are the straight-crested ripples; these include both symmetric and asymmetric profiles. Straight-crested or rectilinear ripples can be traced laterally for many times further than their wavelength. They are oriented perpendicular to the direction of wave or current movement which generates them.

Sinuous ripples show continuous but slightly undulating crest lines.

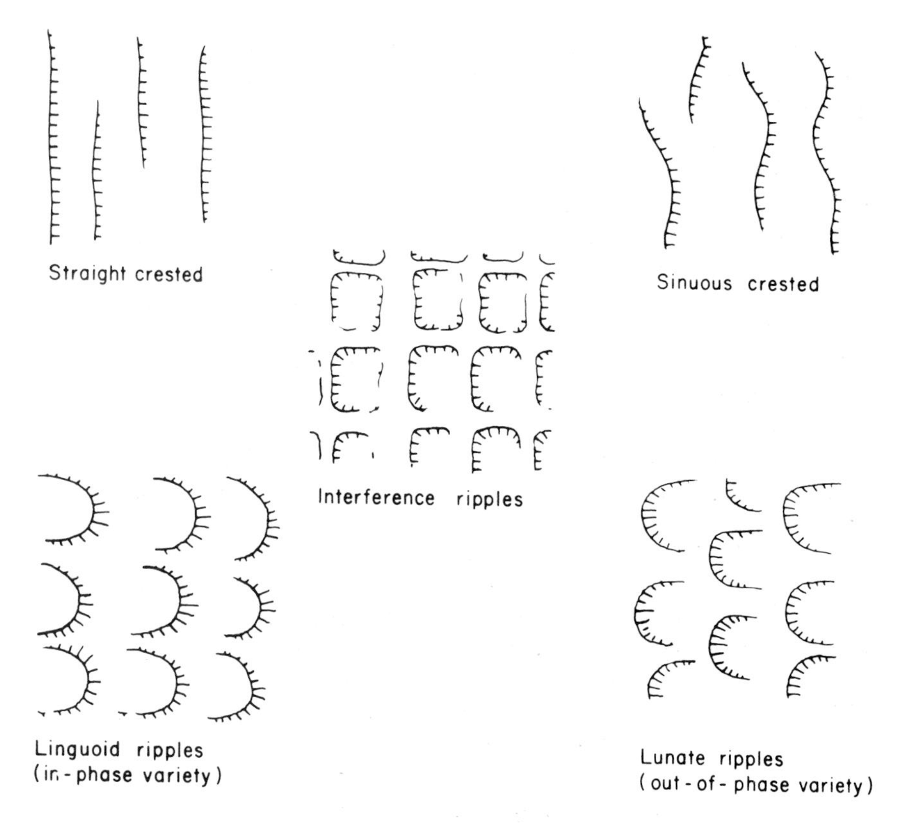

Fig. 112. Nomenclature of rippled bed forms as seen in plan view. (After Allen, 1968a.) Current moves from left to right.

The second main group of ripples, as seen in plan, are those whose crest lengths are generally shorter than their wavelength. These are exclusively asymmetric current ripples. Two important subtypes can be recognized. Lunate ripples have an arcuate crest which is convex up-current. Linguoid ripples have an arcuate crest which is convex down-current.

In plan view, successive linguoid or lunate ripples may be arranged "*en echelon*", out of phase with one another, or in phase, if crests all lie on the same flow axis.

In the same way that trough cross-bedding originates in the migrating hollows of complex dune fields, so does a variety of trough cross-lamination form from migrating complex ripple trains. This structure is picturesquely described as "rib and furrow".

A third main group of ripples can be recognized from their appearance in plan. These are interference ripples which, as their name suggests, consist of two obliquely intersecting sets of ripple crests. Interference ripples result from the modification of one ripple train due to one set of conditions by a later train, generated by waves or currents with a different orientation. "Tadpole nests" is a quaint synonym for interference ripples.

Having examined their morphology, let us now look at the origin of ripples in rather more detail.

It is a matter of observation that ripples do not form in clay nor in coarse sand or gravel. They are restricted to coarse silt and sand with a grain size of less than about 0·6 mm diam. Analysis of traction currents shows that ripple bed forms occur in the lower part of the lower flow regime with a low Froude number (p. 184).

Particular attention has been paid to the way in which ripples are actually formed from a plane bed of sand. It has been suggested that ripple trains develop down-stream from pre-existing irregularities of the sediment substrate (e.g. Southard and Dingler, 1971). An alternative school of thought argues that ripples can form spontaneously on a plane sand bed. Initiation is by random turbulent vortices which scour the first irregularities (e.g. Williams and Kemp, 1971).

Of more interest to geologists is, not so much what initiates ripple formation, but what can be learnt from the resultant structure. One would like to gain information of flow conditions, direction, water depth and environment.

Attempts to relate ripple type to flow conditions have not been noticeably successful. Allen (1968a) has, however, recognized a broad correlation between current ripples and decreasing depth and concomitant increasing velocity. The sequence changes from straight-crested ripples to sinuous crests and so to lunate and linguoid trains. The water depth in which ripples form has no effect on their height or wavelength except at extremely shallow depths. Flow direction, on the other hand, can be very easily determined from ripples, both from their strike and lee face in plan view, and from the dip direction of their internal cross-lamination. This aspect of palaeocurrent analysis is discussed later in the chapter (p. 242).

Ripples occur today in many different environments, ranging from sand dunes, through rivers and deltas to the ocean bed. It has already been pointed out that ripples are closely related to a given set of flow conditions and that these may be encountered in diverse environments. One would not, therefore, expect to find a relationship between ripple morphology and environment, as opposed to process. Nevertheless, Tanner (1967, 1971) has empirically developed several statistical indices which appear to be capable of differentiating ripples from different sedimentary environments.

D. Post-Depositional Sedimentary Structures

The third main group of sedimentary structures are those due to deformation. These may be termed post-depositional because, obviously, they can only form after a sediment has been laid down.

There are a great variety of deformational structures; many are ill-defined and strangely named. They can be arranged, however, into three main groups arbitrarily defined according to whether the sense of movement was dominantly vertical or dominantly lateral, and according to whether the sediment deformed plastically in an unconsolidated state, or whether it was sufficiently consolidated to shear along slide planes (Table XXIV).

These three groups of deformation structures will now be described.

Table XXIV
A classification of post-depositional deformational sedimentary structures

Sense of movement	Structure	Nature of deformation
Dominantly vertical	Load casts and pseudo-nodules, convolute bedding, recumbent foresets, convolute lamination	Plastic (sediments lack shear strength)
Dominantly horizontal	Slumps Slides	Brittle (sediments possess shear strength)

1. Vertical plastic deformational structures

Deformational structures which involve vertical plastic movement of sediment are of two main types. One group occurs within sand beds and may be loosely referred to as quicksand structures. The second group develop at the interfaces of sand overlying mud.

The simplest type of quicksand structure is seen in vertical section as a series of plastic folds. Typically broad flat synclines separate sharp peaked anticlines. The anticlines are sometimes overturned down-current (as shown by cross-bedding in the overlying sand). In plan view the folds often show an elongation perpendicular to current direction. This type of quicksand structure involves the deformation of whole beds of sand up to a metre or more thick (Fig. 113). It is loosely referred to as convolute bedding. This structure is found in many types of sandstone, but is particularly characteristic of fluvial sands (see Selley *et al.*, 1963, and McKee *et al.*, 1967, for ancient and modern examples respectively).

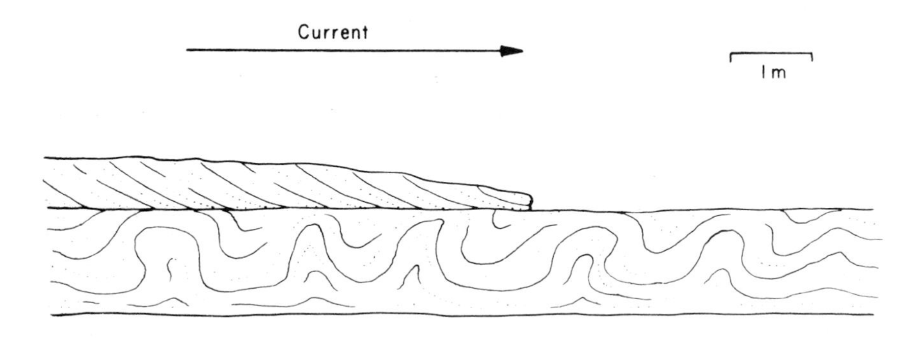

Fig. 113. Convolute bedding due to the expulsion of pore water from loosely packed sand. Torridonian (Pre-Cambrian) fluvial sandstones, Raasay, Scotland.

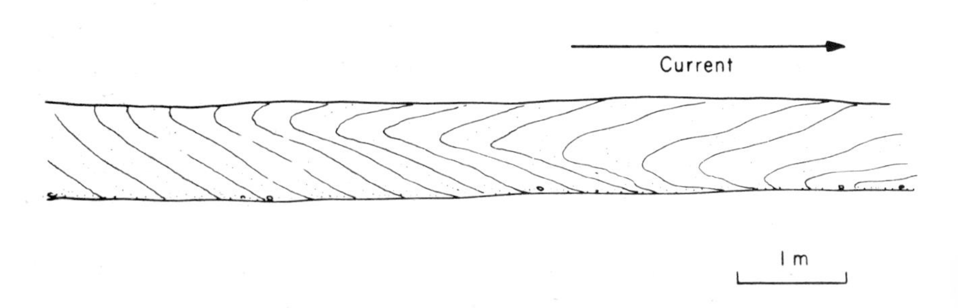

Fig. 114. Recumbent foreset deformation in fluvial sandstone. Cambro-Ordovician, Jordan.

Often associated with convolute bedding is the deformation of cross-bedded sands. The foresets become overturned down-current in the shape of recumbent folds. The axial plane of the fold is commonly tilted down-current in any one set (Fig. 114). Recumbent foresets, like convolute bedding, are found in divers traction deposited sands, and are especially common in the coarse sands of braided alluvium.

Considerable attention has been paid to the origin of convolute bedding and recumbent foresets (e.g. Allen and Banks, 1972). There is widespread agreement that these structures are caused by the vertical passage of water through loosely packed sand. This water may be due to a hydrostatic head of water, as for example on an alluvial fan (e.g. Williams, 1970, p. 409). Alternatively, the water may be derived from the sediment itself. A sand will not compact significantly at the surface, but its grains may be caused to fall into a tighter packing. This results in a decrease in porosity. Excess pore water will be vertically expelled.

Laboratory experiments have shown that this process can indeed generate convolute bedding (Selley, 1969). These experiments showed that the sands could fall into a tighter packing both by vibration and by turbulent eddies in the overlying water.

Convolute bedding has been recorded in modern sediments both as a result of earthquakes and without them (e.g. Barratt, 1966 and McKee *et al.*, 1967, respectively).

The down-current overturning of convolute folds and their association with down-current deformed foresets strongly suggest that powerful currents play a significant part in their genesis. This structure is not restricted to aqueously deposited sediment, however, but also occurs in eolian ones (Doe and Dott, 1980).

On a smaller scale, laminated fine sands and silts also show pene-contemporaneous vertical deformation structures termed convolute lamination. This is similar in geometry to convolute bedding, but occurs in finer-grained sediment on a much smaller scale; generally in beds only a decimetre or so high. Convolute lamination is especially characteristic of turbidite sands, involving deformation of both the laminated and cross-laminated Bouma units. Correlation of fold axes with ripple crests and the presence of deformed intrabed scour surfaces suggests that movement was virtually synchronous with deposition. Convolute lamination probably originates, therefore, by the dewatering of the sediment aided by the shear stresses set up by the turbidity flow itself (see also Davies, 1965; Anketell *et al.*, 1970; Visher and Cunningham, 1981).

Convolute lamination, convolute bedding and recumbent foresets are the three main types of intrabed vertical deformational structures.

A variety of structures develop where sands overlie muds. The mud/sand interface is often deformed in various ways. Most typically irregular-rounded balls of sand depend from the parent sand bed into the mud beneath. These structures are variously termed loadcasts, ball and pillow structures, etc. They are a variety of the broad group of structures termed sole markings or bottom structures. It is important, however, to distinguish deformational bottom structures, like loadcasts, from erosional markings such as grooves and flutes. Sometimes erosional bottom structures become deformed.

In extreme cases the sand lobes may become completely detached from their parent bed above. Similarly thin sand beds may split up along their length to form isolated cakes of sand in mud (Fig. 115). These discrete bodies of sand in mud are termed "pseudonodules" to distinguish them from normal diagenetic nodules (Macar and Antun, 1949).

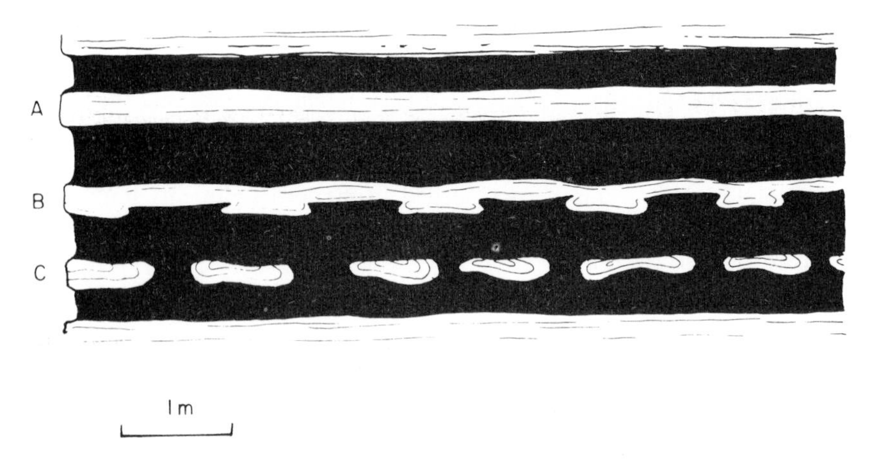

Fig. 115. Vertical load structures in interbedded sands and siltstones. Torridonian (Pre-Cambrian) Fladday. Scotland. Bed A is undeformed, bed B shows well developed load casts on its lower surface. Bed C has split up into discrete pseudonodules.

Load casts and pseudonodules occur at sand/mud interfaces in various environments, both modern and ancient. They are a common feature of turbidite facies, yet they also occur in deltaic and fluvial sediments. There is general agreement that these structures are generated by the differential loading of a waterlogged sand on an unconsolidated mud. They are easy to make in the laboratory (e.g. Kuenen, 1958).

2. Slumps and slides

Slump structures, like the structures previously described, involve the penecontemporaneous plastic deformation of sand and mud. Slump folds, however, commonly show clear evidence of extensive lateral movement in a consistent direction. Slump folds are commonly associated with penecontemporaneous faulting and with major low-angle zones of decollement termed slide planes. Large masses of sediments are laterally displaced along slide surfaces. In rare, but fascinating cases, the top of a slump bed may be covered by volcanoes of sand complete with axial vents and bedded cones (Fig. 116). These formed from sand carried up during dewatering of the slump after it came to rest (Gill and Kuenen, 1958).

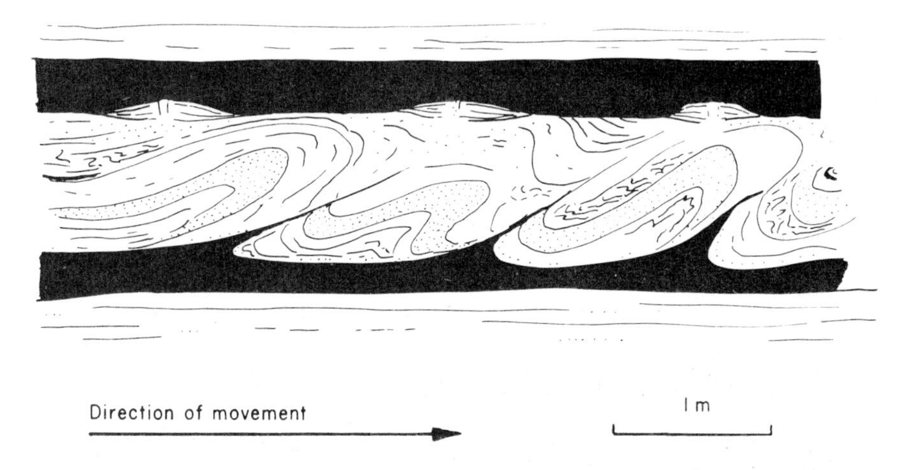

Fig. 116. Slumped beds showing recumbent folds, slide surfaces and sand volcanoes. The presence of the latter demonstrate not only that movement was penecontemporaneous, but also that it occurred at the sediment/water interface before deposition of the overlying shale. Based on examples in the Carboniferous of County Clare, Ireland.

When sliding and slumping were first studied there was considerable controversy over whether these phenomena were tectonic, or whether they were penecontemporaneous soft sediment features. Distinctive criteria for penecontemporaneous movement include the fact that folds and faults are overlain by undisturbed sediment, and their orientations may be unrelated to regional tectonic style and orientation. Disturbed beds may be penetrated by undeformed plant roots or animal burrows and penecontemporaneous faults lack gangue minerals.

Slumps and slides require for their generation the rapid deposition of muddy sediment on an unstable slope. Lateral movement may be initiated by earthquakes, storms or perhaps purely spontaneously. These conditions are best met with on delta fronts in actively subsiding basins. Many case histories of sliding and slumping have been recorded from these situations (e.g. Kuenen, 1948; Klein *et al.*, 1972; Blanc, 1972). Nevertheless, slumping occurs on all scales, from the caving of a river bank to the collapse of a continental margin. For example, large slump masses of Pleistocene sediment have been delineated seismically off the Rockall Bank (North Atlantic) and on the continental slope off the east coast of the North Island of New Zealand (Roberts, 1972 and Lewis, 1971, respectively). In the last of these examples 10–50 m thick beds of Pleistocene sand and silt slumped down slide planes of 1–4°.

On the extreme end of the scale, it has been suggested that the island of Barbados slid down into an ocean basin and was then lifted up again tectonically (Davies, 1971).

This example, right or wrong, demonstrates that extensive lateral sediment displacement is often structurally controlled as it grades into the realms of gravity tectonics (e.g. De Sitter, 1964, pp. 232–256).

E. Miscellaneous Structures

Among the vast number of sedimentary structures which have been described, there are many which do not conveniently fit into the simple tripartite scheme outlined above. These miscellaneous structures include rain prints, salt pseudomorphs and various vertical dike-like structures of diverse morphology and origin. These include desiccation cracks, syneresis cracks, sedimentary boudinage and sand dikes.

Rain prints occur within siltstones and claystones, and where such beds are overlain by very fine sandstones. In plan view, rain prints are circular or ovate if due to wind-blown rain. They are typically gregarious and closely spaced. Raised ridges are present around each print. The whole structure ranges from 2 to 10 mm diam. Rain prints are good indicators of subaerial exposure but are not exclusive to arid climates, though they may have a higher preservation potential in such conditions. Care should be taken to distinguish rain prints from pits formed where sand grains impress soft mud. This is sometimes found where fissile shales contain thin coarse sand laminae. Such sand grain imprints lack the raised rim of rain prints (Fig. 117).

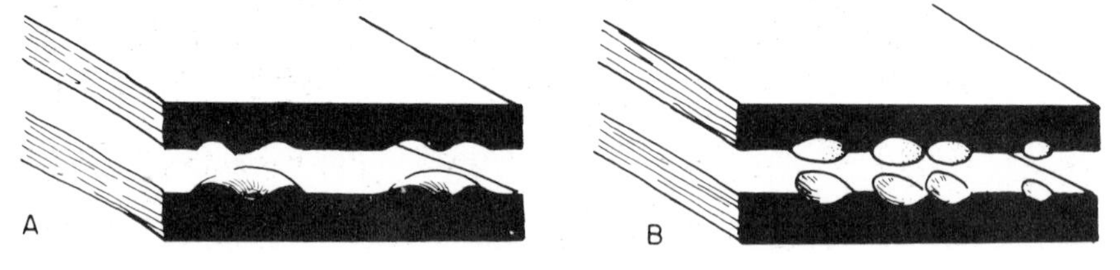

Fig. 117. A: rain prints showing characteristic raised marginal rims. B: pits produced by coarse sand grains on shaley parting.

Salt pseudomorphs occur in similar lithological situations to rain prints. They are typically found where claystones or siltstones are overlain by siltstones or very fine sandstones. Salt pseudomorphs are moulds formed in soft mud by cubic halite crystals. They often show the concave "hopper" habit. The salt crystals grow in mud deposited on the substrate of hypersaline waters. Influx of turbid non-saline water dissolves the salt crystals and buries the mould beneath a new layer of sediment.

A variety of vertical planar structures have been recognized in sediments, these include desiccation cracks, syneresis cracks, sedimentary dikes and Neptunean dikes.

Desiccation cracks, also known as sun cracks, are downward tapering cracks in mud, which are infilled by sand. In plan view they are polygonal. Individual cracks are a centimetre or so wide. Polygons are generally about half a metre across. The cracks may extend down for an equivalent distance (Fig. 119). Picard (1966) has described discontinuous linear desiccation cracks which do not join into polygons, but are oriented parallel to the local palaeoslope. Desiccation cracks are generally taken to indicate subaerial exposure of the sediment surface.

Syneresis cracks are formed in mud by the spontaneous dewatering of clay beneath a body of water (White, 1961). They are distinguishable from

Fig. 118. Desiccation cracks in lacustrine Torridon Group (Pre-Cambrian) shales. Raasay, north-west Scotland.

desiccation cracks by the fact that they are infilled by mud similar or only slightly coarser in grade than that in which they grow. Furthermore, syneresis cracks are generally much smaller than desiccation cracks; typically only one or two millimetres across. Examples have been described

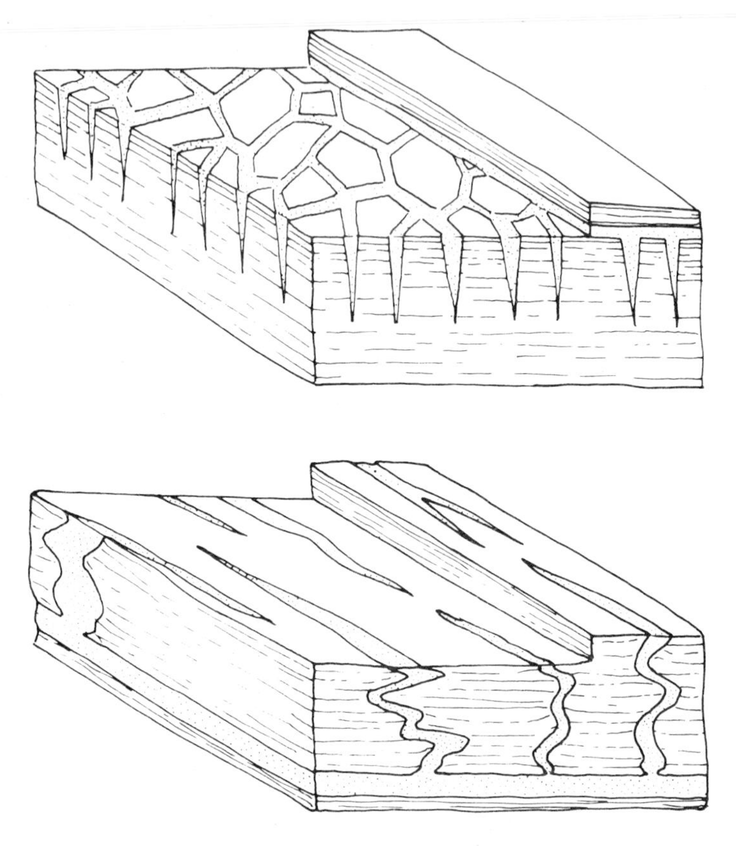

Fig. 119. Upper: desiccation cracks, caused by the contraction of mud to form downward-tapering fissures arranged in polygonal motifs, subsequently infilled by sand. Lower: sandstone dikes, showing ptygmatic contortions due to compaction and attachment to underlying parent sand bed.

from the Torridon Group (Pre-Cambrian) of Scotland (Selley, 1965, p. 373), and from the Devonian Caithness Flags also in Scotland (Donovan and Foster, 1972).

The distinction between subaerial desiccation cracks and subaqueous syneresis cracks is not always easy to make. In particular the huge polygons of modern playas may be due to a combination of subaerial and subaqueous dehydration with complex histories related to Quaternary climatic changes.

Sand dikes are vertical sheets of sand which have been intruded into muds from sand beds beneath. Though they are sometimes polygonally arranged, they can be distinguished from desiccation cracks by their tendency to die out upward and by the fact that they are rooted to the parent sand bed below. Sand dikes are intruded as liquified quicksand into water-saturated mud. Like desiccation cracks, they often show ptygmatic compaction effects (Fig. 119).

A notable example of sand dike intrusion occurs in the Miocene of the Panoche Hills, California. Smyers and Peterson (1971) have described a complex of over 350 sand dikes and sills which intrude the Moreno shale. They range from a decimetre to seven metres in width. Individual dikes are up to a kilometre in length. Earthquakes have been known to generate sand dikes (e.g. Barratt, 1966).

A particular variety of sand dikes is the sedimentary boudinage structure. This is morphologically similar to tectonic boudinage (e.g. De Sitter, 1964, p. 284). Instead of being due to necking of a competent limestone or sandstone in incompetent shale, the converse is true. Sedimentary boudinage typically occurs where interbedded, unconsolidated, water-saturated sands and muds are subjected to tension, as for example adjacent to a slump. Clay beds develop necks and are sometimes divided up into blocks by sand intrusion from above and below (Fig. 120).

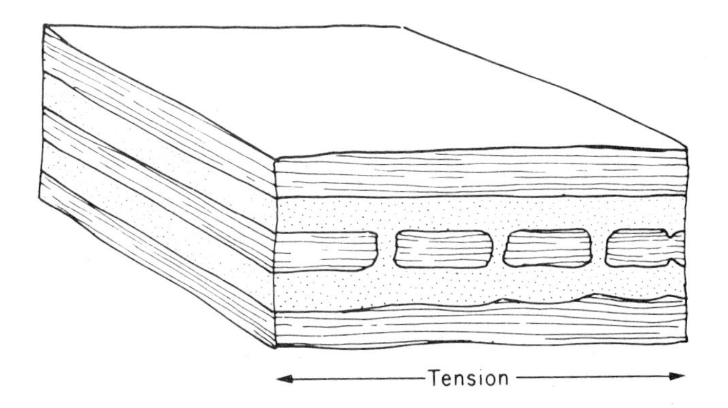

Fig. 120. Sedimentary boudinage. Formed by tensional splitting of clay bed accompanied by quicksand injection along the incipient fissures.

IV. PALAEOCURRENT ANALYSIS

The preceding analysis of sedimentary structures shows that they can be used to determine depositional processes. Since depositional processes occur in several environments, few structures are immediately diagnostic of a specific environment; assemblages of structures are more useful. There is, however, one further use for sedimentary structures. They can indicate the direction of palaeocurrent flow, palaeoslope, palaeogeography and sand-body trend. Palaeocurrent analysis, as this discipline is called, forms an integral part of facies analysis both at outcrop and, using the dipmeter, in subsurface studies. There is an extensive literature on this topic. Potter and Pettijohn (1977) is the definitive text.

The methodology, interpretation and applications of palaeocurrent analysis will now be described in turn.

A. Collection of Palaeocurrent Data

A wide range of sedimentary structures can be used in palaeocurrent analysis. Some structures yield only the sense of current flow, others yield both sense and direction. Examples of the first group include groove marks, channels, washouts and parting lineation. Examples of the second group include pebble imbrication (see p. 36), cross-lamination, cross-bedding, slump folds, flute marks, and the asymmetric profile of ripples.

The measurement of the orientation of sedimentary structures must be done with care. Ideally some kind of areal sampling grid should be used for regional palaeocurrent mapping. In practice, this ideal approach is commonly restricted by limitations of access, exposure and time.

Each sample station will generally consist of a cliff, quarry, stream section or road cut, etc. If it is to be worth anything, palaeocurrent analysis must be integrated with a full sedimentological study. Thus each sample station will also be the location of a measured section, or at least some detailed notes on stratigraphy, lithology, facies and fauna. At each station it is necessary to record structural dip and strike. If it is excessive (greater than about 10°), each measurement must be corrected on a stereographic net. The orientation of the structures will be recorded, including both the azimuth and dip of planar structures which need correction. For linear structures, and planar structures in outcrops of low tectonic dip, only the azimuth need be recorded. At the same time, it is necessary to note the type and scale of the structure and the lithology in which it occurs.

Foreset dip directions from cross-bedding should always be measured in plan view. Dip directions seen in vertical sections should only be recorded as a last resort. There are two reasons for this. First, as pointed out earlier in this chapter, cross-beds do not always dip directly down-current. In troughs and laterally infilled channels, foresets are deposited oblique or perpendicular to current flow. Examination of cross-bedding in plan view gives a clue to the structural arrangement of the foresets. If cross-bedding is measured from vertical faces generally only an apparent dip can be recorded. This may diverge considerably from the true dip direction, especially if there is well-developed jointing. The discrepancy will not be too erroneous, however, as foresets appear horizontal when viewed normal to their dip direction (Fig. 101).

The number of readings which need to be measured at a sample station are a matter for debate and may fortunately be dictated by the size of the exposure. Discussions of the statistics of sampling are given in Miller and Kahn (1962) and Krumbein and Graybill (1965). There is great scope here for statistical gymnastics. As a rule of thumb in unipolar cross-bed systems, as in alluvium, 25 readings are generally sufficient to determine a vector mean with an accuracy of ±30°. This is sufficiently accurate for most purposes. Many more readings may be needed, however, to establish well-defined modes in a section of interbedded facies with different and often polymodal vectors. For example, in shoreline deposits, fluvial channel sands with unimodal down-slope dipping foresets may be interbedded with marine sands with bipolar dipping foresets due to tidal currents unrelated to the palaeoslope.

B. Presentation of Palaeocurrent Data

Palaeocurrent data may be entered in a field note book and subsequently published in tabular form. The azimuths are, however, generally manipulated in some way to make their interpretation easier. The first step involves

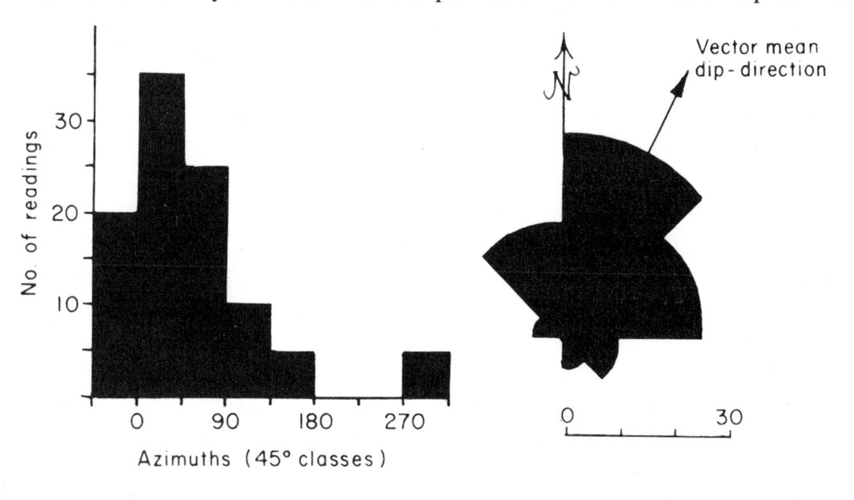

Fig. 121. Presentation of palaeocurrent data. Both methods display the same set of readings.

the removal of tectonic dip on a stereographic net where applicable (Potter and Pettijohn, 1977, p. 371; Schlumberger, 1970). Then the azimuths are divided into class intervals from 0 to 360°. Class intervals of between 30 and 45° are about typical. The data may then be presented on a histogram. More usually, however, a compass rose is used (Fig. 121). This is histogram converted to a circular distribution. It is conventional to indicate the direction of dip of foresets. This is contrary to the convention for wind roses which indicate the direction from which the wind blew. The dominant

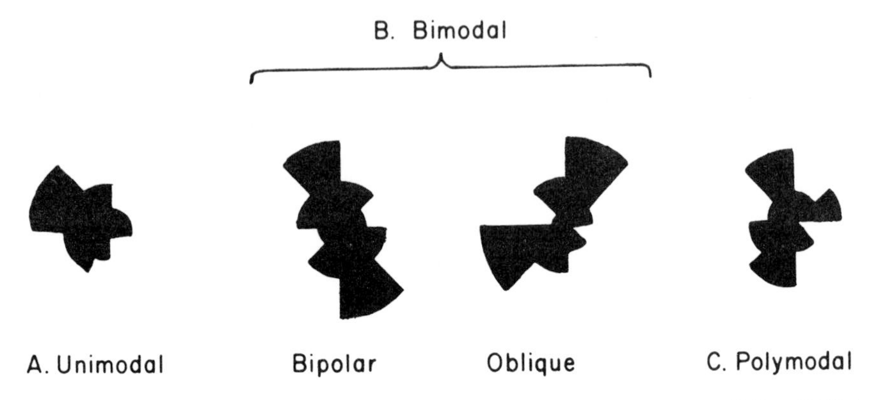

Fig. 122. Various characteristic azimuthal patterns of palaeocurrent data. (From Selley, 1968b.)

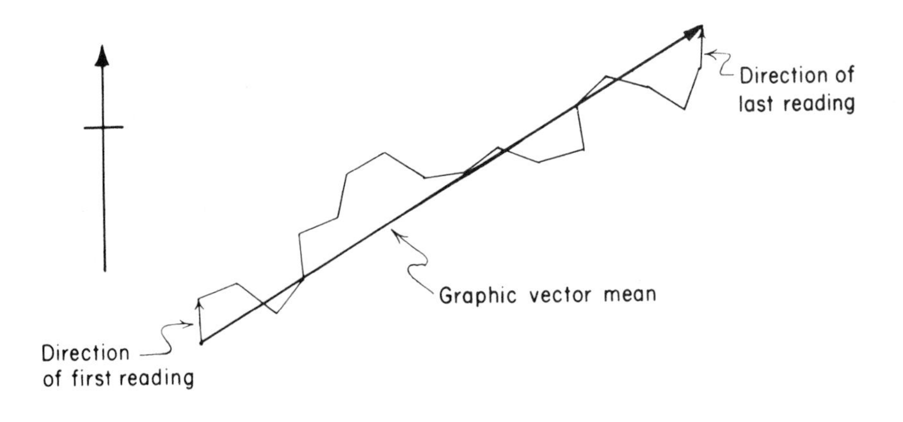

Fig. 123. The graphic method finding the mean vector of a set of palaeocurrent data. (After Reich, 1938 and Raup and Miesch, 1957.)

dip directions, termed modes, are at once apparent from a compass rose. Several types of azimuthal pattern are recognizable in palaeocurrent data (Fig. 122). A vector mean may then be statistically determined for unimodal data. This may be done mathematically or graphically.

A graphic method has been developed by Reiche (1938) and Raup and Miesch (1957). Starting at a point, a unit length is drawn (i.e. 1 cm) on the azimuth of the first reading. A unit length is then drawn along the azimuth of the second reading starting at the end of the first. This process is continued until all the readings have been plotted. A line connecting the point of origin to the distal end of the last reading records the graphical vector mean (Fig. 123). This is a quick simple method for innumerate geologists. It is particularly useful for highlighting dip switches (abrupt vertical variations in dip direction) on measured sections or dip meter runs of bore holes.

A more sophisticated way of calculating the vector mean is by the following formula (Harbaugh and Merriam, 1968, p. 42):

$$X_v = \arctan \left[\frac{\sum_{i=1}^{n} n_i \sin X_i}{\sum_{i=1}^{n} n_i \cos X_i} \right]$$

where X_v is the directional vector mean, n is the total number of observations, n_i is the number of observations in each frequency class and X_i is the midpoint azimuth in the ith class interval.

More simply, the humble arithmetic mean may be calculated by adding all the azimuths together and dividing by the total number of observations. This does not work if the azimuths are dispersed about the 360° point as this is then likely to yield a mean direction of about 180°, the exact opposite of the true mean. The arithmetic means of such sets of data may be calculated by using a false origin. Ninety degrees, for example, are added to all the data. The azimuths are summed and divided by the total number of readings, as before. Subtraction of 90° from the result then yields the true arithmetic mean.

Additional statistical methods are available for measuring the amount of dispersion of the data around the vector mean (Potter and Pettijohn, 1977, p. 374; Harbaugh and Merriam, 1968, p. 42).

These techniques are only applicable to unimodal distribution of azimuthal data. They may not be used on bimodal or polymodal data. In such instances it may be safest to present the data as compass roses (Tanner, 1959).

Considerable attention has been paid to the degree of scatter of palaeocurrent data, and to the calculation of statistical variance. This might give an insight to the sinuosity of fluvial channels and to the differentiation of unidirectional continental and polymodal marine current systems. One very good compilation of palaeocurrent data has been published by Long and Young (1978) who found that the statistical variance of fluvial palaeocurrent data was less than 4000, and of marine data tended to be above that figure.

Palaeocurrent data can be used as an element of regional facies mapping. Where there are sufficient sample points of unimodal data, their vector means may be plotted and contoured. The contours are dimensionless isolines which record the regional palaeostrike. The azimuth vectors, hopefully aided by facies analysis, indicate the palaeodip. Regional palaeocurrent maps may be subjected to mathematical smoothing techniques such as trend surface analysis.

C. Interpretation of Palaeocurrent Data

Palaeocurrent analysis involves several stages before the data are actually interpreted, viz.

(a) measurement of structures;
(b) *deduction* of palaeocurrent;
(c) manipulation of palaeocurrent data;
(d) *deduction* of palaeoslope.

The two deductive phases of the exercise deserve special attention. Considering the deduction of palaeocurrent direction from sedimentary structures, it has already been pointed out how foresets are sometimes oblique or perpendicular to current flow; antidunes actually point up-current. Palaeocurrents must thus be deduced carefully from the structures actually recorded. The sedimentary structures should be studied in the field and their genesis considered before measurement commences. Many published studies today include compass roses of palaeocurrents without making clear whether these are actual measured structural orientations or deduced flow directions.

A further point of deducing palaeocurrents from structures concerns the weighting to be given to different structures. A ripple reflects a much more local and smaller current flow than a dune. A dune, in turn, reflects a smaller flow than a channel. A channel may itself meander and deviate from the regional topographic slope. We may thus think of these structures as the hierarchical members of a total flow system (Allen, 1966). Thus when measuring palaeocurrent data a channel axis is immensely more significant than a few cross-bed orientations, and these should count for more than an equivalent number of cross-laminae. Few geologists have attempted to address the problem of weighting sedimentary structures of different rank (Iriondo, 1973). It is, however, reassuring to find that measurements of cross-beds in modern channels do give mean dip directions which correspond to the channel axis (e.g. Potter and Pettijohn, 1977, p. 103; Smith, 1972).

This leads on to the second main problem of palaeocurrent interpretation, namely the relationship between palaeocurrent and palaeoslope. In certain environments the flow systems are slope-controlled, in others they are not. In the first case, palaeocurrent analysis can give valuable information on palaeogeography and basin evolution. In the second case, it cannot.

Palaeocurrents are slope-controlled in fluvial, deltaic and (most) turbidite environments. Palaeocurrents are not related to slope in eolian and marine shoreline environments. Klein (1967) has reviewed the relationship between sedimentary structures, palaeocurrents and palaeoslopes in modern deposits. Selley (1968b) has defined a number of regional palaeocurrent models which have been recognized in ancient sedimentary deposits. Each major depositional environment is characterized by a particular palaeocurrent model (Table XXV).

Figure 124 summarizes a typical example of palaeocurrent analysis in a regional study of complex shoreline deposits of diverse facies.

On a still broader scale, palaeocurrents can indicate the age of formation of structural features. In particular, they may show whether a palaeohigh

Table XXV
A classification of some palaeocurrent patterns

Environment	Local current vector	Regional pattern
Alluvial — braided	Unimodal, low variability	Often fan-shaped
Alluvial — meandering	Unimodal, high variability	Slope-controlled often centripetal basin fill
Eolian	Uni-, bi- or polymodal	May swing round over hundreds of kilometres around high pressure systems
Deltaic	Unimodal	Regionally radiating
Shorelines and shelves	Bimodal (due to tidal currents), sometimes unipolar or polymodal	Generally consistently oriented on-shore, off-shore, or long-shore
Marine turbidite	Unimodal (some exceptions noted, see p. 188)	Fan-shaped, or, on a larger scale, trending into or along trough axes

Based on data in Selley (1968b).

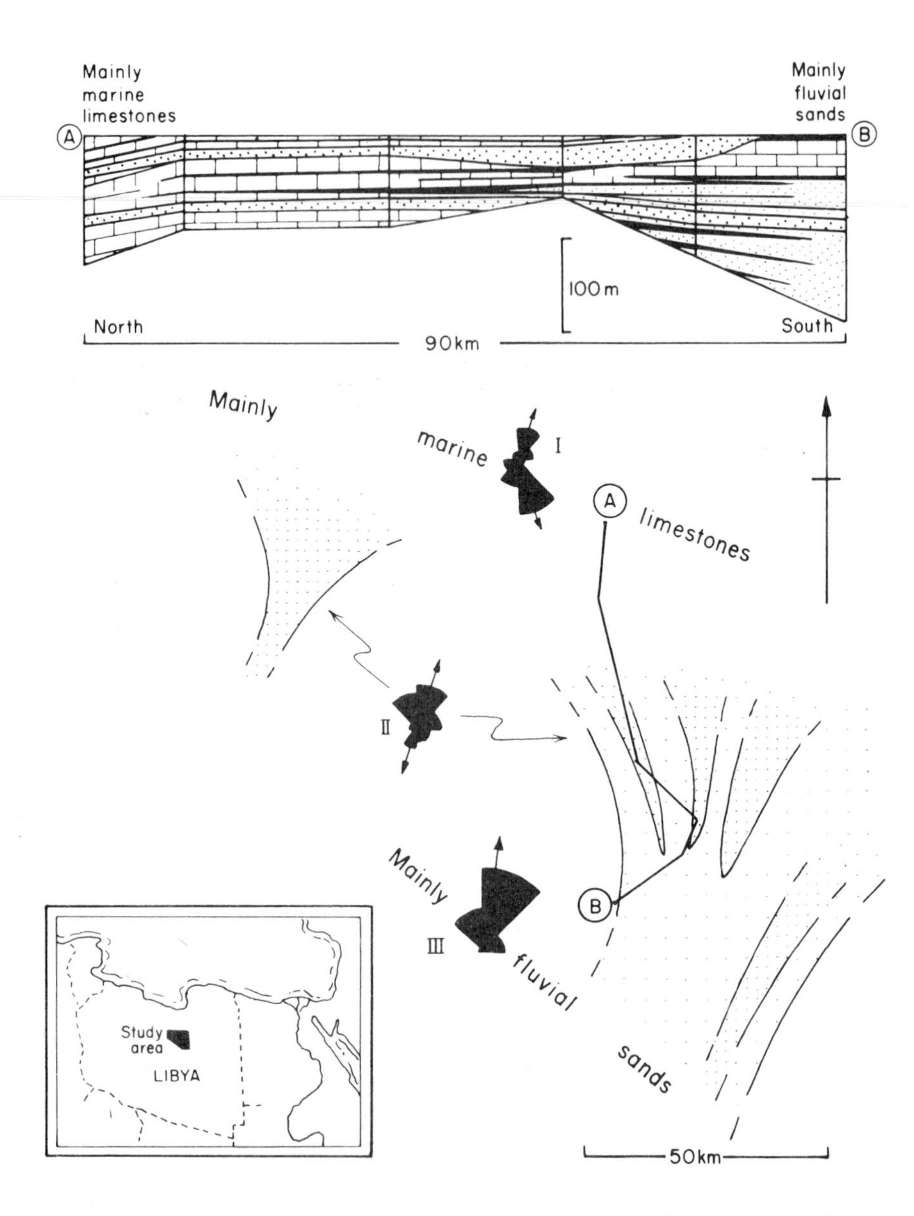

Fig. 124. An example of palaeocurrent analysis integrated with a palaeoenvironmental study of a Libyan Miocene shoreline. (From Selley, 1968c.) In the north of the area marine calcarenites show a bipolar azimuthal pattern of foreset dips (I). This reflects tidal currents with a net on-shore component. In the south fluvial sands show unipolar north (seaward)-dipping foresets (III). Radiating estuarine channel complexes show bipolar cross-bedding dips (II), suggesting tidal currents with the ebb current dominant.

was active during sedimentation or whether it rose after deposition of the sediments which drape it (Fig. 125). Similarly, palaeocurrent analysis can distinguish syndepositional from post-depositional sedimentary basins. In the former, palaeocurrents converge on the centre of the basin. In post-depositional (tectonic) basins, palaeocurrents sweep across the basin with a more or less uniform palaeostrike (Fig. 126). The South Wales Pennsylvanian basin shows centripetal palaeocurrents indicative of its syndepositional origin (Bluck and Kelling, 1963). The Illinois basin provides

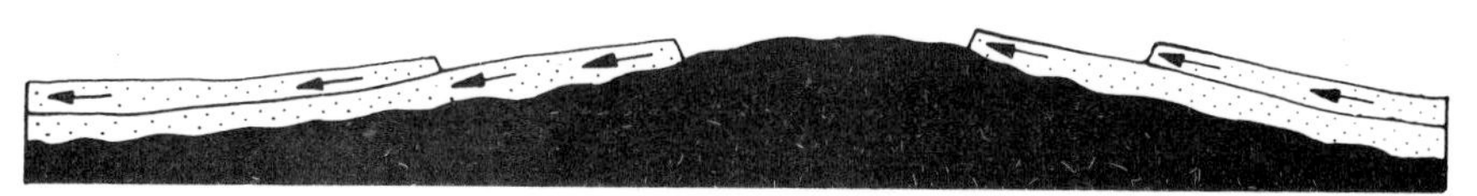

Fig. 125. Where palaeocurrents reflect palaeoslope, as in fluvial environments, they provide clues as to the age of regional structural arches. Palaeocurrents move in a constant direction across post-depositional arches (A), but diverge from axes of syndepositional uplift (B).

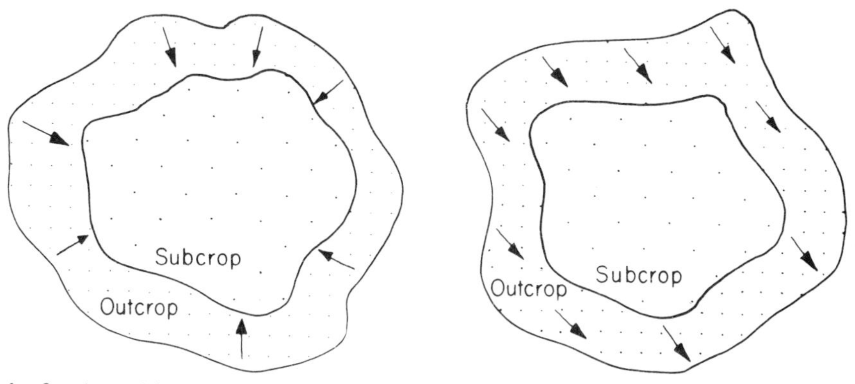

Fig. 126. Where palaeocurrents reflect palaeoslope, as in fluvial deposits, they provide clues as to the age of formation of sedimentary basins. In syndepositional basins (A) they converge on what was the topographic centre of the basin. In post-depositional basins, however, (B), palaeocurrents show no relation to present structural morphology.

a good example of a post-depositional tectonic basin with a regional consistent palaeostrike (Potter *et al.*, 1958).

Interpretive studies such as these can only be made from palaeocurrents deduced from sediments deposited in slope-controlled environments. These examples do show how palaeocurrent analysis can be an important and integral feature of regional facies analysis.

V. REFERENCES

Allen, J. R. L. (1963). The classification of cross-stratified units, with notes on their origin. *Sedimentology* **2**, 93–114.

Allen, J. R. L. (1964). Primary current lineation in the Lower Old Red Sandstone (Devonian), Anglo-Welsh basin. *Sedimentology* **3**, 89–108.

Allen, J. R. L. (1966). On bedforms and palaeocurrents. *Sedimentology* **6**, 153–190.

Allen, J. R. L. (1968a). On criteria for the continuance of flute marks, and their implications. *Geologie Mijnb.* **47**, 3–16.

Allen, J. R. L. (1968). "Current Ripples" North-Holland, Amsterdam. 433pp.

Allen, J. R. L. (1969). Some recent advances in the physics of sedimentation. *Proc. geol. Ass.* **80**, 1–42.

Allen, J. R. L. (1970). "Physical Processes of Sedimentation" Allen and Unwin, London. 248pp.

Allen, J. R. L. (1971). Transverse erosional marks of mud and rock: their physical basis and geological significance. *Sedimentary Geol.* **5**, 167–385.

Allen, J. R. L. (1979). A model for the interpretation of wave ripple-marks using their wavelength, textural composition, and shape. *J. geol. Soc. Lond.* **136**, 673–682.

Allen, J. R. L. (1980). Sandwaves: a model of origin and internal structure. *Sedimentary Geology* **26**, 281–328.

Allen, J. R. L. and Banks, N. L. (1972). An interpretation and analysis of recumbent-folded deformed cross-bedding. *Sedimentology* **19**, No. 3/4.

Anketell, J. M., Gegla, J. and Dzulinski, S. (1970). On the deformational structures in system with reversed density gradients. *Ann. Soc. geol. Pologne.* **15**, 3–29.

Bagnold, R. A. (1954). "The Physics of Blown Sand and Desert Dunes" Methuen, London. 265pp.

Bagnold, R. A. (1979). Sediment transport by wind and water. *Nordic Hydrology* **10**, 309–322.

Barratt, P. J. (1966). Effects of the 1964 Alaskan earthquake on some shallow water sediments in Prince William Soud, S. E. Alaska. *J. sedim. Petrol.* **36**, 992–1006.

Bigarella, J. J. (1972). Eolian environments—their characteristics, recognition and importance. Recognition of Ancient-Sedimentary Environments (J. K. Rigby and W. K. Hamblin, Eds). *Spec. Publs Soc. econ. Palaeont. Miner., Tulsa* **16**, 12–62.

Blanc, J. J. (1972). "Slumpings" et figures sedimentaires dans le Cretace superieur du bassin du Beausset, France. *Sedimentary Geol.* **7**, 47–64.

Bluck, B. H. and Kelling, G. (1963). Channels from the Upper Carboniferous coal measures of South Wales. *Sedimentology* **2**, 29–53.

Bouma, A. H. (1962). "Sedimentology of Some Flysch Deposits" Elsevier, Amsterdam. 168pp.

Bromley, R. G. (1975). Trace fossils at omission surfaces. *In* "The Study of Trace Fossils" (R. W. Frey, Ed.). Springer-Verlag, Berlin. 97–120.

Bromley, R. G. and Asgaard, U. (1972). Freshwater *Cruziana* from the Upper Triassic of Jameson Land, East Greenland. *Grønl. geol. Undersogelse Rapp.* **49**, 7–13.

Bucher, W. H. (1919). On ripples and related sedimentary surface forms and their palaeogeographic interpretation. *Am. J. Sci.* **47**, 149–210, 241–269.

Campbell, C. V. (1967). Lamina, laminaset, bed and bedset. *Sedimentology* **8**, 7–26.

Conybeare, C. E. B. and Crook, K. A. W. (1968). "Manual of Sedimentary Structures" *Bull. Bur. Miner. Resour. Geol. Geophys. Aust.* **12**, 327pp.

Crimes, T. P. and Harper, J. C. (Eds) (1970). "Trace Fossils" Lpool Geol. Soc. 547pp.

Crimes, T. P. and Harper, J. C. (1977). "Trace Fossils" Vol. 2. Seel House Press, Liverpool. 351pp.

Darwin, G. H. (1883). On the formation of ripple-marks in sand. *Proc. R. Soc.* **36**, 18–43.

Davies, H. G. (1965). Convolute lamination and other structures from the Lower Coal Measures of Yorkshire. *Sedimentology* **5**, 305–326.

Davies, S. N. (1971). Barbados: a major submarine gravity slide. *Bull. geol. Soc. Am.* **82**, 2593–2602.

De Sitter, L. U. (1964). "Structural Geology" McGraw-Hill, London. 551pp.

Doe, T. W. and Dott, R. J., Jr (1980). Genetic significance of deformed cross-bedding—with examples from the Navajo and Weber Sandstones of Utah. *J. sedim. Petrol.* **50**, 793–812.

Donovan, R. N. and Foster, R. J. (1972). Subaqueous shrinkage cracks from the Caithness Flagstone Series (Middle Devonian) of Northeast Scotland. *J. sedim. Petrol.* **42**, 309–317.

Dzulinski, S. and Sanders, J. E. (1962). Current marks on firm mud bottoms. Trans. *Conn. Acad. Arts Sci.* **42**, 57–96.

Dzulinski, S. and Slaczka, A. (1959). Directional structures and sedimentation of the Krosno beds (Carpathian flysch). *Ann. soc. geol. Polska* **28**, 205–260.

Dzulinski, S. and Walton, E. K. (1965). "Sedimentary Features of Flysch and Greywacke" Elsevier, Amsterdam. 274pp.

Frazier, D. E. and Osanik, A. (1961). Point-bar deposits. Old River Locksite, Louisiana. *Trans. Gulf-Cst Ass. geol. Socs* **11**, 127–137.

Frey, R. (Ed.) (1975). "The Study of Trace Fossils" Springer-Verlag, Berlin. 562pp.

Friend, P. F. and Moody-Stuart, M. (1972). Sedimentation of the Wood Bay Formation (Devonian) of Spitsbergen: regional analysis of a late orogenic basin. Norsk Polarinstituut Skrift. No. 157, Oslo, 77pp.

Gill, W. D. and Kuenen, P. H. (1958). Sand volcanoes on slumps in the Carboniferous of County Clare, Ireland. *Q. Jl geol. Soc. Lond.* **113**, 441–460.

Gubler, Y. (Ed.) (1966). "Essai de Nomenclature et de Caracterisation des Principales Structures Sedimentaires" Editions Technip, Paris. 291pp.

Hamblin, W. K. (1962). X-ray radiography in the study of structures in homogenous sediments. *J. sedim. Petrol.* **32**, 201–210.

Hand, B. M., Middleton, G. V. and Skipper, K. (1972). Antidune cross-stratification in a turbidite sequence, Cloridorme Formation, Gaspe, Quebec. *Sedimentology* **18**, 135–138.

Hantzschel, W. (1975). Trace fossils and problematica. *In* "Treatise on Invertebrate Paleontology" (R. C. Moore, Ed.), Part W, p. W269. Geol. Soc. Am. New York.

Harbaugh, J. W. and Merriam, D. F. (1968). "Computer Applications in Stratigraphic Analysis" John Wiley, New York. 282pp.

Harms, J. C. and Fahnestock, R. K. (1965). Stratification, bedforms and flow phenomena (with an example from the Rio Grande). *In* "Primary Sedimentary Structures and their Hydrodynamic Interpretation" (G. V. Middleton, Ed.). *Spec. Publs Soc. econ. Palaeont. Miner., Tulsa* **12**, 84–115.

Harms, J. C., MacKenzie, D. B. and McCubbin, D. G. (1963). Stratification in modern sands of the Red River, Louisiana. *J. Geol.* **71**, 566–580.

Heckel, P. H. (1972). Recognition of ancient shallow marine environments. Recognition of Ancient Sedimentary Environments (J. K. Rigby and W. K. Hamlin, Eds). *Spec. Publs Soc. econ. Palaeont. Miner., Tulsa* **16**, 226–286.

Hulsemann, J. (1955). Grossrippeln und Schragschichtungsgefuge im Nordsee-Watt und in der Molasse. *Senckenberg leth.* **36**, 359–388.

Imbrie, J. and Buchanan, H. (1965). Sedimentary structures in Modern Carbonate sands of the Bahamas. *In* "Primary Sedimentary Structures and Their Hydrodynamic Interpretation" (G. Middleton, Ed.). *Spec. Publs Soc. econ. Palaeont. Miner., Tulsa* **12**, 149–172.

Iriondo, H. H. (1973). Volume factor in paleocurrent analysis. *Bull. Am. Ass. Petrol. Geol.* **57**, 1341–1342.

Jopling, A. V. (1965). Hydraulic factors controlling the shape of laboratory deltas. *J. sedim Petrol.* **35**, 777–791.

Jopling, A. V. and Walker, R. G. (1968). Morphology and origin of ripple-drift cross-lamination, with examples from the Pleistocene of Massachusetts. *J. sedim. Petrol.* **38**, 971–984.

Kindle, E. M. (1917). Recent and fossil ripple marks. *Mus. Bull. Can. geol. Surv.* **25**, 1–56.

Klein, G. de Vries (1967). Paleocurrent analysis in relation to modern marine sediment dispersal patterns. *Bull. Am. Ass. Petrol. Geol.* **51**, 366–382.

Klein, G. de Vries, de Melo, U. and Favera, J. C. D. (1972). Subaqueous gravity processes on the front of Cretaceous deltas, Reconcavo Basin, Brazil. *Bull. geol. Soc. Am.* **83**, 1469–1492.

Krumbein, W. C. and Graybill, F. A. (1965). "An Introduction to Statistical Methods in Geology" McGraw-Hill, New York. 475pp.

Kuenen, P. H. (1948). Slumping in the Carboniferous rocks of Pembrokeshire. *Q. Jl geol. Soc. Lond.* **104**, 365–385.

Kuenen, P. H. (1958). Experiments in geology. *Trans. geol. Soc. Glasg.* **23**, 1–28.

Lewis, K. B. (1971). Slumping on a continental slope inclined at 1°–4°. *Sedimentology* **16**, 97–110.

Long, D. G. F. and Young, G. M. (1978). Dispersion of cross-stratification as a potential tool in the interpretation of Proterozoic arenites. *J. sedim. Petrol.* **48**, 857–862.

Lyell, Sir C. (1865). "Elements of Geology" (6th Edition) John Murray, London. 794pp.

Macar, P. and Antun, P. (1949). Pseudonodules et glissements sous-aquatiques dans l'Emsian inferior de l'Oesling. *Ann. Soc. geol. Belg.* **73**, 121–150.

Martinsson, A. (1965). Aspects of a Middle Cambrian Thanatotope in Oland. *Geol. För. Stockh. Förh.* **87**, 181–230.

McBride, E. F. and Hayes, M. O. (1962). Dune cross-bedding on Mustang Island, Texas. *Bull. Am. Ass. Petrol. Geol.* **46**, 546–551.

McKee, E. D. and Weir, G. W. (1953). Terminology for stratification and cross-stratification. *Bull. geol. Soc. Am.* **64**, 381–390.

McKee, E. D., Crosby, E. J. and Berryhill, H. L. (1967). Flood deposits, Bijou Creek, Colorado, June 1965. *J. sedim. Petrol.* **37**, 829–851.
Miller, R. L. and Kahn, J. S. (1962). "Statistical Analysis in the Geological Sciences" John Wiley, New York. 483pp.
Otto, G. H. (1935). The sedimentation unit and its use in field sampling. *J. Geol.* **46**, 569–582.
Payne, T. G. (1942). Stratigraphical analysis and environmental reconstruction. *Bull. Am. Ass. Petrol. Geol.* **26**, 1697–1770.
Pettijohn, F. J. and Potter, P. E. (1964). "Atlas and Glossary of Primary Sedimentary Structures" Springer-Verlag, Berlin. 370pp.
Picard, M. D. (1966). Oriented, linear-shrinkage cracks in Green River Formation (Eocene), Raven Ridge area, Uinta basin, Utah, *J. sedim. Petrol.* **36**, 1050–1057.
Potter, P. E. and Pettijohn, F. J. (1977). "Paleocurrents and Basin Analysis" (2nd Edition) Springer-Verlag, Berlin. 425pp.
Potter, P. E., Nosow, E., Smith, N. W., Swann, D. H. and Walker, F. H. (1958). Chester cross-bedding and sandstone trends in Illinois basin. *Bull. Am. Ass. Petrol. Geol.* **42**, 1013–1046.
Raup, O. B. and Miesch, A. T. (1957). A new method for obtaining significant average directional measurements in cross-stratification studies. *J. sedim. Petrol.* **27**, 313–321.
Reiche, P. (1938). An analysis of cross-lamination of the Coconino sandstone. *J. Geol.* **44**, 905–932.
Reineck, H. E. (1961). Sediment bewegungen an kleinrippeln im watt. *Senckenberg leth.* **42**, 51–67.
Reineck, H. E. and Wunderlich, F. (1968). Classification and origin of flaser and lenticular bedding. *Sedimentology* **11**, 99–104.
Rittenhouse, G. (1972). Cross-bedding dip as a measure of sandstone compaction. *J. sedim. Petrol.* **42**, 682–683.
Roberts, D. G. (1972). Slumping on the eastern margin of the Rockall Bank, North Atlantic Ocean. *Mar. geol.* **13**, 225–237.
Rodriguez, J. and Gutschick, R. C. (1970). Late Devonian—early Mississippian ichnofossils from Western Montana and Northern Utah. *In* "Trace Fossils" (T. P. Crimes and J. C. Harper, Eds.), 407–438. Lpool Geol. Soc.
Rust, B. R. (1978). A classification of alluvial channel systems. *In* "Fluvial Sedimentology" (A. D. Miall, Ed.), 187–198. Can. Soc. Pet. Geol. Calgary.
Schlumberger Ltd. (1970). "Fundamentals of Dipmeter Interpretation" Schlumberger, New York. 145pp.
Schumm, S. A. (1969). River metamorphosis. *J. Hydraul. Div. Am. Soc. civ. Engrs* **95**, 255–273.
Seilacher, A. (1964). Biogenic sedimentary structures. *In* "Approaches to Paleoecology" (J. Imbrie and N. D. Newell, Eds), 296–315. John Wiley, New York.
Seilacher, A. (1967). Bathymetry of trace fossils. *Mar. geol.* **5**, 413–428.
Selley, R. C. (1965). Diagnostic characters of fluviatile sediments in the Pre-Cambrian rocks of Scotland. *J. sedim. Petrol.* **35**, 366–380.
Selley, R. C. (1967). Palaeocurrents and sediment transport in the Sirte basin, Libya. *J. Geol.* **75**, 215–223.
Selley, R. C. (1968a). Facies profile and other new methods of graphic data presentation: application in a quantitative study of Libyan Tertiary shore-line deposits. *J. sedim. Petrol.* **38**, 363–372.
Selley, R. C. (1968b). A classification of palaeocurrent models. *J. Geol.* **76**, 99–110.

Selley, R. C. (1968c). Nearshore marine and continental sediments of the Sirte basin, Libya. *Q. Jl geol. Soc. Lond.* **124**, 419–460.
Selley, R. C. (1969). Torridonian alluvium and quicksands. *Scott. J. geol.* **5**, 328–346.
Selley, R. C. (1970). Ichnology of Palaeozoic sandstones in the southern desert of Jordan: a study of trace fossils in their sedimentologic context. *In* "Trace Fossils" (T. P. Crimes and J. C. Harper, Eds), 477–488. Lpool Geol. Soc.
Selley, R. C., Sutton, J., Shearman, D. J. and Watson, J. (1963). Some underwater disturbances in the Torridonian of Skye and Raasay. *Geol. Mag.* **100**, 224–243.
Skipper, K. (1971). Antidune cross-stratification in a turbidite sequence, Cloridorme Formation, Gaspe, Quebec. *Sedimentology* **17**, 51–68.
Smith, N. D. (1972). Some sedimentological aspects of planar cross-stratification in a sandy braided river. *J. sedim. Petrol.* **42**, 624–634.
Smyers, N. B. and Peterson, G. L. (1971). Sandstone dikes and sills in the Moreno Shale, Panoche Hills, California. *Bull. geol. Soc. Am.* **82**, 3201–
Sorby, H. C. (1859). On the structures produced by the currents present during the deposition of stratified rocks. *The Geologist* **2**, 137–149.
Southard, J. B. and Dingler, J. R. (1971). Flume study of ripple propagation behind mounds on flat sandbeds. *Sedimentology* **16**, 251–263.
Tanner, W. F. (1959). The importance of modes in cross-bedding data. *J. sedim. Petrol.* **29**, 211–226.
Tanner, W. F. (1967). Ripple mark indices and their uses. *Sedimentology* **9**, 89–104.
Tanner, W. F. (1971). Numerical estimates of ancient waves, water depth and fetch. *Sedimentology* **16**, 71–88.
Terwindt, J. H. J. and Breusers, H. N. C. (1972). Experiments on the origin of flaser, lenticular and sand–clay alternating bedding. *Sedimentology* **19**, 85–98.
Visher, G. S. and Cunningham, R. D. (1981). Convolute laminations—a theoretical analysis: examples of a Pennsylvanian sandstone. *Sed. Geol.* **28**, 175–188.
White, G. (1961). Colloid phenomena in the sedimentation of argillaceous rocks. *J. sedim. Petrol.* **31**, No. 4, 560–565.
Williams, G. E. (1970). Origin of disturbed bedding in Torridon Group Sandstones. *Scott. J. geol.* **6**, 409–410.
Williams, P. B. and Kemp, P. H. (1971). Initiation of ripples on flat sediment beds. *J. Hydraul. Div. Am. Soc. civ. Engrs* **97** (HY4) Proc. Paper 8042, 502–522.

8 Environments and Facies

I. SEDIMENTARY ENVIRONMENTS

A. Environments Defined

For the geologist the modern earth's surface is his laboratory. Here we can monitor the processes which generate sediments, and here we can study the deposits which are their end product. By applying the principle of uniformitarianism ("the key to the past is in the present"), we may detect the origin of ancient sedimentary rocks. For this reason many geologists, and scientists of other disciplines, have intensively studied modern sedimentary processes and products.

The surface of the earth can be classified by geomorphologists into distinctive physiographic units, such as mountain ranges, sand deserts and deltas. Similarly oceanographers can define morphologic types of sea floor, such as continental shelves, submarine fans and abyssal plains.

The most cursory of such studies shows that there are a finite number of physiographic types. For example, one encounters lakes and deltas on most continents of the globe; submarine fans and coral reefs are spread wide across the hemispheres.

From this observation it follows that the surface of the earth may be classified into different sedimentary realms or environments. A sedimentary environment has been defined as "a part of the earth's surface which is physically, chemically and biologically distinct from adjacent areas" (Selley, 1970, p. 1). As already pointed out, sand deserts, deltas and submarine fans are examples of these different sedimentary environments.

The three defining parameters listed above are numerous and complex.

The physical parameters of a sedimentary environment include the velocity, direction and variation of wind, waves, and flowing water; they include the climate and weather of the environment in all their subtle variations of temperature, rainfall, snowfall and humidity.

The chemical parameters of an environment include the composition of the waters which cover a subaqueous sedimentary environment; they include the geochemistry of the rocks in the catchment area of a terrestrial environment.

The biological parameters of an environment comprise both fauna and flora. On land these may have major effects on sedimentary processes. Over-grazing, defoliation, deforestation and over-cultivation of soils by animals can cause catastrophic increases in rates of erosion in one area accompanied by accelerated rates of deposition elsewhere. Conversely, the colonization of deserts by a new flora has a moderating effect on sedimentary processes.

In marine environments many of the lowliest forms of life are important both because their skeletons can contribute to rock formation, and because their presence in water can change its equilibrium, resulting in the precipitation of chemical sediments (see p. 125). The morphology and history of organic reefs in particular are inextricably related to the ecology of their biota.

This brief review of the physical, chemical and biological parameters which define a sedimentary environment should be sufficient to demonstrate how numerous and complex these variables are.

B. Environments of Erosion, Equilibrium and Deposition

The classification of sedimentary environments by their physical, chemical and biological parameters will be returned to shortly. Now let us look at environments from a slightly different view-point. Examination of the modern earth shows that there are sedimentary environments of net erosion, environments of equilibrium and environments of net deposition.

Sedimentary environments of net erosion are typically terrestrial, and consist largely of the mountainous areas of the world. In such erosional environments weathering is often intensive and erosion is rapid. Locally sedimentation may take place from glacial, mud flow and flash flood processes. Due to renewed erosion, however, such deposits are ephemeral and soil profiles have little time to develop on either bedrock or sediment. Erosional sedimentary environments also occur on cliffed coastlines and, under the sea, in submarine canyons and on current-scoured shelves. It is, however, in these shoreline and submarine situations that the products of deposition dominate the processes of erosion. Subaqueous deposits must take up some 90% of the world's sedimentary cover. Probably some 60% of this total volume is composed of submarine and shoreline deposits. It appears that depositional sedimentary environments are predominantly subaqueous.

To environments of erosion and of deposition must be added a third category which may be termed sedimentary environments of equilibrium.

These are surfaces of the earth, both on the land and under the sea, which for long periods of time are neither sites of erosion nor yet of deposition. Because of this stability such environments often experience intense chemical alteration of the substrate. On land, environments of equilibrium are represented by the great peneplanes of the continental interiors (King, 1962). These, in the case of Central Africa at least, are parts of the earth's surface which have been open to the sky for millions of years. Prolonged exposure to the elements is responsible for the development of weathering profiles and soil formation in the rocks which immediately underlie an environment of equilibrium. Laterite and bauxite horizons are the products of certain specific climatic conditions in conjunction with suitable rock substrates (p. 57). They may be regarded as the products of sedimentary environments of equilibrium.

Environments of equilibrium can also be recognized beneath the sea. Extensive areas, both of the continental shelves and of the abyssal plains, are subjected to currents which are powerful enough to remove any sediment which may have settled by suspension, yet are too weak to erode the substrate. These scoured surfaces are susceptible to chemical reactions with sea water leading to the formation of manganese crusts, to phosphatization and other diagenetic changes (Mero, 1965). In the geological column submarine sedimentary environments of equilibrium are represented by hardgrounds (p. 140). These are mineralized surfaces, generally within limestones, often intensely bored and overlain by a thin conglomerate composed of clasts of the substrate. Examples come from shelf environments such as the Cretaceous chalk of northern Europe (Jefferies, 1963) and from Jurassic pelagic deposits of the Alps (Fischer and Garrison, 1967).

Table XXVI summarizes these concepts of sedimentary environments of erosion, equilibrium and deposition.

It is readily apparent that sedimentary geology is concerned primarily with depositional environments. These must be regarded as a particular type of sedimentary environment.

It is important to make this distinction when interpreting the depositional environment of an ancient sedimentary rock. What one observes is not just the result of the depositional process which deposited the rock. This will be shown unequivocally by the sedimentary structures. The sediment itself and its fossils may have originated to a large extent in erosional or equilibrial environments contiguous to that in which the sediment was actually deposited. Consider, for example, an ancient braided river channel. Its cross-bedding will indicate the direction, force and nature of the depositing current. The composition and texture of the sand itself, however, may largely be inherited from an erosional sedimentary environment in the source area. A fossilized tree trunk enclosed in the channel sand testifies to the coeval existence of an, albeit temporary, environment of equilibrium.

Table XXVI
To illustrate the concept of sedimentary environments of erosion, deposition and equilibrium

			Erosional	Equilibrial	Depositional
Land	Subaerial		Dominant	Development of peneplanes, soils, laterites and bauxites	Rare (eolian and glacial)
Land	Subaqueous		Localized	Unknown?	Localized (fluvial and lacustrine)
Sea	Subaqueous		Rare	Development of "hard-grounds", often nodular and mineralized	Dominant

C. Environments Classified

Since the earliest days of geological studies it has been found convenient to classify sedimentary environments into various groups and subgroups. Such a classification provides a useful formal framework on which to base more detailed analyses of specific environmental types. Table XXVII is an example of the classical classification of environments. Variants of this type will be found in many text books (e.g. Twenhofel, 1926, p. 784; Krumbein and Sloss, 1959, p. 196; Pettijohn, 1956, p. 633; Dunbar and Rodgers, 1957; Blatt *et al.*, 1980).

This sort of all-embracing scheme, while excellent for modern environmental studies, is limited in its application to ancient sedimentary rocks. There are two reasons for this. First, this is a classification of sedimentary environments and, as already shown, only a limited number of these are quantitatively significant depositional environments. Thus rocks of spelean, glacial and abyssal origin are very rare in the geological column. A second reason why this classification is difficult to apply to ancient sediments is because it is extremely hard to determine the depth of water in which ancient marine deposits originated. It is generally possible to recognize the relative position of a sequence of facies with respect to a shoreline, but it is often extremely hard to equate these with absolute depths (Hallam, 1967,

Table XXVII
An example of the classical type of classification of sedimentary environments

Continental	Terrestrial	Desert
		Glacial
	Aqueous	Fluvial
		Paludal (swamp)
		Lacustrine
		Cave (Spelean)
Transitional		Deltaic
		Estuarine
		Lagoonal
		Littoral (intertidal)
Marine		Reef
		Neritic (between low tide and 183 m)
		Bathyal (between 183 and 1830 m)
		Abyssal (deeper than 1830 m)

p. 330). For this reason a classification of marine environments into depth-defined neritic, bathyal and abyssal realms is hard to apply to ancient sediments.

In a review of the problems of environmental classifications, Crosby (1972) cites a new version compiled by Shepard and McKee. This is particularly useful because its hierarchical structure allows the inclusions of many minor subenvironments which have been identified during the boom in modern sediment studies over the last 15 years (Table XXVIII). Crosby (ibid.) has also prepared a useful classification of marine environments based on water depth, water circulation and energy level. These three parameters define the limits of a number of depositional environments which have been commonly recognized in sedimentary rocks.

Finally Table XXIX is a classification of major depositional sedimentary environments. This tabulates only those environments which generate quantitatively significant deposits and can be confidently identified in ancient sediments. This scheme recognizes the three major environmental types of continental, transitional shoreline and marine.

Within the continental environments no room is found for the rare glacial and cave deposits. Swamp deposits are omitted; they can generally be considered as subenvironments of fluvial, lacustrine and shoreline deposits.

The classification of shorelines lays stress on whether they are linear-barrier shorelines or lobate deltaic ones. Though estuaries are clearly recognizable at the present day they are extremely hard to recognize in ancient sediments and their parameters are ill-defined. Cliffed coastlines are omitted as they are essentially environments of non-deposition. Tidal flats,

Table XXVIII
Listing of environments of deposition
by Shepard and McKee, in Crosby, 1972, pp. 9–10

Terrestrial	Landslide	
	Talus	
	Alluvial fans and plains	
	River channels	
	Flood plains	
	Glacial moraine	
	Outwash plains	
	Dunes	Unidirectional wind types
		Bidirection wind types
		Multidirectional wind types
Lacustrine	Playas	
	Salt lakes	
	Deep lakes	
Delta	On-shore	Distributary channel
		Levée
		Marsh and swamp
		Interdistributary
		Beach
	Off-shore	Channel and levée extensions
		Distributary mouth bar
		Delta front platform
		Prodelta slope
Beach	Back-shore	
	Berm	
	Fore-shore	
Near-shore zone		
Off-shore zone		
Barrier	Beach	
	Dunefield	
	Barrier flat	
	Washover fan	
	Inlet	
Bar (submerged)	Longshore bar	
	Bay bar	
Tidal-flat area	Salt marsh	
	Tidal flat	
	Tidal channel	
Lagoon	Hypersaline	
	Brackish	
	Fresh	
Estuary	Shallow	
	Deep	
Continental shelf		
Epicontinental sea		

continued

Table XXVIII—cont'd.

Deep intracontinental depression	Trough Basin
Continental borderland	Basin Trough
Continental slope	
Deep sea	Deep-sea fan Abyssal plain Marine areas marginal to glaciers
Reef	Linear Patch Fringing

Table XXIX
A classification of depositional sedimentary environments

Continental	Fanglomerate Fluviatile Lacustrine Eolian		Braided Meandering
Shorelines	Lobate (deltaic) Linear (barrier)		Terrigenous Mixed carbonate–terrigenous Carbonate
Marine	Reef Shelf Turbidite Pelagic	Terrigeneous Carbonate	

This table tabulates only those environments which have generated large volumes of ancient sediments. From Selley, 1978, Fig. 1.2.

tidal channels, lagoons, salt marshes and barrier bars are all considered as subenvironments of deltaic or linear shorelines.

The classification of marine environments shuns all attempts to recognize depth zones. Shelf deposits can be recognized by a combination of lithologic, palaeontologic and sedimentary features as well as their structural setting. Terrigenous and carbonate shelves are easily differentiated. Reefs, in the broadest sense of the word, are recognizable to most geologists (saving the geosemanticists). Logically most reefs are a variety of the carbonate shelf environment, but their abundance and geologic importance justifies giving them independent status. Turbidites are, strictly speaking, the product of a particular type of sedimentary process which occurs in many environments. Large volumes of turbidites, however, are generally found to have been deposited at the feet of deltas and continental slopes. Pelagic deposits are those literally "of the open sea". These are largely fine-grained chemical deposits, argillaceous, calcareous or siliceous. Sediments like these originated away from terrestrial influence, but this may have been due either to great depth or distance from land. The designation pelagic evades this dilemma.

The classification of depositional environments shown in Table XXIX has certain limitations. Nevertheless, it forms a useful framework for the discussion of the main sedimentary models which are described later in this chapter.

II. SEDIMENTARY FACIES

Having examined the concept of sedimentary environments and their classification, let us now discuss their ancient products.

Geologists first began to study sedimentary rocks in some detail in the early part of the last century. William Smith in 1815 published his "Geological Map of England and Wales, with Part of Scotland". The principles on which this map was based were that strata occurred in sequences of diverse rock types; that these strata had a lateral continuity which could be mapped, and that they were characterized by different assemblages of fossils.

As the discipline of stratigraphic palaeontology progressed, it became apparent that some fossils appeared to be restricted to certain geological time spans, while others were long ranging but appeared to occur in certain rock types.

The implications of these facts were realized by Prevost in 1838. He proposed the name "formation" for lithostratigraphic units. By using biostratigraphy he demonstrated that different "formations" were formed in the same geological "epoch", and that similar formations could occur in different epochs.

Almost simultaneously, Gressly (1838), working in the Alps, reached similar conclusions. He coined the name "facies" for units of rock which were characterized by similar lithological and palaeontological criteria. His original definition (as translated by Teichert, 1958) reads:

> To begin with, two principal facts characterize the sum total of the modifications which I call facies or aspects of a stratigraphic unit: one is that a certain lithological aspect of a stratigraphic unit is linked everywhere with the same

palaeontological assemblage; the other is that from such an assemblage fossil genera and species common in other facies are invariably excluded.

Gressly took the word "facies" from the writings of Steno (1669) and, though he was not aware of the distinction of zonal and facies fossils, he formulated a number of laws which govern the vertical and lateral transitions of facies.

By the middle of the nineteenth century, it was established that bodies of rock could be defined and mapped by a distinctive combination of lithologic and palaeontologic criteria. These rock units were called "formations" or "facies" by different geologists. Over the years, however, these two terms assumed different meanings. "Formation" became ever more rigorously defined, while "facies" became used in ever broader meanings.

In 1865, Lyell was still writing:

> The term "formation" expresses in geology any assemblage of rocks which have some character in common, whether of origin, age, or composition. Thus we speak of stratified and unstratified, fresh-water and marine, aqueous and volcanic, ancient and modern, metalliferous and non-metalliferous formations.

In modern usage this definition better fits facies than formation.

Over the years, however, the term formation became more and more restricted to the description of discrete mappable rock units. This concept was finally embalmed in Article 6 of the Code of Stratigraphic Nomenclature of the American Commission on Stratigraphic Nomenclature, viz.

> The formation is the fundamental unit in rock stratigraphic classification. A formation is a body of rock characterized by lithologic homogeneity. It is prevailingly, but not necessarily, tabular and is mappable at the earth's surface or traceable in the subsurface.

Similarly, the Geological Society of London states:

> The formation is the basic practical division in lithostratigraphical classification. It should possess some degree of internal lithological homogeneity, or distinctive lithological features that constitute a form of unity in comparison with adjacent strata. (Geol. Soc. Lond. Recommendations on Stratigraphical Usage, 1969.)

Thus Prevosts' original term used for describing a rock mass by lithology and palaeontology has been restricted to a lithostratigraphic unit.

Meanwhile, Gressly's term "facies" became used in ever vaguer ways. In particular it ceased to be used to define rocks purely by lithology and palaeontology. The palaeogeographic and tectonic situation of a rock also became one of the defining parameters (e.g. Stille, 1924). Thus the terms geosynclinal, orogenic and shelf facies became used. The concept of facies was also applied to metamorphic rocks (Eskola, 1915).

Recognition of the breadth of usage of facies is shown by Krumbein and Sloss's terms "lithofacies", "biofacies" and "tectofacies" (1959). "The expressions of variation in lithologic aspect are lithofacies and the expressions of variation in biologic aspect as biofacies" (ibid. p. 268). "Tectofacies are defined as the laterally varying tectonic aspect of a stratigraphic unit" (ibid. p. 383).

Further extensive of the use of the term facies came when environmental connotations were added. As early as 1879, Mojsisovics wrote "Following Gressly and Oppel, one now customarily applies the term facies to deposits formed under different environmental conditions" (translation in Teichert, 1958). As the analysis of ancient sedimentary environments progressed this approach to facies became increasingly common until the same body of rock could be termed "flysch" facies, geosynclinal facies, or turbidite facies. Only the first of these three usages is descriptive and consistent with the original definition. The terms geosynclinal and turbidite give a genetic connotation to the term, describing its interpreted tectonic situation and depositional process. Teichert (1958), Erben (1964) and Reading (1978) have reviewed the origin of the facies concept and its subsequent growth and diversification.

It seems most useful to restrict the term facies to its original usage. This text uses facies as a descriptive term as stated by Moore (1949): "Sedimentary facies is defined as any areally restricted part of a designated stratigraphic unit which exhibits characters significantly different from

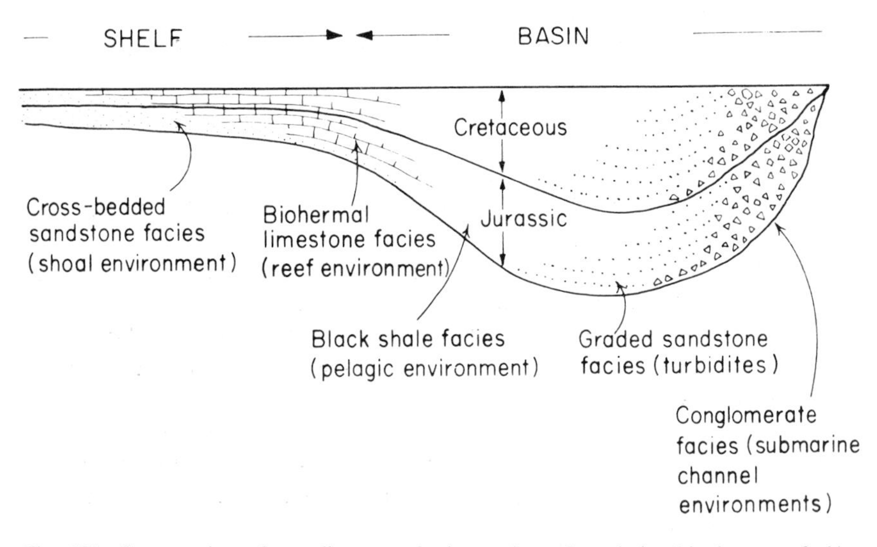

Fig. 127. Cross-section of a sedimentary basin to show the relationship between facies, environments and time.

those of other parts of the unit.'' A further refinement of this definition is that a facies has five defining parameters, viz. geometry, lithology, palaeontology, sedimentary structures and palaeocurrent pattern (Selley, 1970, p. 1).

Thus one may talk of a red bed facies, an evaporite facies or a flysch facies. Terms like shelf facies or geosynclinal facies are unnecessary and should not be used. Similarly, terms such as fluvial facies or turbidite facies are regarded as inadmissible; they are genetic and conflict with the original definition. It may be correct, however, to make statements to the effect that ''this pebbly sand facies was deposited in a fluvial environment and this flysch facies was deposited by turbidity currents''.

Figure 127 illustrates the differences between facies, and environments.

III. SEDIMENTARY MODELS

A. The Model Concept

The academic geologist has always been concerned with interpreting geologic data. The industrial geologist has to go one step further and make predictions based on these interpretations. One of the most useful tools in both these exercises is the concept of the sedimentary model. This is due to a concatenation of the concepts of environments and facies which were discussed in the two preceding sections.

The concept of the sedimentary model is based on two main observations and one major interpretation. These may be dogmatically stated as follows:

Observation 1. There are on the earth's surface today a finite number of sedimentary environments.

(*Qualifying remarks*: detailed examination shows no two similar environments to be identical. Environments show both abrupt and gradational lateral transitions.)

Observation 2. There are a finite number of sedimentary facies which reoccur in time and space in the geological record.

(*Qualifying remarks*: detailed examination shows no two similar facies to be identical. Facies show both abrupt and gradational lateral and vertical transitions.)

Interpretation: The parameters of ancient sedimentary facies of unknown origin can be matched with modern deposits whose environments are known. Thus may the depositional environments of ancient sedimentary facies be discovered.

Conclusion: ''There are, and always have been, a finite number of sedimentary environments which deposit characteristic sedimentary facies, these may be classified into various ideal systems or models" (Selley, 1970, p. 213).

These ideas have been implicit, though seldom so baldly stated, in most, if not all facies analyses.

Potter in particular has discussed sedimentary and facies models (apparently as interchangeable terms). Potter and Pettijohn (1963, p. 226) write:

> A sedimentary model in essence describes a recurring pattern of sedimentation

and again (ibid, p. 228)

> the fundamental assumption of the model concept is that there is a close relationship between the arrangement of major sedimentation in a basin and direction structures in as much as both are a product of a common dispersal pattern.

Additional discussion of models, sedimentary of facies, have been given by Potter (1959), Visher (1965), Selley (1978), Pettijohn *et al.* (1972, p. 523), Walker (1976, 1979), Curtis (1978) and Blatt *et al.* (1978).

The next part of this chapter is devoted to the description of the major sedimentary models which can be defined. The classification of these models is based on the classification of depositional environments given in Table XXIX. Each major depositional environment is correlative with a sedimentary model. More detailed accounts of these will be found in books by Reading (1978), Selley (1978), Laporte (1979) and Klein (1980).

B. Some Models Described

1. Piedmont fanglomerates

Between mountains and adjacent lowlands it is often possible to define a distinct belt known as the piedmont zone (literally from the French: mountain foot).

At the foot of the mountains, valleys debouch their sediments into the plains. The piedmont zone is characterized by small alluvial cones which are fed by gullies and built out onto pediment surfaces cut subhorizontally into the bedrock. Larger alluvial fans are fed by complex valley systems and grade out into the deposits of the alluvial plain (Fig. 128).

Depositional gradients in the piedmont zone are steep, up to 30° in marginal screes, but diminish radially down-fan. This change in gradient correlates with changes in process and sediment type.

Alluvial cones and the heads of alluvial valleys are characterized by boulder beds and conglomerates deposited by gravity slides from the adjacent mountain sides. These grade down the fan into conglomerates and massive or crudely-bedded argillaceous pebbly sandstones and siltstones.

Deposits of this type, termed "diamictites", originate from mud flows (p. 199). They in turn grade down-fan into poorly sorted massive or flat-bedded pebbly sandstones. They sometimes show irregular scours and silt laminae. These beds are deposited from flash floods, and pass down-slope into the alluvium of braided channel systems.

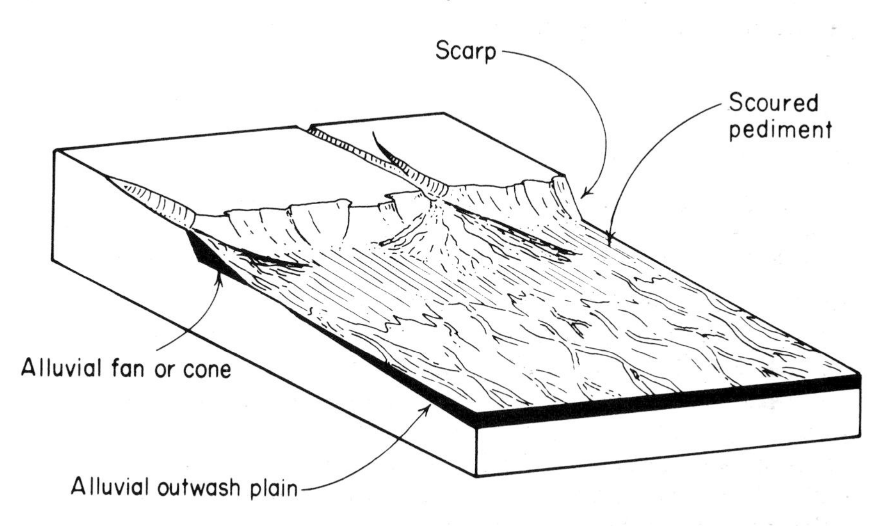

Fig. 128. Sketch to illustrate the morphology of the piedmont zone. Coarse sands and gravels are deposited by landslides, mud flows and flash floods on the alluvial fans. Sands with minor silts are deposed in braided channel systems by ephemeral floods on the outwash plains.

Deposits of the piedmont zone are characterized, therefore, by extremely coarse grain size and poor sorting, by massive or subhorizontal bedding and an absence of fossils. The term "fanglomerate" is applied to deposits of the piedmont zone, aptly denoting their lithology and geometry (Lawson, 1913).

The previous account has been drawn largely from studies of modern piedmont zones, especially in the Rocky Mountains, by Blackwelder (1928), Blissenbach (1954), Bluck (1964), Denny (1965), Van Houten (1977) and Trowbridge (1911).

Modern piedmont zone deposits are widespread around mountain chains from the Arctic to the Equator.

Piedmont zone deposits have been frequently identified in ancient sediments (Fig. 129). They are especially characteristic of fault-bounded intracratonic rifts. Notable examples occur in the Devonian of the Midland Valley of Scotland (Bluck, 1967) and in the Connecticut Triassic trough of eastern North America (Klein, 1962). In such situations great thicknesses of fanglomerates can form due to repeated synsedimentary movement along fault scarps.

A second common setting for piedmont zone deposits is as thin laterally extensive veneers above major basal unconformities of thick continental sequences. Examples of this type have been documented from the Torridon Group (Pre-Cambrian) of north-west Scotland by Williams (1969) and from the sub-Cambrian unconformity of Jordan (Selley, 1972).

There are many other instances of ancient fanglomerate piedmont deposits.

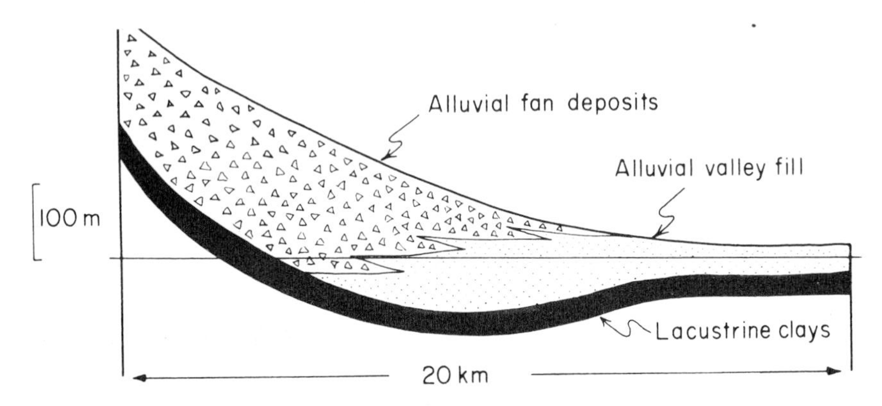

Fig. 129. Cross-section of Pleistocene alluvial fan complex deposited against the Sierra Nevada. (After Magleby and Klein, 1965.)

2. *The fluvial model*

The processes and deposits of modern rivers have been intensively studied for a number of very good reasons. Many of the largest concentrations of populations of the modern world lie on major alluvial valleys such as the Ganges, the Indus, the Nile and the Mississippi. It is important, therefore, to study alluvial processes and deposits because of the way in which they influence settlement patterns, farming, irrigation, water supply, communications and pollution. Important accounts of modern river systems have been given by Chorley (1969), Gregory (1977) and Schumm (1977).

(i) Fluvial processes and modern alluvium

Studies of modern alluvial deposits show that they can be classified into a number of subfacies. Each subfacies can be defined by its geometry, by the type of sediment, biota, and the type and orientation of its sedimentary structures. It can be seen today that each of these subfacies is formed in a different physiographic subenvironment of the alluvial system. Table XXX classifies these different alluvial subenvironments.

Table XXX
A hierarchical classification of alluvial subenvironments

Channel	Active	Channel floor Channel bar Channel bank or point bar
	Abandoned channel or ox-bow lake	
Overbank	Levée	
	Crevasse splay	
	Flood basin	Pond Swamp and marsh peats

Based on Shantser (1951) and Allen (1965).

Figure 130 illustrates the physiography and mutual relationships of these subenvironments. The characteristic features of the subfacies of each subenvironment will now be described and their genesis explained in the light of fluvial processes.

The banks of a river are inherently unstable due to the erosive power of the current. This is particularly so where rivers flow through their own detritus. This instability shows itself both by sudden switching of channels from place to place and by the gentle lateral erosion of channel walls. This process deserves discussion in some detail. In certain circumstances it is an inherent property of river channels to meander sinuously across their flood plain. As the water flows round a bend the current velocity increases on the outer bank of the curve and decreases on the inner bank. This leads to erosion of the outer bank, to form a subvertical cliff. On the inner part of the meander a slackening of current velocity allows sedimentation of bed load, and the formation of a gently sloping bar profile. On the point bar of major rivers subaqueous dunes are present which migrate down-current round the bend, depositing cross-bedded sands. On the river bed in the centre of the channel cross-sectional profile remains about constant. A lag gravel of intraformational and extraformational clasts may be present together with abraded bones, teeth, shells and water-logged drift wood.

As these processes continue through time the channel meanders sideways to deposit a characteristic sequence of grain size and sedimentary structures. At the base of the sequence is a scoured intraformational erosion surface

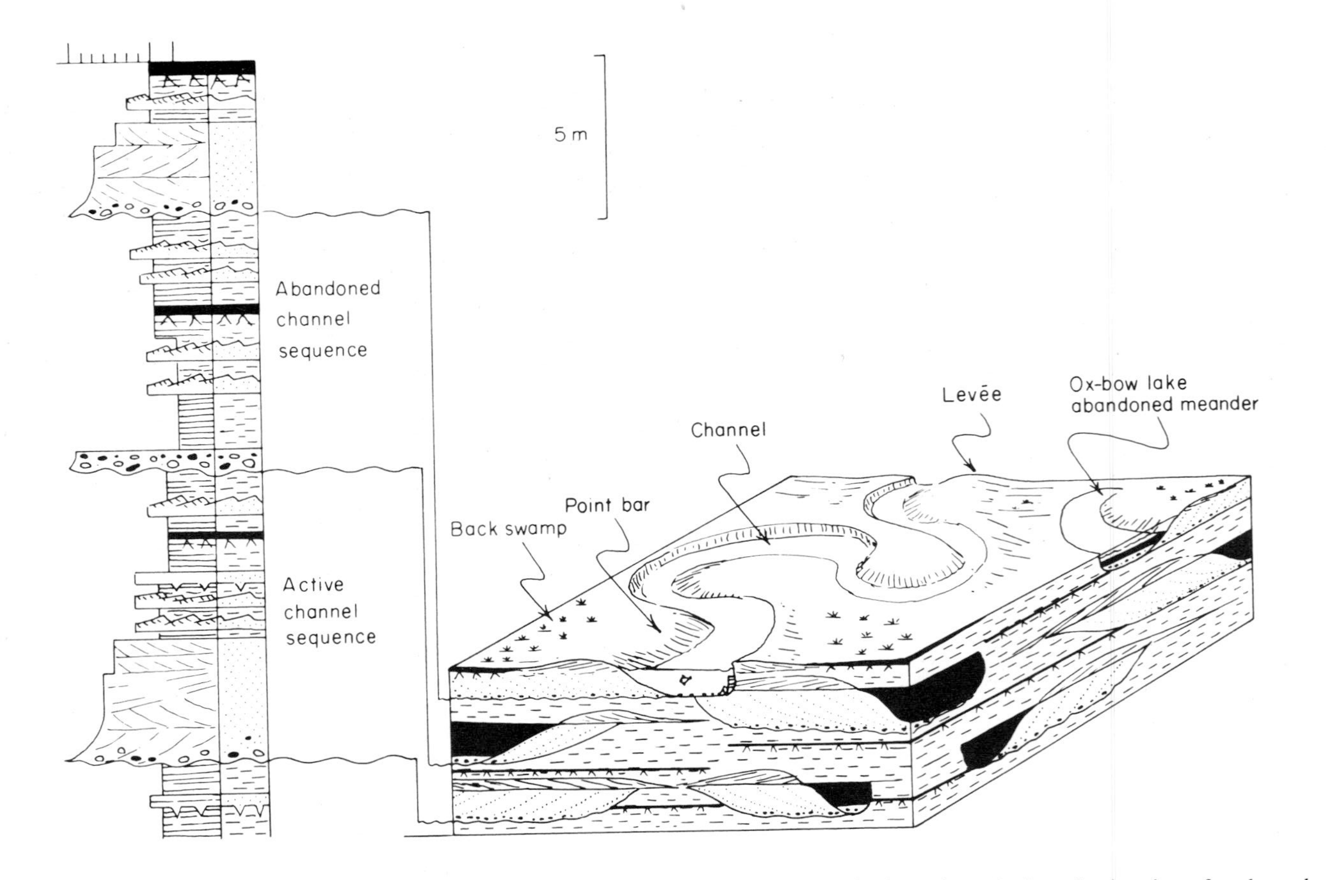

Fig. 130. Physiography and facies of an alluvial flood plain cut by meandering channels. This shows how the lateral migration of a channel generates an upward fining grain size profile on its inner convex bank. (After Visher, 1965.)

bevelled across older alluvium or bedrock. This is overlain by a channel lag conglomerate whose composition has already been described. This may be a veneer only one clast thick, or it may occur in crudely-bedded or cross-bedded sequences measurable in tens of metres or more. Above this unit come the cross-bedded sand bar deposits.

Some studies of modern and ancient point bar sequences record a vertical decline in grain size and set height. This reflects the progressive lateral decline in current velocity from the channel floor across the point bar up to the inner bank of the meander. Vertical fining of grain size will not be present if the source of detritus did not contain a broad enough spectrum of grain sizes.

The rate of discharge in a river channel is seldom constant. Diminishing discharge will result in the river shrinking within its own major channel to find its way in a braided pattern through the bars which it deposited at full flood. An increase in discharge, by contrast, causes a rise in river level until it bursts its banks. On flowing over the channel lip, current velocity may diminish; thus depositing layers of sediment which decrease in grain size away from the lip. These levées may build the banks up higher and higher on either side of the channel. They separate the channel from low-lying flood basins on either side of the alluvial plain.

Flood basin deposits are generally fine-grained sands, clays and silts. They are interlaminated, cross-laminated and characteristically desiccation cracked. Flood basin deposits are often burrowed, frequently pierced by plant roots and, under suitably waterlogged conditions, may become peat-forming swamps and marshes.

This assemblage of levée, flood basin and swamp sediments are collectively referred to as overbank deposits to distinguish them from the assemblage of channel deposits. One further type of overbank deposit to mention are crevasse splays. Rivers at bank-full scour channels through the levées termed crevasses; lobes of sands and fines are deposited where these debouch into the flood basins. These crevasse splays are analogous in origin, though smaller in scale, to the lobes of deltas.

Returning now to processes of channel sedimentation it will be recalled that the classification in Table XXX divides channels into active and abandoned types. The sedimentary processes of channels just described may be abruptly terminated by switching or avulsion of the channel course. This can happen in several ways. A sinuous river channel can ultimately meander back on itself to short circuit its flow by necking, as it is termed. Another type of channel diversion occurs when a river has raised itself up above the flood plain by levée building to the extent that its floor is above the surface of the flood plain. The development of a crevasse in the levée may become so wide and deep that even when the flood diminishes to the normal discharge the river still pours through the breach and no longer flows down the old raised river bed. Channel diversion of this type can often occur on a huge scale with resultant catastrophic flooding and loss of life.

These two types of channel switching are characteristic of meandering river systems.

In low-sinuosity braided stream networks, by contrast, channel abandonment occurs by channel bars enlarging until they block a course and cause it to divert. Alternatively the headward erosion of channel gullies in soft detritus results in capture of a previously active channel course.

In both meandering and braided river systems these processes cause an abrupt change from active channel sedimentation to abandoned channel sedimentation. In some situations the lower end of the abandoned channel may still open out as a backwater into the main channel. After a time, however, this may become blocked to form an isolated ox-bow lake, as it is picturesquely termed, lying within the flood plain away from the present active channel. Abandoned channel deposits are similar to those of flood basins. They are laminated, cross-laminated, desiccation-cracked, burrowed fine sands, silts and clays, which are occasionally pierced by rootlets and interbedded with peats. Abandoned channel deposits are distinguishable from flood basin deposits, however, by their channel-shaped geometry and by the fact that they abruptly pass down into channel-floor lag gravels instead of passing down gradually into cross-bedded point bar sands.

This review of the processes and products of modern alluvial sedimentation is based on a wide range of publications. Particularly important studies have been documented by Shantser (1951), Sundborg (1956), Leopold *et al.* (1964), Allen (1965), Collinson (1978) and Miall (1978).

Studies of modern rivers show that they are of diverse types. It is possible to arrange these in a continuous spectrum which grades from piedmont fanglomerates to braided alluvial fans to meandering river valleys, and so to deltaic deposits at the distal end (Fig. 131). The differences between braided and meandering channel systems and their respective alluvial types will now be described.

(ii) The alluvium of braided rivers

Braided channel systems are characterized by a network of constantly shifting low-sinuosity anastomozing courses. Modern braided river systems occur on alluvial fans in semi-arid and arid climates, along many mountain fronts, the edges of ice caps and the snouts of glaciers. They are characterized by relatively higher gradients and coarser sediments than meandering river systems. Discharge is seasonal and ephemeral in glacial and mountainous realms, and still more sporadic in desert climates. With their coarse grain size and erratic flow, braided channels are generally overloaded with sediment. Continuous formation of channel bars causes thalwegs to continuously diverge until they meet up with another channel course. Braided alluvial plains thus consist essentially of a network of

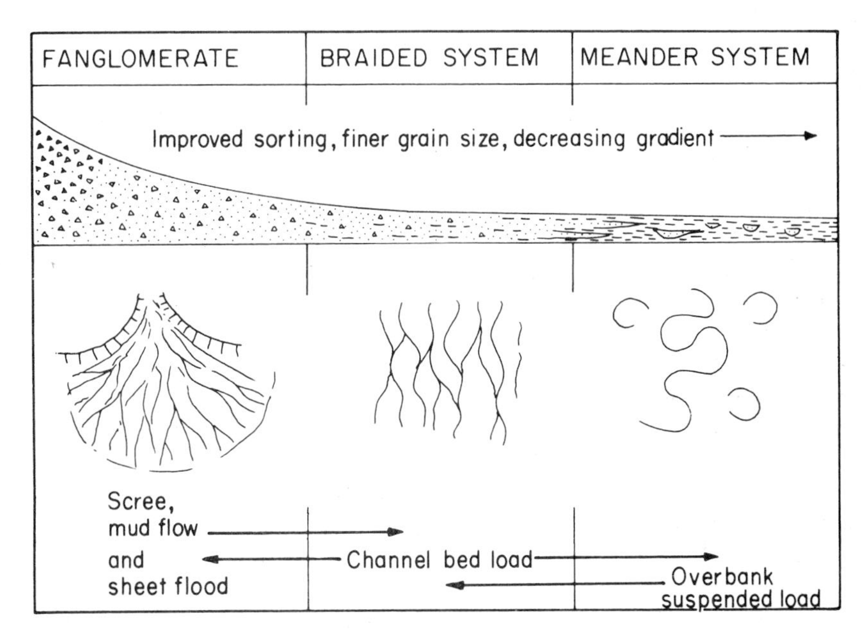

Fig. 131. Diagram to illustrate the transitional relationship between piedmont fans, braided and meandering alluvial systems and facies.

channels with no clearly defined overbank terrain. Fine suspended load sediment can only come to rest in rare abandoned channels and in the pools of active channels when a flood abates. The alluvium of braided river systems consists, therefore, largely of channel lag gravels and of cross-bedded channel bar and braid bar sands. Laminated, cross-laminated and desiccation-cracked fine sands and silts occur in rare shoe-string forms infilling abandoned channels (Fig. 132).

Accounts of mountain front, glacial and arid braided alluvial deposits have been given by Doeglas (1962), Rust (1972) and Williams (1971) respectively. Based on these and other studies Miall (1977) and Rust (1978) have established depositional models for braided channel systems.

(iii) The alluvium of meandering rivers

With increasing distance from source, the gradient of river profiles lessens, grain size decreases, channels diminish in number on the flood plain and

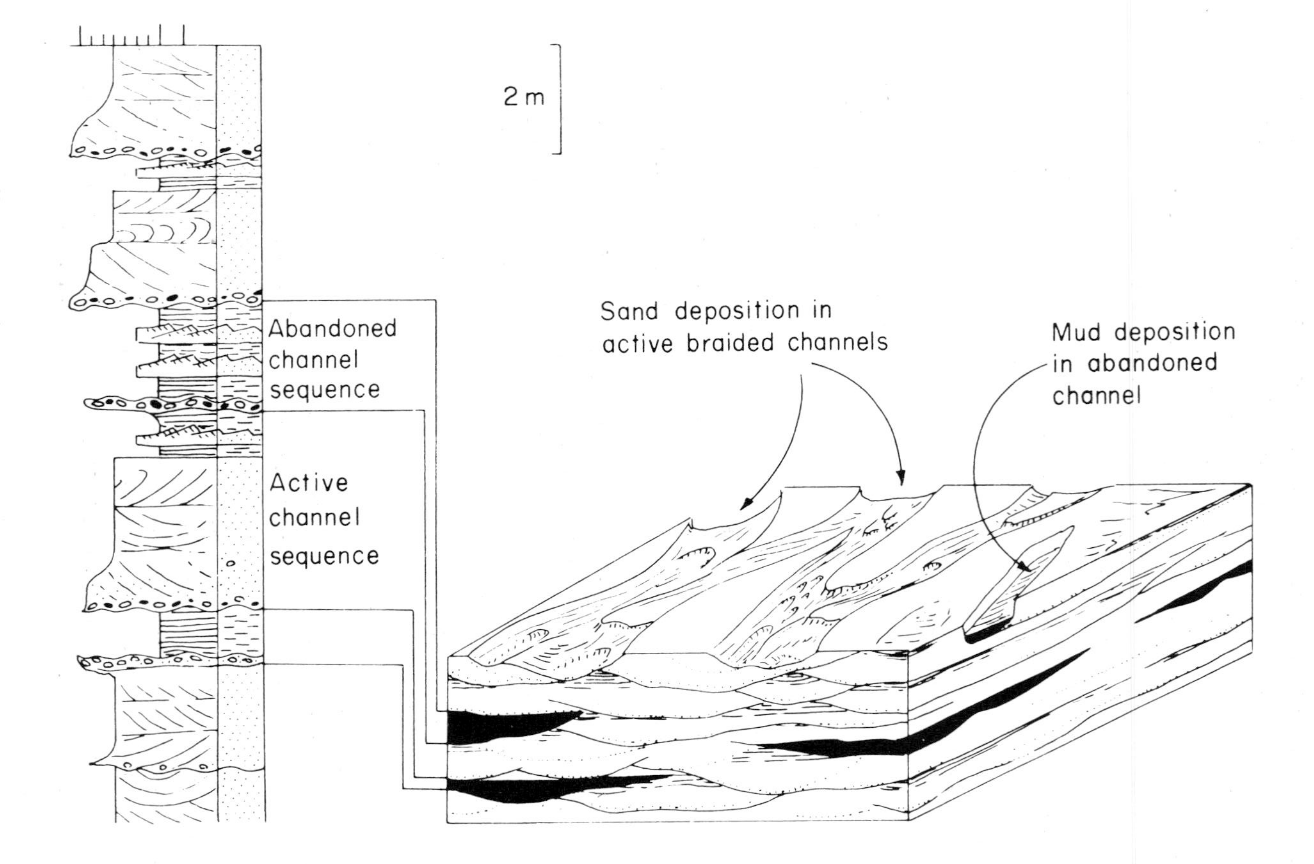

Fig. 132. Physiography and facies of a braided alluvial channel system. Sedimentation occurs almost entirely in the rapidly shifting complex of channels. Silts are rarely deposited in abandoned channels. A flood plain is absent.

increase in sinuosity. Thus braided alluvial plains can change down-stream into broad flood plains traversed by meandering rivers. The resultant alluvium shows the whole suite of active channel, abandoned channel and overbank facies. There is, therefore, a much higher proportion of silt and clay and far less sand and gravel than in the alluvium of braided rivers. Sand and gravel makes up less than 10% of some modern alluvial valley fills. A further important difference is that the erratic switching of braided river channels generates a haphazard vertical sequences of subfacies. The more sedate lateral migration of meandering river channels generates a regular fining-upward motif of grain size correlative with a characteristic suite of sedimentary structures.

(iv) Cyclicity of fluvial deposits

Studies of ancient alluvial facies show that those upward-fining cycles are repeated over great thicknesses of strata. Classic examples of this phenomenon are described from the Devonian Old Red Sandstone of the North Atlantic margins of the Appalachians, South Wales and Spitzbergen (Allen and Friend, 1968; Allen, 1964 and Friend, 1965 respectively). The cycles vary in average thickness from one or two up to 20 m.

The genesis of the fining-upward sequence is implicit in the sedimentary model of the laterally migrating channel. The cause of its repetition has been a subject for debate.

Four fining-up cycles in the Quaternary alluvium of the Mississippi River Valley are correlative with eustatic sea level changes during the ice age (Turnbull *et al.*, 1950). There is also a strong presumption that erratic subsidence of basins bounded by active faults may cause repeated regrading of flood plains, thus generating thick piles of alluvium with upward-fining cyclothems. Cyclic climatic changes, especially precipitation, may effect run-off and river discharge patterns. These can cause pulsating fluctuations in sediment input which, again, may be responsible for the repetition of the upward-fining channel motif.

Beerbower (1964) suggested, however, that it is unnecessary to invoke such external processes as the cause of repeated alluvial cycles. They may be simply explained by the to and fro lateral migration of a river channel across its flood plain coupled with a gradual isostatic adjustment of the basin floor in response to the weight of sediment. Individual cyclothems should be basin-wide if due to external climatic or tectonic causes (termed allocyclic by Beerbower). By contrast, cyclothems due to the lateral meandering of channels, interrupted by abandonment, superimposed on gradual subsidence (termed auto-cyclic by Beerbower) may be expected to be of only local extent.

It is interesting to note that studies of the small-scale facies changes of ancient alluvium sometimes show uncorrelatable sections spaced only a few hundred metres apart (Kazmi, 1964; Friend and Moody-Stuart, 1972, Fig. 21). Extensive palaeostrike-trending cliffs in the Morrison Formation of Colorado also show that upward-fining cycles occur repeatedly at any one point, but are of extremely local extension (Fig. 133).

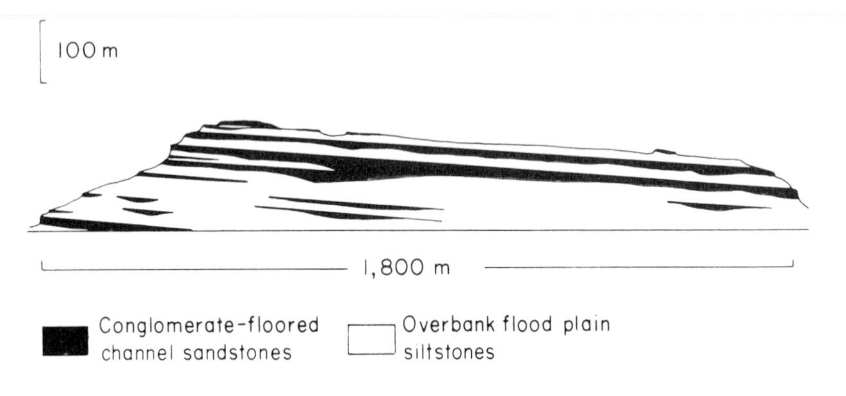

Fig. 133. A cliff in the fluvial Morrison Formation, Slick Rock, Colorado. A measured section at any point in that cliff would record a series of upward-fining cyclothems. Individual channel sequences have only limited lateral extent, however, suggesting an absence of any external (allocyclic) controlling mechanism. (From Shawe *et al.*, 1968.)

Ancient sedimentary facies comparable to modern alluvium are widespread in time and place. Classic examples of facies attributable to braided alluvial channel systems are provided by the Torridon Group (Pre-Cambrian) of north-west Scotland, and by the Cambro-Ordovician sandstones of the Sahara and Arabia (Williams, 1969; Bennacef *et al.*, 1971; Selley, 1972).

Notable examples of facies attributable to deposition from meandering river systems occur in Devonian rocks of the North Atlantic margins as already mentioned.

3. *Eolian*

The processes morphology and structures of wind sedimentation are described on p. 195. This account showed that many data have been gathered from modern eolian dune fields. This knowledge has been used in attempts to recognize ancient dune deposits in the geologic record, notably by Glennie (1970), Bigarella (1972) and McKee (1979). Three main rock series have been attributed to eolian sedimentation. The first of these includes formations which range from Pennsylvanian to Jurassic in age, and outcrop in the Colorado plateau of the USA. These have been extensively documented and reviews have been given by Opdyke (1961), Poole (1964) and Selley (1978), pp. 52–58).

Eolian sedimentation has also been proposed for Permo-Trias rocks in the North Sea basin of Europe. Accounts based on outcrop studies in England have been given by Shotton (1937), Laming (1966) and Thompson (1969). An eolian origin has been postulated for part of the gas-bearing Permian Rotliegende in the southern North Sea (Glennie, 1972), see Fig. 134.

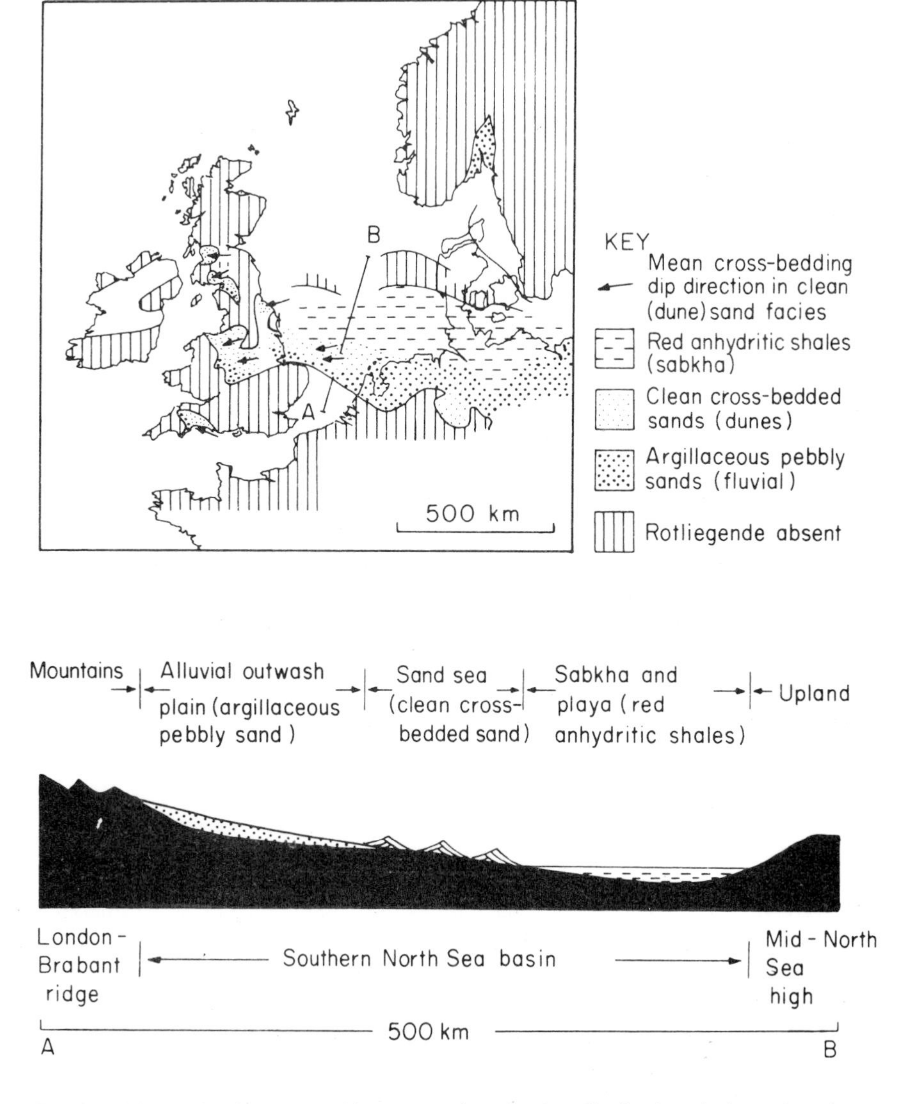

Fig. 134. Map and palaeogeographic cross-section showing distribution, facies and environment of the Rotliegende (Permian) Sandstones of the North Sea basin. (After Glennie, 1972.)

The Late Jurassic—Early Cretaceous Botucatu Sandstone of South America has also been attributed to an eolian origin (Bigarella and Salamuni, 1961; Bigarella, 1979).

Despite these wide ranging studies no viable eolian sedimentary model has emerged. There are perhaps two main reasons for this. First of all, eolian deposits are perhaps one of the hardest of all types to recognize in ancient sediments. This is because no criteria so far proposed seem to be exclusive to eolian deposits. An eolian origin for a facies can only be postulated when one can prove an absence of criteria suggesting aqueous sedimentation. This is not easy and the wind-blown origin of the classic "eolian" deposits have not gone unchallenged. Visher (1971), Freeman and Visher (1975), Stanley *et al.* (1971) and Jordan (1969) all questioned an eolian origin for many of the Colorado Plateau sandstone formations. Pryor (1971) has questioned the previously proposed eolian origin of the Permian Yellow Sands of north-eastern England. In both cases it has been suggested that the maturity, good sorting and high foresets of these facies are due to sedimentation from subaqueous dunes in a shelf sea. In this context it is significant that these groups of rock lie in similar settings. They both occur associated with red beds, evaporites and carbonate facies. This assemblage suggests a palaeogeography in which fluvial deposits passed laterally into sabkha and marine-shelf environments. By analogy with modern shorelines, eolian dunes and submarine sand shoals would both be expected and, with repeated marine advances and retreats, could be so reworked as to make it impossible to detect the last process which actually deposited them.

It can be seen, therefore, that one reason why an eolian sedimentary model has not emerged is because ancient eolian deposits are hard to identify.

A second reason for this lack of a viable model is that modern sand seas are very largely environments of erosion or of equilibrium. Areas of actual net eolian sand sedimentation are rare and unstudied. Most published work describes isolated dunes which migrate across deflation surfaces. This raises the questions: do eolian deposits have a low preservation potential? Eolian processes may play a large part in determining the texture, shape and sorting of sediments, but are these sands actually laid to rest in the geological column by sporadic catastrophic floods rather than by the wind that shapes them?

These may be some of the reasons why a viable eolian sedimentary model has not yet been defined.

4. Lacustrine models

Lacustrine deposits, unlike eolian beds, are easy to recognize in the geological record. Diagnostic criteria include evidence of aqueous

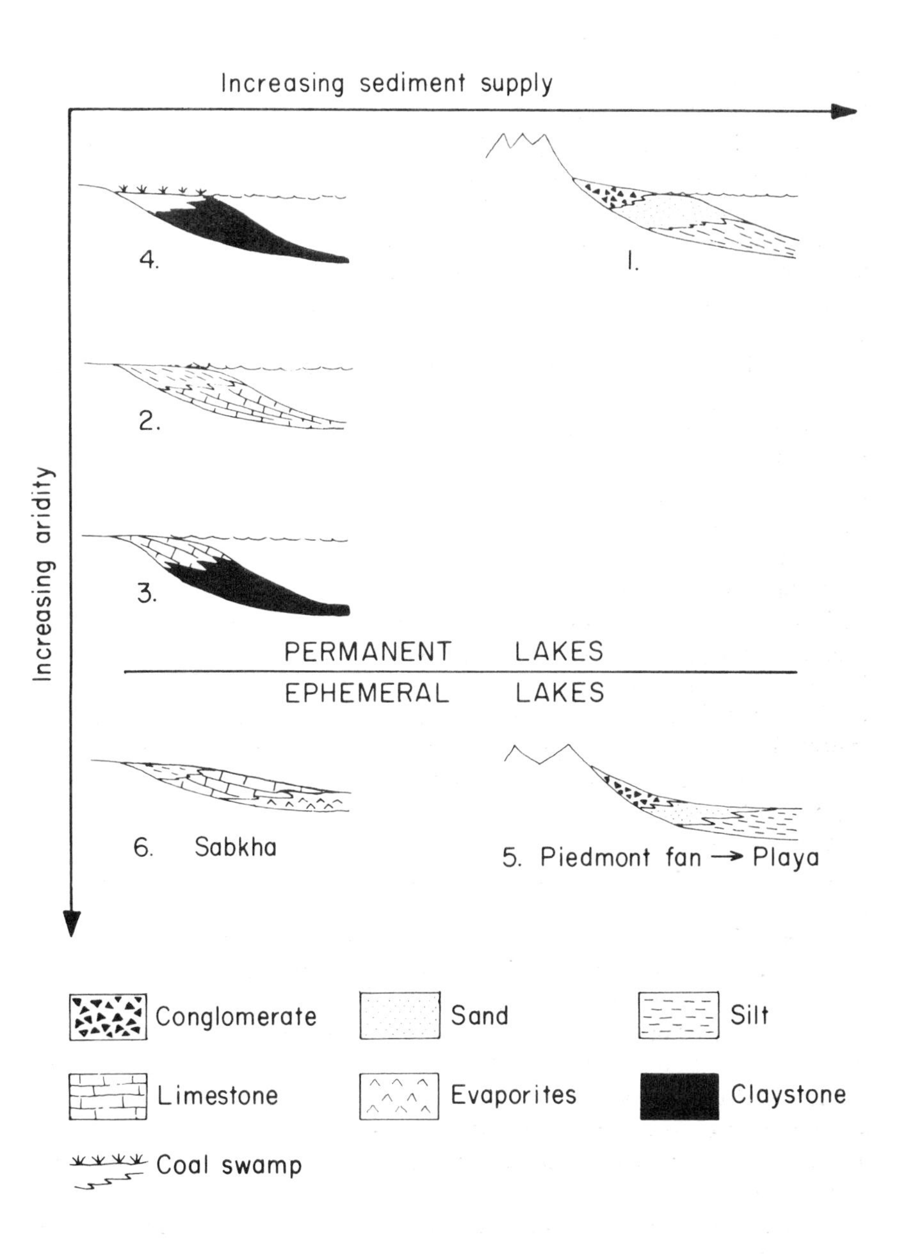

Fig. 135. Lacustrine sedimentary models. This combines Visher's (1965) concept of regressive lake infilling with Kukal's (1971) classification of lakes based on centripetal lithological variation. Two additional models are included for playa lakes and sabkhas in arid ephemeral lake basins.

deposition coupled by an absence of marine biota and the presence of fresh-water fossils. Lacustrine deposits are generally fine-grained and laminated (Matter and Tucker, 1978).

Attempts to define lacustrine sedimentary models have been made by Visher (1965), Kukal (1971) and Picard and High (1972a, 1979). These attempts reach only a broad consensus. Visher and Picard and High worked essentially from the evidence of ancient lacustrine deposits. They defined an ideal model showing an upward-coarsening grain size profile. Picard and High write (ibid. p. 115): "Inasmuch as all lakes are ultimately filled, regression dominates the history of a lake". Visher reached a similar conclusion comparing the lacustrine sedimentary sequence to that produced by a regressive marine shoreline.

According to these concepts an ideal lake sedimentary model generates a sequence which commences with laminated fine sediments deposited in the deep lake centre. As the lake is infilled, marginal fluvial, deltaic and swamp environments encroach inwards. The laminated fines may grade up through turbidite sands, such as are known from Lake Zug and Lake Meade, into beach sands, cross-bedded fluviodeltaic channel sands and marsh peats.

By contrast to this simple model, Kukal (1971) defined four different lake types, based on the areal distribution of different kinds of sediment in modern lakes. By combining the classic geological approach of looking at lakes vertically with the bird's-eye view of a Recent sedimentologist, it is possible to define the sedimentary models shown in Fig. 135. In addition to Kukal's four classes of lakes, this scheme introduces two models for the ephemeral lakes of arid regions. The total of six models have been arranged according to the degree of aridity and of topographic relief that may be expected to generate the various lacustrine types.

(i) Terrigenous permanent lakes

Kukal's first type of lake is the permanent variety which is infilled by terrigenous sediment (Fig. 135). These are commonly found today in mountainous terrain where high precipitation and sediment run-off are combined. Numerous examples may be cited, such as Lake Constance and Lake Zug in the Alps and Lake Titicaca in the Andes. Such lakes correspond to the typical lake model of Visher in that marginal fluvio-deltaic sands prograde into the lake to bury finer sediment settled out from suspension. Because of their location in mountainous terrains, the preservation of such lakes in the geological record must be slim.

(ii) Autochthonous permanent lakes

Kukal's second class of lake occurs in low-lying terrain in temperate and warm humid climates. Minor amounts of fines are brought in by rivers.

Carbonate sedimentation occurs away from the river mouths, both around the lake shores and in the deeper part of the lake. This takes the form of charophyte calcareous algal marls in the deeper parts of the lake. Around the shores, various freshwater mollusc shells break down to calcarenites, where there is sufficient wave action. This type of sedimentation occurs in the north German lakes such as Lake Schonau and the Great Ploner Lake, and in lakes of southern Canada (Fig. 135.2).

In Kukal's third type of lake, sapropelites in the centre are ringed by carbonate shoreline sediments of algal and molluscan origin (Fig. 135.3).

Good examples of permanent lakes with extensive autochthonous sedimentation occurred in the Tertiary basins of the Rocky Mountain foothills. One of the best known is Lake Uinta which, during the Eocene, covered some 23 000 km^2 of Utah and Colorado, depositing over 2000 m of diverse sediment types. These included marginal deltaic sands and coal swamps, which prograded basinward over varved oil shales of the Green River Formation. These regressive phases alternated cyclically with transgressions when carbonate marls in the basin centre passed into skeletal sands and algal oolites around the margin. Intensively studied, because of oil fields in the marginal facies, key papers on Lake Uinta include those by Bradley (1948), Picard (1967), Picard and High (1968, 1972b) and Eugster and Surdam (1973).

Oil shales such as occur in the Green River Formation are found also in Permo-Carboniferous lakes of New South Wales and in the Carboniferous Lake Cadell in the Midland Valley of Scotland (Greensmith, 1968).

Kukal's fourth type of lake consists of marginal marshes which prograde centripetally to overlie organic muds (sapropelites) deposited in the central part of the lake basins (Fig. 135.4). Examples of this type of sedimentation occur in the cold wet gytta lakes of northern Canada and northern Europe and Asia. While sapropelites and oil shales are common lacustrine deposits they seldom occur unassociated with other rock types in ancient lake sediments.

(iii) Ephemeral lakes

Modern lakes which only exist for intermittant spans of time include the "playas" of the North American desert, the Kavirs of Iran and the great inland drainage basins of the Australian interior. In such desert climates precipitation is very erratic. Though rain may only come once every few years, when it does fall it pours down. In the 1967 rainstorms of the Australian Lake Eyre basin, overnight falls of 150 mm were recorded (Williams, 1971). Downpours such as these lead to the formation of ephemeral inland lakes.

Intermontane playa lakes, such as those of the North American Desert, are rimmed by piedmont alluvial fans whose sedimentology has already been described (Fig. 135.5). The width of this marginal facies can be very variable. In playas such as Qa Saleb and Qa Disi in the Southern Desert of Jordan, the mud flats locally impinge on sheer cliffs hundreds of metres high.

The typical deposit of modern playa lakes is a red-brown mudstone containing varying amounts of clay, silt and disseminated carbonate. Scattered wind-blown sand grains are not uncommon.

Because of their tendency to evaporation, ephemeral lakes can also deposit evaporite minerals. Hence the soda flats of the North American desert and the inland sabkhas of the Saharan and Arabian deserts (Fig. 135.6).

Ancient lakes analogous to the playas of modern arid climates have been well documented, notably from Triassic rocks of Europe and North America.

Specific examples include the Popo Agie member of the Chugwater Formation in Wyoming (Picard and High, 1968), the Lockatong Formation of the Newark Group of New Jersey and Pennsylvania (Van Houten, 1964), and the red Keuper marls and associated evaporites of north-west Europe.

To summarize the foregoing discussion, it is apparent that a well-defined lacustrine sedimentary model has not been established. Ancient lacustrine deposits can generally be recognized with some confidence, and it can be argued that lake sediments tend to reflect an upward-coarsening regressive sequence. This model will not bear up under close scrutiny because the diverse environmental parameters of lakes can generate a wide range of allochthonous and autochthonous lithologies. Furthermore, the susceptibility of lakes to climatic changes can cause fluctuating lake levels. These are reflected in transgressive/regressive cycles which disrupt any overall regressive migration of lake facies.

5. Deltaic models

The term "delta", the Greek character Δ, was used to describe the mouth of the Nile by Heroditus nearly 2500 years ago. This term is still used by geographers and geologists alike. A modern definition cites a delta as "the subaerial and submerged contiguous sediment mass deposited in a body of water (ocean or lake) primarily by the action of a river" (Moore and Asquith, 1971, p. 2563). This definition, though broadly correlative with the original meaning of Heroditus, lays no stress on a triangular geometry. Not all deltas, as presently defined, possess this feature.

(i) Processes in a model delta

Reduced to its simples elements, a delta is formed where a jet of sediment-laden water intrudes a body of standing water (Fig. 136).

Current velocity diminishes radially from the jet mouth, depositing sediment whose settling velocities allow grain size to diminish radially from the jet mouth. Sedimentation around the jet mouth builds up to the air/water interface, but the force of the jet maintains a scoured channel out through the sediment. Ridges on either side of the distributary channel are termed "levées". As sedimentation continues, the delta progrades out into the standing body of water. Three main morphological units appear. The

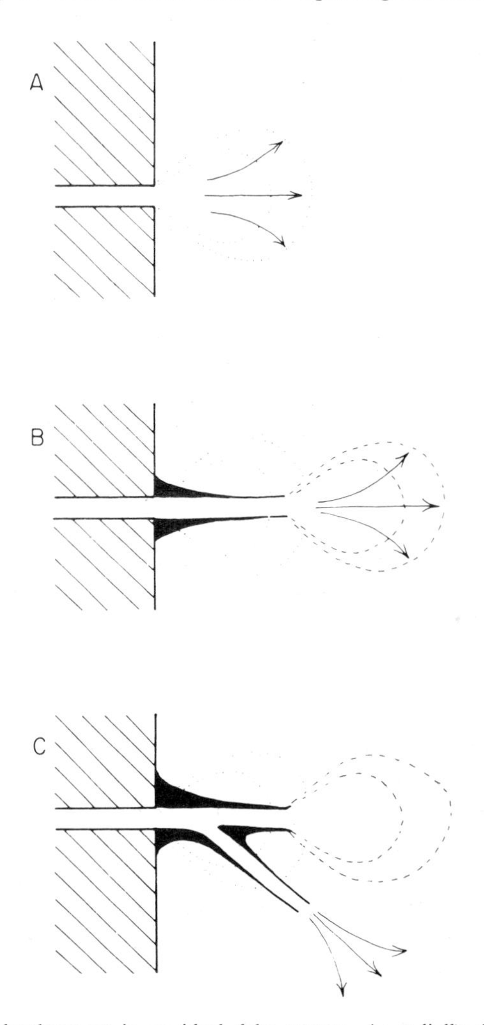

Fig. 136. Stages of development in an ideal delta system. A: radially decreasing current velocites from jet mouth, deposit concentric arcs of sand, silt and clay. B: delta progrades, forcing a channel through marginal levées. C: channel mouth chokes, levée ruptures and a new delta builds out from the crevasse.

delta platform is the subhorizontal surface nearest the jet mouth. It is basically composed of sand and is traversed by the distributary channel and its flanking levées. The delta platform grades away from the source into the delta slope on which finer sands and silts come to rest, and this in turn passes down into the pro-delta area on which clay settles out of suspension. A vertical section through the apex of a delta should thus reveal a gradual vertical increase in grain size. At the base the pro-delta clays grade up through delta silts into sands of the delta platform. Classically, these three elements have been termed the bottomset, foreset and topset respectively (Fig. 137).

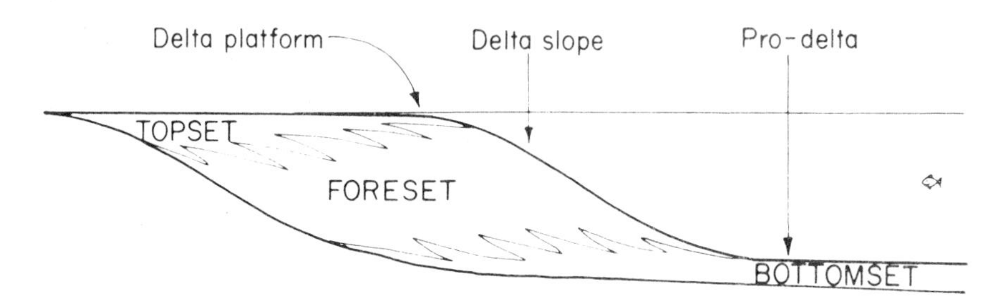

Fig. 137. Nomenclature of a delta profile.

Eventually a distributary channel becomes so long that its mouth becomes choked with sediment. At a point of weakness the levée bursts and a new distributary system is established. The abandoned distributary is choked by suspended sediment, and the whole abandoned lobe sinks beneath the water as it compacts.

Theoretically, this process may continue indefinitely as the distributaries switch from side to side from their original point of sediment input. This ideal delta model consists of a series of interdigitating lobes, each one showing a gradual upward increase in grain size and a decrease in grain size from its point of origin.

This ideal model will now be contrasted with modern deltas.

(ii) Modern delta systems

Without doubt the Mississippi is one of the most intensely studied modern deltas. Some of the key papers include those of Fisk (1955), Coleman and Gagliano (1965), Kolb and Van Lopik (1966) and Gould (1970).

The Mississippi delta bears a close relation to the ideal model (Fig. 138). A series of seven separate Quarternary delta lobes can be mapped (Fig. 139). Topset, foreset and bottomset sedimentary facies can be recognized, each with a characteristic suite of lithologies, biota and sedimentary structures. Though the modern Mississippi compares well with the previously derived "ideal model" it is in fact a very dangerous analogue to use to interpret ancient deltas. It is unusual for two reasons. It has a far higher ratio of mud to sand than most deltas, ancient or modern.

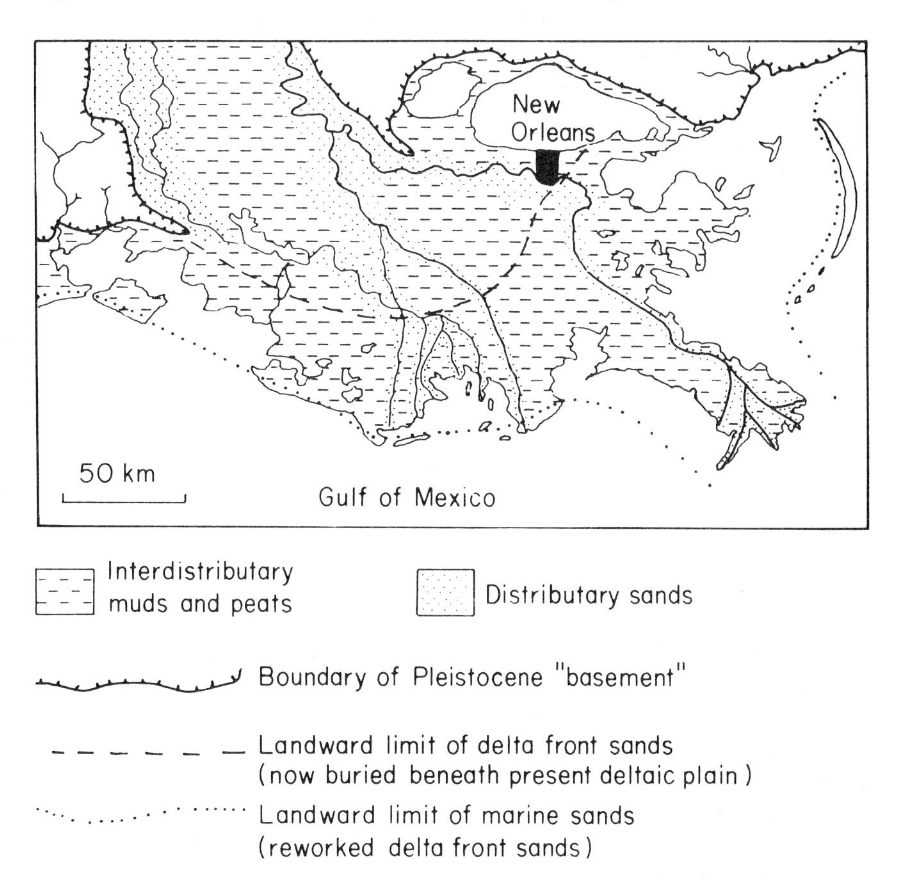

Fig. 138. Map of the modern Mississippi delta showing distribution of major sand facies.

Furthermore the Mississippi builds into a sheltered marine embayment of low tidal range. Thus the marine processes which redistribute the alluvial sediments of most deltas are largely absent from the Mississippi.

Only about 25% of the Mississippi's load is sand, the rest is silt and clay. This means that a very small amount of each delta lobe is sand. Almost all the sand load is deposited at the mouths of the distributary channels in what are called bar-finger sands. As the distributary extends seawards, so the bar-finger sands take on a linear geometry. Switching of the delta means that these bar-finger sands generate an overall pattern of radiating shoe-strings, analogous to the fingers of a hand.

In the older lobes of the Mississippi the main locus of sand deposition was the seaward edge of the delta platform, where it is possible to define an

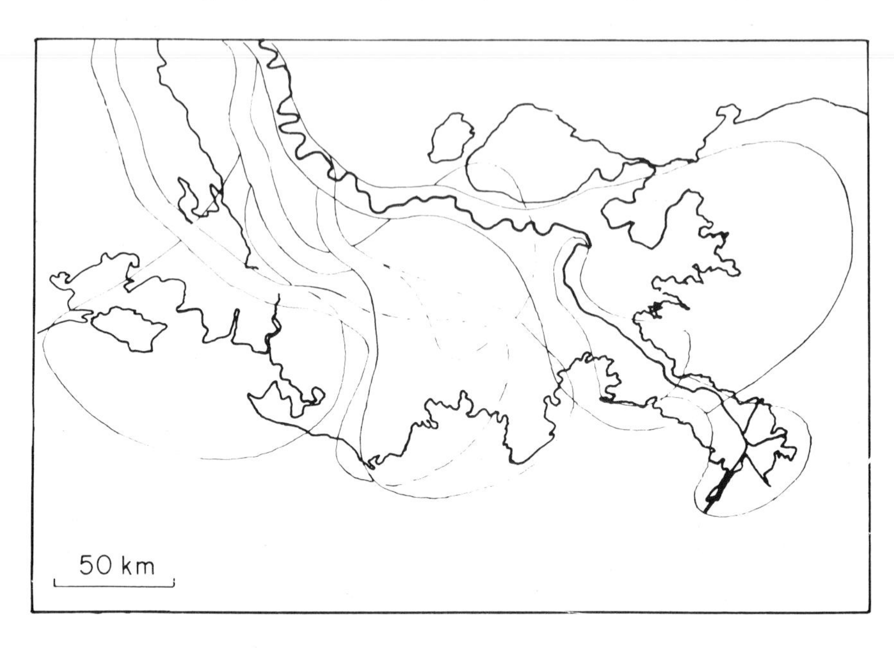

Fig. 139. Map showing the distribution of the post-Pleistocene lobes of the modern Mississippi delta. (After Kolb and Van Lopik, 1966.)

arcuate belt of delta-front sheet sands which were deposited at the mouths of distributary channels (Fig. 140). The delta-front sheet sands pass shorewards into, and are overlain by, silts, clays and peats deposited in levée, interdistributary and swamp environments. Seaward progradation of the delta has thus generated an upward-coarsening sequence ranging from marine clays of the pro-delta up into the delta-front sheet sands. These are in turn overlain by a suite of brackish and non-marine fine-grained facies dissected by radiating distributary channel sands.

To the east of the present active mouth of the Mississippi lies the arcuate archipelago of the Chandeleur islands. These mark the edge of the abandoned lobe of the old St Bernard delta. It is clear that when the sediment supply of a delta is cut off then it is extremely susceptible to reworking by marine influences, both tides and waves. As the delta lobe compacts and subsides, the sea transgresses it and reworks its upper part. Fine sediment tends to be winnowed out and settle in deeper quieter water.

The sands are reworked and redeposited as a transgressive marine blanket-sand which may thus unconformably overstep the various facies of the delta.

Instead of consisting of a simple upward-coarsening sequence, examination of the modern Mississippi shows an upper fine-grained

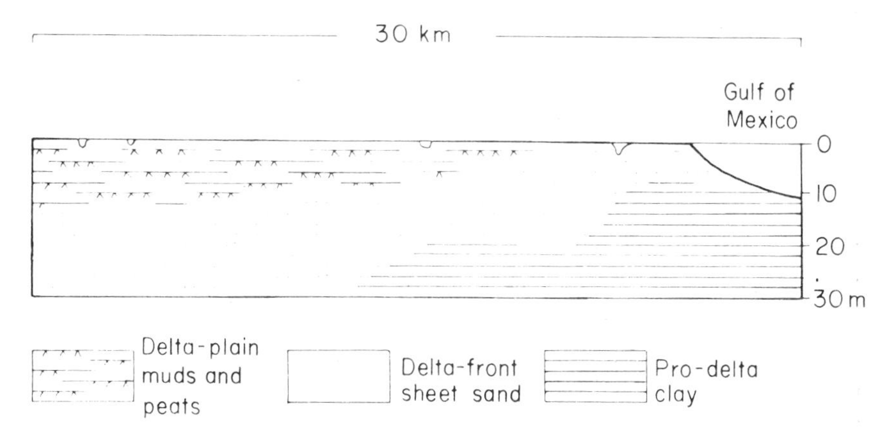

Fig. 140. Section of the Lafourche subdelta of the modern Mississippi, showing main locus of sand deposition as a sheet at the delta front. This is gradually buried beneath a prograding sheet of delta-plain muds and peats deposited in interdistributary bays and swamps, with minor distributary channel sand bodies. (After Gould, 1970.)

non-marine unit in a delta sequence. Furthermore, while a delta is, of its very nature, a regressive prograding prism, each lobe may contain both constructive regressive and destructive transgressive phases.

The bird-foot Mississippi type of delta, with radiating distributary channel networks, is rare world-wide, and is seen more commonly in lakes rather than seas (e.g. the St Clair River delta of Canada (Pezzetta, 1973)). There may be two reasons for this. First, not all rivers carry as much fine material as the Mississippi. Gravel deltas such as those of the Arctic, and of steep desert coasts, such as the eastern side of the Gulf of Aquaba, retain the basic form of alluvial cones. They can be followed from mountain front to sea bed with no differentiation into the diverse subenvironments described from the muddy Mississippi.

The second main qualifying factor of deltaic geometry, other than sediment type, is the relative importance of marine and fluvial influences. The Mississippi maintains a bird-foot geometry because of its relatively sheltered position in the Gulf of Mexico. The tidal range of the northern part of the Gulf of Mexico is low (less than a metre) so tidal currents are relatively insignificant. Prevailing winds are from the north and east so the

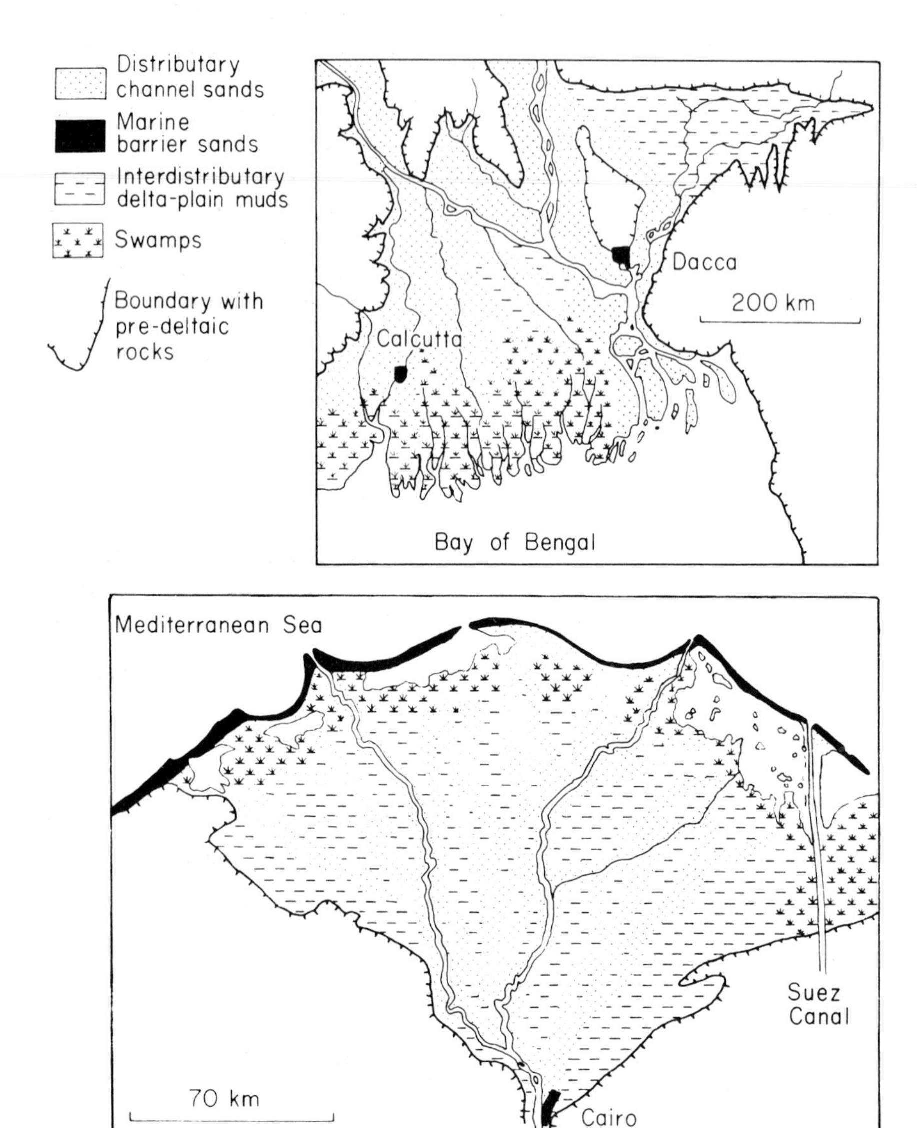

Fig. 141. Map showing the distribution of modern deltaic facies on the Ganges–Brahmaputra delta (top) and the Nile delta (bottom). The first is an example of a tide-dominated delta with braided seaward-trending channel sands. The Nile is an example of a wave-dominated delta with extensive fringing barrier sand bodies. (Based on Morgan, 1970, Coleman *et al.*, 1970, and Wright and Coleman, 1973).

fetch of waves is short, apart from those due to the occasional hurricane.

More exposed deltas, such as those of the Nile and Niger, show smoothed arcuate coastlines (Fig. 141). This is because sand is no sooner deposited at a distributary mouth than it is reworked by the sea and redeposited along the delta front, often in the form of barrier islands. Marine influence, however, may take the form of wave action, on exposed coasts, or of tidal scour on coasts with high tidal ranges. In the Bay of Bengal tidal ranges vary from 3 to 5 m. Largely because of this, deltas of the Bay of Bengal are quite different from that of the Mississippi. These deltas include those of the Ganges-Brahmaputra, the Klang, Langat and Mekong (Morgan, 1970; Coleman *et al.*, 1970). The scouring effect of strong tidal currents redistributes fluvial sediment into broad tidal flats where extensive mangrove swamp development acts as an additional sediment trap. The distributary channels themselves are wide, deep, straight, braided estuaries (Fig. 141).

The preceding analysis shows that deltas are composed of a series of upward- and landward-coarsening clastic lobes. These lobes, essentially regressive, may contain fine-grained topsets, and embody both constructive and destructive phases. Furthermore, the environments and facies of deltas vary widely according to the relative importance of fluvial, tidal and wave processes.

Table XXXI attempts to classify the different delta types which are recognizable.

Table XXXI
A classification of deltas based on the dominant processes

Dominant process		Environments	Sand facies	Example
Fluvial		Radiating bird-foot distributary/levée systems	Radiating mouth-bar sands	Mississippi
Marine	Waves	Distributaries truncated by barrier sands	Arcuate delta-front barrier sands	Nile and Niger
	Tides	Extensive tidal flats and scoured braided estuaries	Delta-front sheet sand	Mekong and Ganges–Brahmaputra

Based on Fisher *et al.* (1969), Morgan (1970) and Wright and Coleman (1973).

(iii) Ancient deltaic deposits

Many sedimentary facies have been attributed to deltaic deposition. Not only are ancient deltaic deposits common in the geological record, they are also very important economically because they host most of the world's coal reserves and a significant part of its oil and gas. Rapid sedimentation of organic-rich sediment in a dominantly reducing environment is conducive to the metamorphism of coal from peat and for the generation of oil and gas.

Useful accounts of deltaic sedimentation, ancient and modern, have been edited by Shirley and Ragsdale (1966), Morgan and Shaver (1970), Fisher *et al.* (1969) and Broussard (1975).

In a general way it is possible to distinguish fluvial-dominant from marine-dominant deltas. Notable examples of cyclic deltaic sediments have been described from the coal bearing Pennsylvanian strata of the Illinois basin e.g. Potter (9162) and Wanless *et al.* (1970). Detailed mapping of sediment increments between regionally widespread coal marker horizons shows radiating shoe-string sand bodies analogous to those of the modern Mississippi (Fig. 142). Sedimentology and field relationships of these sands suggest that they are channel sands, rather than bar-finger sands. Thus, though the Pennsylvanian sands may be different in genesis from those of the modern Mississippi, it is reasonable to suppose that these were laid down in fluvially-dominated deltas.

Examples of marine-dominant destructive deltas are not hard to find either. In the upper Cretaceous a broad sea-way stretched along the eastern edge of the rising Rocky Mountains from the Canadian Arctic to the Gulf of Mexico. In response to the uplift of the Laramide orogeny, vast quantities of detritus were brought into this sea-way from the west. This was deposited in a wide range of environments ranging from the piedmont fanglomerates of the Mesaverde Group to the off-shore marine shales of the Pierre and Lewis Formations. Transitional shoreline deposits were laid down in a great diversity of environments. Both linear shoreline and lobate shoreline deposits have been recognized. The deltaic deposits seem to have been deposited in marine dominant deltas analogous to the Niger and the Nile. Notable accounts of these beds have been given by Weimer (1970) and Asquith (1970). The Castlegate Sandstone of Utah provides a good specific example (Fig. 143). This commenced as a constructive delta in which fluvial sands passed seawards through delta-platform coals, sands and shales into a facies of thinly interbedded fine sands and muds, laminated, carbonaceous, and burrowed towards the top. This latter assemblage is interpretable as the deposit of the delta front and slope. There is evidence at the top of the Castlegate Sandstone, however, that marine influences began to dominate.

The coaly delta platform sediments are separated from the slope beds by a coarse, clean, well-sorted, upward coarsening, sand unit. This has gently seaward-dipping foresets and steep, shoreward-dipping foresets. These characteristics suggest that this was a shoal sand body, formed by reworking the delta during a constructive marine dominant phase.

These examples show that it is possible to recognize both fluvial-dominant and marine-dominant deltas in ancient sediments.

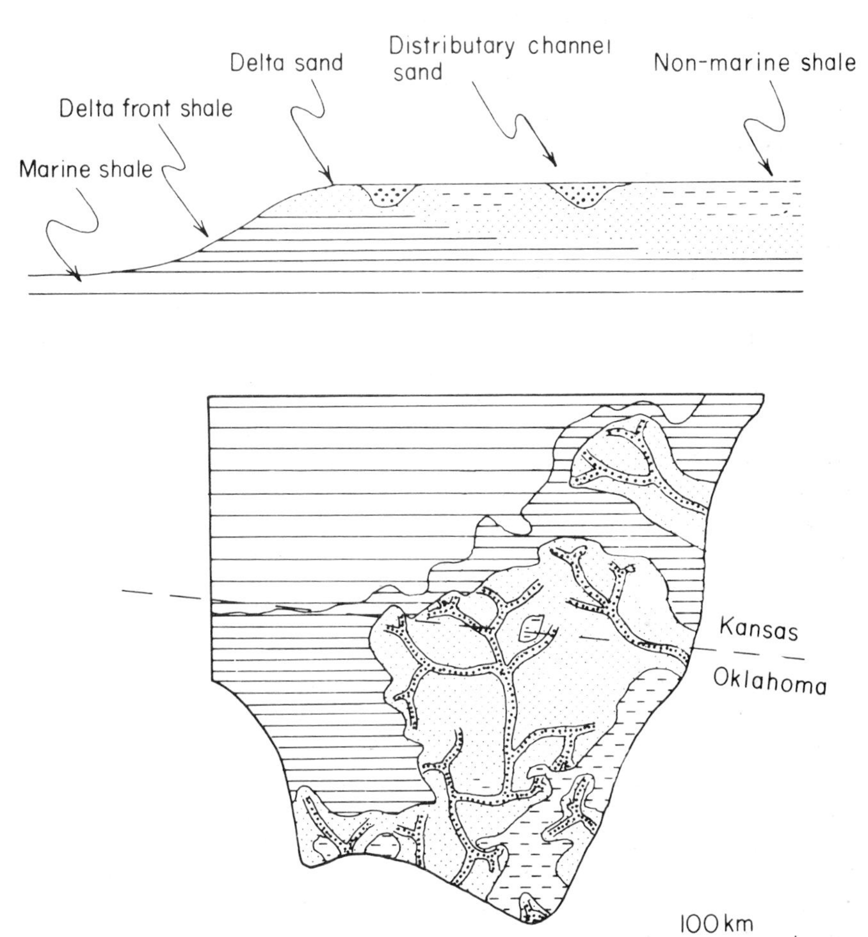

Fig. 142. Environmental and facies map of the deltaic sediments of the Noxie Sandstone member, Kansas City Group and contemporary strata. (From Horne in Wanless *et al.*, 1970.) This map shows the progradation of a series of deltaic lobes from a general south-easterly direction. The geometry of the distributary sands, and the apparent absence of marine delta-front sand shoals, suggests that this was a fluvially dominated delta with little tidal or wave action.

A further distinction of some importance can be made, however. The two specific deltas just described graded seaward into open marine muds. Others, such as the Pennsylvanian deltas of the Appalachian plateau, and of the Yoredale series of northern England, prograded across carbonate platforms (e.g. Ferm, 1970; Moore, 1959).

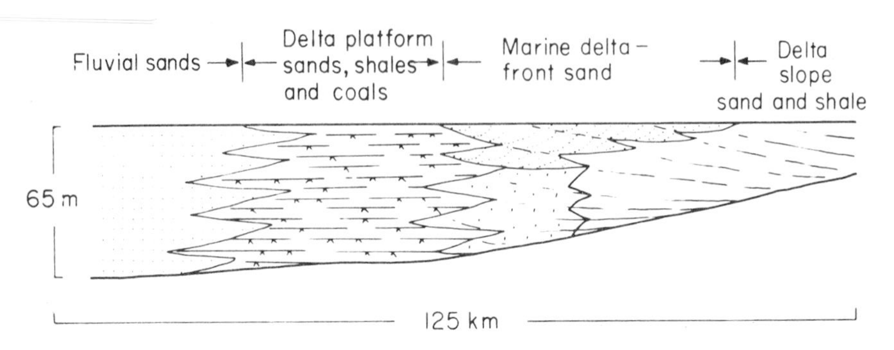

Fig. 143. Cross-section of the Castlegate Sandstone (Cretaceous), Utah, from De Graaff (1972). The low-angle seaward dips and high-angle landward dips of the delta-front sand, together with its distribution, suggest that it was formed by marine reworking of the delta front sands.

There is, however, yet another important type of delta which generates turbidite sands at its foot. The development of slumps and slides is known from modern deltas, and there is evidence that these transport sand by turbidity currents onto the basin floor. This has been described from the Mississippi, Fraser and Niger deltas (Shepard, 1963, pp. 494 and 500; Burke, 1972). Great depth of both water and sediment make it hard to study modern delta-front turbidites. Many ancient examples have been described however. Notable case histories have been documented from the Carboniferous of England (Walker, 1966; De Raaf *et al.*, 1964), from the Ordovician rocks of the Appalachians (Horowitz, 1966), from the Coaledo Formation of Oregon (Dott, 1966), and from the Tertiary–Recent wedge of the Niger delta (Fig. 144).

To conclude: at its simplest the delta process generates upward-coarsening lobes of sediment which grade from marine muds, upwards and shorewards into diverse non-marine sands, muds and, often, coals. This simple model may be modified by marine destructive influences. Furthermore, if the delta slope was sufficiently unstable to slide and slump, then redeposited turbidite sands may be present at the delta foot.

Recognition of these diverse deltaic models is critical to the effective exploitation of hydrocarbons from ancient deltas. Sand reservoirs in fluvial-dominated deltas are radiating shoe-strings on the delta platform.

Marine-dominated deltas tend to have arcuate motifs of shoal sands. Additional reservoir sands may be present in the submarine canyons and fans of high-slope deltas (see p. 379).

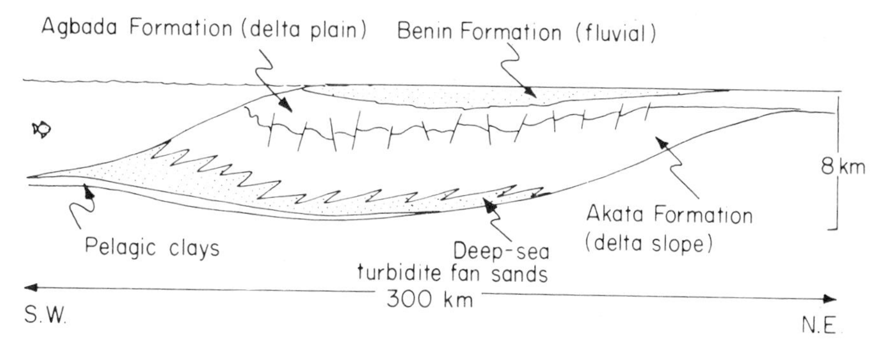

Fig. 144. Cross-section of the Tertiary–Recent sediment prism of the Niger delta showing position of turbidite fan sands at the foot of the delta slope. (From Burke, 1972.)

6. Linear barrier coasts

With increasing marine influence deltas pass gradually into linear barrier islands. All stages of the transition are present in modern shorelines and in ancient rocks; yet the two end members are sufficiently distinctive to be clearly definable models.

Modern deltas which pass along the shore into linear barrier-island coastlines can be seen on either side of the Mississippi delta in the Gulf of Mexico, on the Dutch coast to the north-east of the Rhine delta, on the Sinai coast to the east of the Nile delta and on the coast to the west of the Niger delta.

In addition, there are many modern barrier coasts which are unrelated to any major delta. Examples include much of the east coast of North America between New Jersey and Florida, parts of the north German and Polish coasts and the Younghusband peninsula of South Australia. These coasts do show, nevertheless, considerable input of sediment from rivers draining their hinterlands.

Coastlines have been classified according to their tidal range. Microtidal coasts have a tidal range of less than 2 m. Mesotidal coasts have tidal ranges between 2 and 4 m. Macrotidal coasts have tidal ranges in excess of 4 m (Davies, 1973). Barrier islands are generally best developed where tidal range is relatively low.

(i) Modern barrier coasts

Recent barrier islands and the processes which form them have been intensely studied and debated (Davis, 1978; Swift and Palmer, 1978).

There is considerable controversy over the genesis of modern barrier islands. The two major mechanisms proposed are the progressive up-building of off-shore bars and the submergence of coastal beaches and dune belts. One of the problems of studying modern bars is that we live in an ice age, albeit an interglacial, and eustatic shoreline changes have occurred several times in the last million years. Modern shorelines show both raised and drowned beach features and many modern barrier islands contain cores of older sediments. This is, therefore, not the best point in geological time to study barrier island formation. It seems reasonable to postulate a polygenetic origin for barrier islands (Schwarz, 1971); additional data and discussions will be found in King (1972), Steers (1971), Guilcher (1970) and Hoyt (1967).

Some of the best studied of modern barrier island complexes include the Gulf Coast of west Texas in general, and Padre and Galveston islands in particular (Shepard, 1960; Bernard *et al.*, 1962). Of the barrier coastlines of the eastern coast of North America, Sapelo island is one of the most studied (Hoyt *et al.*, 1964; Hoyt and Henry, 1967). Outside North America, the Dutch and German barrier coasts are some of the best known (Van Straaten, 1965; Horn, 1965).

Coupling studies of these modern coasts with their ancient analogues, a well-defined barrier island sedimentary model has been put together by Visher (1965), Potter (1967), Shelton (1967), Davies *et al.* (1971) and Elliott (1978).

Essentially, a barrier island is a linear sand body exposed at high tide which runs parallel to the coast, separating the open sea from sheltered bays, lagoons and tidal flats. In most modern barrier coasts, two high-energy environments alternate laterally with two low-energy environments. On the landward side is a fluvial coastal plain of sands, silts, clay and peats. This grades seawards; generally through salt marsh deposits, into quiet-water tidal flats and lagoons. The sediments of these facies consist of laminated cross-laminated and flaser-bedded fine sands, silts and clays. This zone is characterized by intense bioturbation, by shell beds, often of oysters and mussels, and by upward fining tidal creek sequences (e.g. Evans, 1965; Ginsburg, 1975).

The barrier island itself consists of a number of distinctive physiographic units. On its landward side there may be a complex of wash-over fans and barrier flats formed of sand flushed over the barrier island during storms. The crest of the barrier island is often formed of wind-blown sand dunes, though these are sometimes stabilized by soil profiles and vegetation. This crestal area passes seawards through a beach zone to the open sea.

The barrier island sands are generally mature and well sorted. Attempts to distinguish beach, dune and river sands by granulometry are described elsewhere (p. 19). Internally the beach deposits are horizontally or subhorizontally bedded with gentle seaward dips. Trough and planar foresets are also present in subordinate amounts. Barrier islands are sometime cross-cut by tidal channels: these may possess subaqueous deltas on both their landward and seaward sides (Armstrong-Price, 1963). Cross-bedded channel sequences are formed from the lateral accretion of these channels; such deposits may compose a considerable part of the barrier sand body that is actually preserved in the geological record (Hoyt and Henry, 1967).

Seawards, the beach deposits of the barrier island pass out into open marine environments. Many modern barrier sands, particularly those of the North Sea, pass out into scoured marine shelves, which are essentially environments of erosion or equilibrium. In the coasts of Nigeria and of the Gulf of Mexico, by contrast, barrier islands pass off-shore into deeper water. Here fine sediment may settle out of suspension. Detailed seaward traverses from such barrier beaches reveal a gradual decrease in grain size from the high-energy surf zone to the lower energy environment beneath the limit of wave and tidal current-action. Bioturbated muddy flaser-bedded sands and silts below low-water pass seaward into the laminated muds of the open marine environment.

The subenvironments and the facies generated by an ideal sedimentary model for a prograding barrier island complex are shown in Fig. 145. As one would expect, the resultant sedimentary sequence is very similar to that generated by a wave-dominant delta. The main differences are that the deltaic sequence contains distributary channel sands within the lagoon and tidal flat sequence, and in the upper part of the sand sheet.

(ii) Ancient barrier coasts

A large number of ancient sedimentary facies have been attributed to deposition in bar environments. Examples are given in the previously cited papers by Visher and Shelton, and Davies *et al.* Particularly well documented case histories come from the Cretaceous Eagle Sandstone and Muddy Sandstone of Montana (Shelton, 1965; Berg and Davies, 1968), the Bisti oil field of New Mexico (Sabins, 1963), and the Red Fork Sandstone of the Enid embayment, Oklahoma (Withrow, 1968).

These examples, taken from both surface and subsurface studies, have a number of common features. They all tend to show an upward-coarsening grain size profile. The bases of the sand bodies are transitional with open marine shales. The tops of the sands are abrupt, often marked by a break in sedimentation, and overlain by marine or non-marine shales. Internally these sand bodies are generally bioturbated in their lower parts and massive or subhorizontally bedded towards the top. The most diagnostic feature of

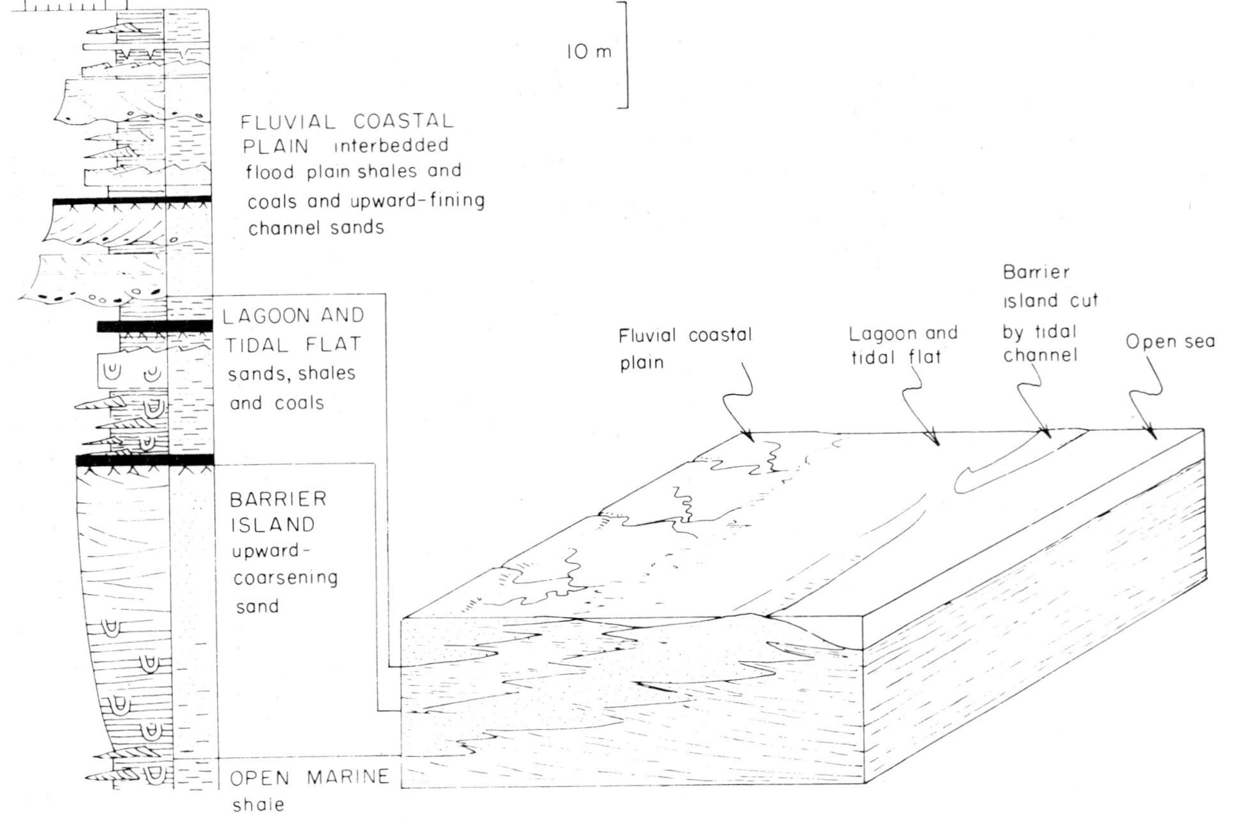

Fig. 145. Sedimentary environments, facies and vertical profile produced by a seaward-prograding barrier island complex.

all, however, is that these sand bodies are linear shoe-strings running parallel to the local shoreline.

Modern coastal geomorphologists rigorously define coastal sand bodies into barrier islands, off-shore bars, spits and tombolas. It is easy to make these distinctions of modern sand bodies. They are not easy to make of ancient sand bodies because their spatial relationships are seldom sufficiently clear. Most geologists are content to label a particular unit as a bar sand. Additional refinements of its degree of exposure or geomorphology are largely academic.

Lateral progradation of barrier bars can generate not just a shoe-string, but a sheet sand body. This requires a critical balance between sediment input and subsidence. Nevertheless, examples of barrier sheet sands have been described, notably from the Cretaceous of the Rocky Mountain foothills and from the Gulf Coast of Mexico Tertiary. Descriptions of Rocky Mountain bar sheet sands have been given by Weimer (1961) and Hollenshead and Pritchard (1961), and Asquith (1970). Descriptions of the Gulf Coast sand bodies have been given by Boyd and Dyer (1966) and Burke (1958).

These studies show that the sand sheets occur not just as regressive sand sheets which pass down into marine and up into non-marine facies, but also as transgressive sands which pass up into marine facies. The transgressive sand bodies tend to be less well-developed, however, and the transgressions are more often represented by surfaces of erosion than by depositional units. Close examination of the transgressive sand bodies shows that they are actually composed of a series of upward-coarsening shoe-string units which are vertically arranged *en echelon* (e.g. Asquith, 1970, Fig. 34). This implies that deposition of sand during a marine transgression actually takes place only during still-stands of the sea when bars may prograde seaward (Fig. 146).

Barrier shoreline coasts often show, therefore, a series of transgressive and regressive cycles. Unlike the deltaic sedimentary model, the barrier model lacks a built-in cycle generator. Such cycles are generally attributed to external causes such as tectonic movement and eustatic changes (Tanner, 1968).

In conclusion it can be seen that a clearly defined barrier island sedimentary model can be recognized in modern shorelines.

Ancient bar sands are often recognizable by their sequence of grain size and sedimentary structures. There can be little doubt of the bar sand origin of a shoe-string enclosed in shale, trending parallel to the palaeoshoreline. It is hard, and often irrelevant, to prove whether such a sand body was an off-shore bar or a barrier island.

Sand sheets which vertically separate marine shales from non-marine sediments can be formed by both prograding marine-dominated deltas and by barrier island coasts.

Detailed studies of such sand bodies show that bar sands form an integral part of the overall prograding sediment prism and that a division of the shoreline into deltaic or barrier models is irrelevant and may be misleading. It is more important to make accurate environmental diagnoses of discrete

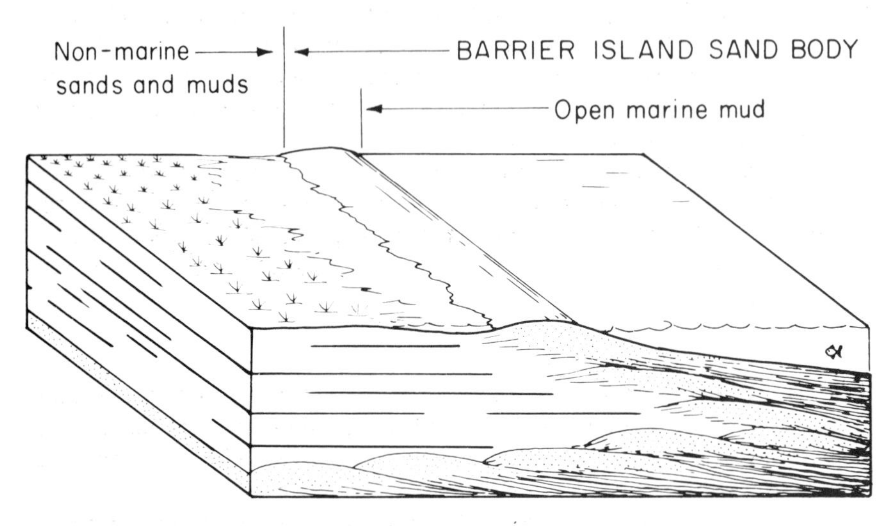

Fig. 146. Regressive/transgressive barrier sand bodies may be composed of a multitude of discrete prograding sand bar increments.

sand lenses to better predict their geometry. Weber's (1971) classic unravelling of the barrier bar and tidal channel sands on the Tertiary topset of the Niger delta illustrates this point.

7. A shelf sea model

A distinctive sedimentary model can be recognized for the deposits of continental shelves (Swift *et al.*, 1973). In the same way that modern coasts are in some ways unsuitable for studying the evolution of barrier deposits, so are modern shelves not ideal for study as a key to their ancient counterparts. Modern continental shelves include both environments of erosion, equilibrium and deposition. These have inherited features gained when the shelves were largely exposed due to low sea levels in the glacial maxima. The present-day sediments of the continental shelves were largely laid down in fluvial, periglacial and lagoonal environments. They have subsequently been reworked since the last postglacial rise in sea level. It has been estimated that some 70% of the continental shelves of the world are covered by these "relic" sediments (Emery, 1968).

Bearing this idea in mind, modern shelf deposits are of three kinds: those in which the Pleistocene sediments have been reworked so as to be in equilibrium with the present hydrodynamic environment; those which are truly "relic" and have been essentially unaltered by the present regime; and those which are now being buried beneath a mud blanket (Curray, 1965; Klein, 1977; Johnson, 1978).

Continental shelf deposits in equilibrium with the present hydrodynamic environment tend to be shallow and inshore while the unmodified relict deposits are largely preserved in the deeper waters near the continental margins. Burial of relict Pleistocene deposits beneath mud occurs near the mouths of major rivers in areas of low current velocities.

These facies types contrast with the concept of a graded shelf; that is a shelf whose sedimentary cover is in equilibrium with its hydrodynamic regime and, ideally, one in which sediment grade fines seaward (Swift, 1969). The Bering Sea has been cited as one example of such a modern graded shelf (Sharma, 1972).

(i) Modern shelf deposits

The erratic behaviour of sea level in the last million years makes it difficult to use modern continental shelves to study the evolution of deposits through time. Nevertheless the processes and structures of individual environments can be studied and applied to ancient sedimentary facies.

There are three main environments on modern shelves: an in-shore zone of sand bars and tidal flats, an open marine high-energy environment and an open marine low-energy environment. In the ideal situation of a gentle uniformly-dipping shelf, these three zones would trend parallel to the shore in linear belts. In practice, because of the complex tectonic and glacial histories of many modern shelves, these zones are erratically arranged.

The in-shore zone consists largely of tidal flat deposits which form an integral part of barrier coasts. On the basis of the studies previously cited, it is possible to demonstrate an overall upward-fining prograding sequence for in-shore tidal flat deposits. The thicknesses of the individual units of this sequence are controlled by the tidal range. It has been argued that the palaeotidal range of ancient tidal flat sequences may be measured from them (Klein, 1971). On terrigenous coasts, the tidal flat deposits grade up from quartzose sands, through clays, into marsh peats. Along the shorelines of arid coasts, salt marshes or "sabkhas" (as they are known in Arabic) develop. These show a broadly comparable upward-fining sequence, but the basal unit is generally a skeletal carbonate sand. This grades up into carbonate muds, often dolomitized, with nodular layers of gypsum or anhydrite (Fig. 147). Thin peats of algal stromatolitic origin may be present (Evans *et al.*, 1969). Ancient cyclic deposits of this sabkha type have been described by Wood and Wolfe (1969). Modern and ancient sabkhas have been reviewed by Till (1978) and Wilson (1975, p. 297).

The relatively sheltered in-shore environments of tidal flats are generally protected from the open sea by a high-energy zone. This may be a barrier island, as already described. Alternatively rock reefs and subaqueous sand shoals may provide protection. The high-energy shelf environment may be one of erosion, where a marine terrace is cut across bedrock; it may be an environment of equilibrium where sand shoals migrate to and fro by tidal scour; rarely is it an environment of deposition.

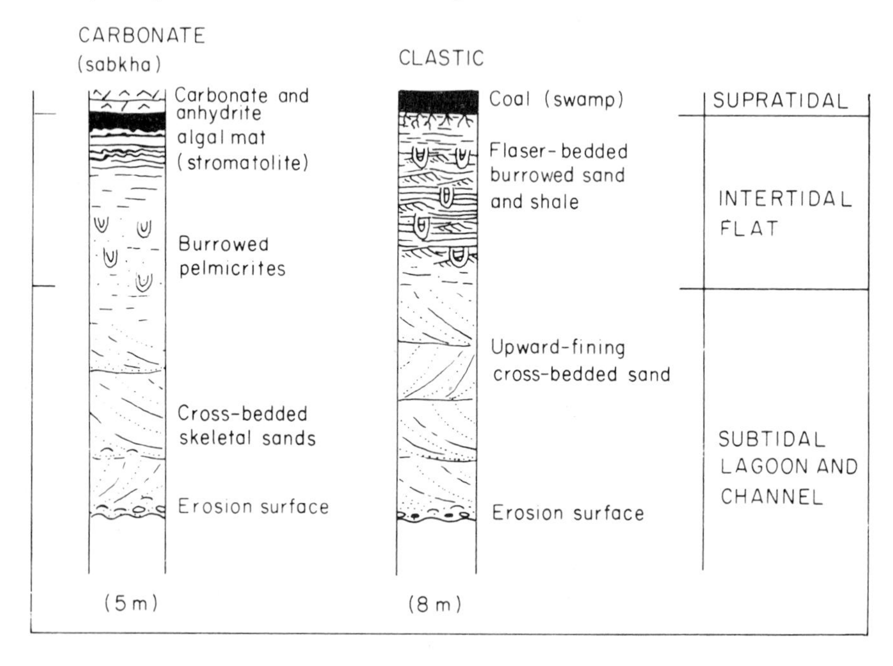

Fig. 147. Comparative profiles of tidal flat sequences in arid sabkhas (producing carbonates) and clastic sediments, based on the coasts of Abu Dhabi and the Dutch Wadden Sea. (From Evans, 1970.) The thickness of the sequences are related to tidal range.

Modern high-energy shelves have been intensively studied (Swift *et al.*, 1972; Stanley and Swift, 1976). The continental shelf of north-west Europe is one of the best known (e.g. Stride, 1963; Kenyon and Stride, 1970; Belderson *et al.*, 1971; Banner *et al.*, 1979). Much of this shelf is floored by Pleistocene glacial and fluvioglacial deposits. Three major zones can be recognized, one predominantly gravel floored, the second sand, and the third mud (Fig. 148).

The gravel floored parts of the shelf are those subjected to the strongest tidal currents. They are essentially areas of erosion from which sand and mud have been winnowed to leave a lag gravel deposit. These gravel sea beds are traversed by ephemeral sand ribbons up to 2·5 km long and 100 m wide, which are aligned parallel to the axis of tidal flow.

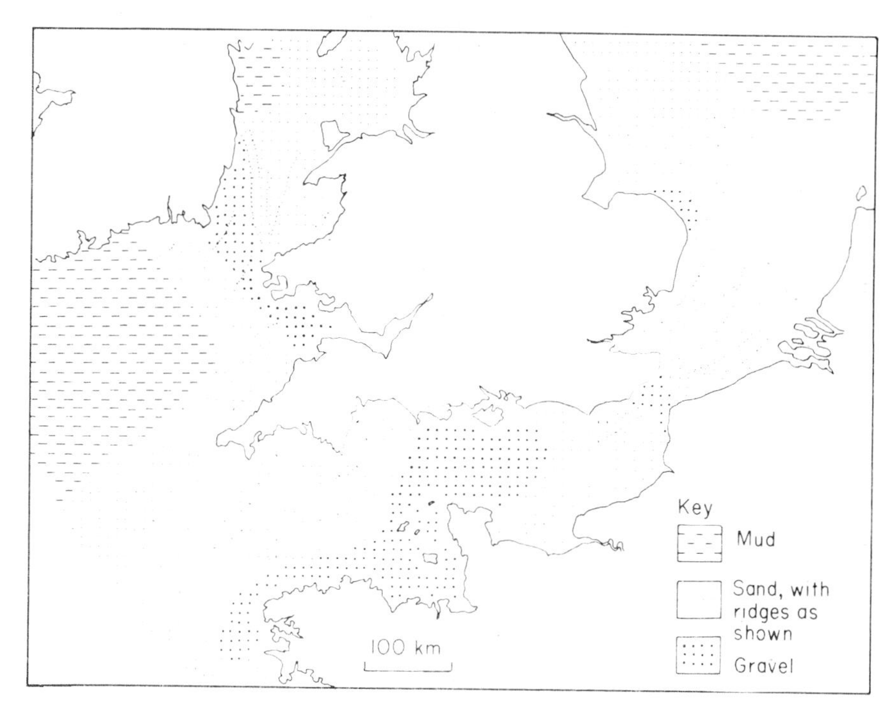

Fig. 148. Map showing the distribution of sediment types off the southern coast of England. Distribution is not correlative with water depth, being also controlled by tidal current flow and inherited glacial sedimentary pattern. (From Stride, 1963.)

The sand floored parts of the shelf are essentially environments of equilibrium. Much sediment is moved to and fro, but there is little net sand deposition. The dominant bed form of these areas is sand waves. These are large underwater dunes with heights of up to 20 m and wavelengths of up to 1 km. The surface of these sand bodies are modified by smaller dunes and ripples.

Acoustic and sparker surveys show low-angle bedding within these sand bodies. Coring reveals cross-bedded sets of shelly sand. Detailed studies of these sand bodies show a complex relationship between external morphology and internal structure (Houbolt, 1968). There is no correlation between dune height and water depth (Stride, 1970). The complex morphology and structure reflect the response of the sand waves to the ever-changing tidal flow regime.

Studies of analogous smaller scale in-shore sand bodies have been carried out where they are exposed at low tide. Bipolar cross-bedding has been observed dipping in the two opposing tidal current directions (Hulsemann, 1955; Reineck, 1963, 1971). Ancient analogues of these tidal sand bodies have been recognized by Narayan (1970), Reineck (1971), De Raaf and Boersma (1971) and Swett *et al.* (1971).

The third sediment type of the north-west Europe shelf are the mud patches. These occur in areas where the current velocities are sufficiently low for mud to settle out from suspension. As Fig. 148 shows, these are not only areas of deeper water.

The European shelf is comparable to other modern terrigenous shelves of the world which have been studied. Shelf-sea tidal sand bodies have also been described from the north-west Atlantic shelf and from the Malacca Straits (e.g. Jordan, 1962; Keller and Richards, 1967).

Modern carbonate shelves show many similar features to terrigenous shelves. In the large gulf between Arabia and Iran it is possible to define zones of mud, sand and gravel sediments analogous to those of the North Atlantic shelves (Fig. 149) (Purser, 1973). The in-shore tidal flat deposits are of the sabkha type previously described.

The morphology, genesis and structure of oolitic and skeletal sands have been described from the Bahamas platform (Newell and Rigby, 1957; Purdy, 1961; 1963; Imbrie and Buchanan, 1965; Ball, 1967).

This review of modern shelf deposits shows that three main realms can be recognized: depositional areas of mud sedimentation, equilibrial or

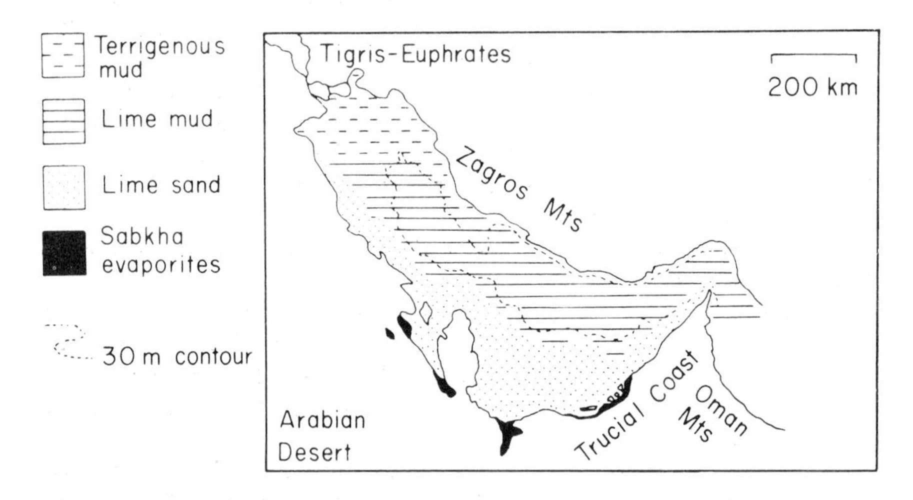

Fig. 149. Map showing the present-day sediment distribution of the gulf between Arabia and Iran (from Emery, 1956). Compare with Fig. 150.

slow-depositional high-energy sand shoals, and equilibrial or erosive areas of gravel and bedrock. These zones can be distinguished on both terrigenous and carbonate shelves.

Rapid regressions and transgressions across modern shelves in the last million years make it hard to define a spatial model for their facies. This can be made, however, from the study of older shelves which experienced less eccentric sea level changes.

(ii) Ancient shelt deposits

A general theory of shelf sea sedimentation has been put forward, based largely on the study of ancient rocks (Shaw, 1964; Irwin, 1965; Heckel, 1972). The thesis on which this is based states that in quiescent tectonic epochs of the past there were broad stable subhorizontal shelves with gradients of less than one in a thousand. These gently sloping surfaces are intersected by two horizontal surfaces of great significance: sea level and effective wave base.

The intersections of these surfaces with the sea bed define three sedimentary environments. In the deepest part of the shelf, below effective wave base, fine-grained mud settles out of suspension. Resultant sedimentary facies are laminated shales and calcilutites, sometimes with chert bands, and a biota of sparse well-preserved macrofossils and pelagic foraminifera.

Up-slope of the point at which effective wave base impinges on the sea bed, is a high-energy environment. Because of the gentle gradient of the shelf, this belt may be tens of kilometres wide. This is a zone of shoals and bars. The resultant sedimentary facies include biogenic reefs cross-bedded oolites, skeletal and mature quartz sands. To the lee of this high-energy belt is a sheltered zone which may stretch for hundreds of kilometres to the shoreline. This low-energy environment generates pelmicrites, micrites, dolomicrites and evaporites in the lagoons, tidal flats and sabkhas of arid carbonate realms. Clays, sands and peats form in the analogous environments of terrigenous realms (Fig. 150).

Regressions and transgressions allow the three facies belts to migrate to and fro over each other in a cyclic manner.

The epeiric shelf sea model was first defined from the carbonate Palaeozoic deposits of the Williston basin, North America. Its simplicity has an appeal which has lead the model to be applied to terrigenous shelf deposits such as those of the Lower Palaeozoic of the Arabian Shield (Selley, 1970, p. 145).

The facies and processes of epeiric seas can be studied on modern continental shelves, but, because of the eccentric behaviour of recent sea levels, the mutual relationships of shelf facies are better studied in old

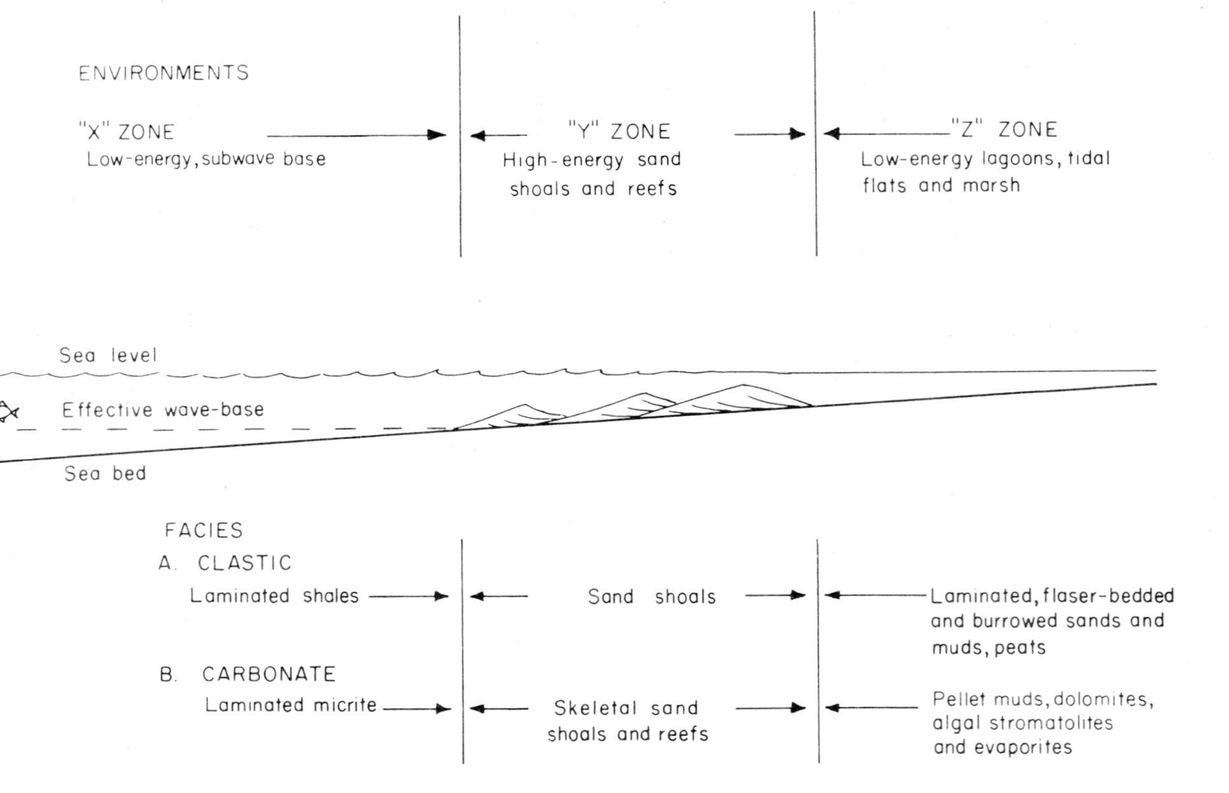

Fig. 150. Epeiric sea shelf sedimentary model. (Based on Irwin, 1965.)

rocks. A more sophisticated analysis of carbonate shelf facies models has been given by Wilson (1975).

8. Reefs

A "reef" has been defined as "a chain of rocks at, or near the surface of water: a shoal or bank" (Chambers Dictionary, 1972). This lay definition places no stress on the organic origin, or on the wave-resistant potential of a reef.

To geologists, however, reefs have, in a vague sort of way, always meant wave-resistant organic structures, built largely of corals.

Reefs, their genesis and diagenesis, are important because reef rocks are often mineralized or host hydrocarbon reservoirs. These aspects are discussed elsewhere (pp. 383 and 394). The following discussion is concerned mainly with the morphology and nomenclature of reefal carbonates.

(i) Present day reefs

Modern organic reefs have been intensely studied both by biologists, and by geologists interested in carbonate sedimentation (e.g. Jones and Endean, 1973).

The morphology, structure and ecology of modern reefs are well known. It is accepted that organic reefs can form in a wide range of water depths, temperatures and salinities. Reefs of calcareous algae and molluscs can grow in lakes. Marine reefs can be made of almost any of the sedentary lime-secreting invertebrates. Corals can grow in deep, cold water (Teichert, 1958; Maksimova, 1972). Nevertheless, the majority of modern reefs occur in warm, clear, shallow sea water and they contain coralline frame-building organisms in significant amounts.

Modern coral reefs are divisible into a number of distinct physiographic units. Each unit is characterized by distinctive sediment types and biota. There are considerable variations in the species and genera of reefs, yet a well-defined ecological zonation characterizes the subenvironments of any specific example (Fig. 151). As Fig. 152 shows there are three main morphological elements to a reef: the fore-reef, the reef flat and the back-reef. Additional terms are recognized for lower order features.

The fore-reef, or reef talus, grades seawards into deeper water toward the open sea. This zone slopes away from the reef front with a decreasing gradient. It is composed largely of detritus broken off the reef and is, therefore, a kind of subaqueous scree. Boulders of reef rock may be present at the foot of the reef front, but grain size decreases from calcirudite, through calcarenite to calcilutite down-slope. The talus is itself colonized by corals, calcareous algae and other invertebrates which tend to bind the scree

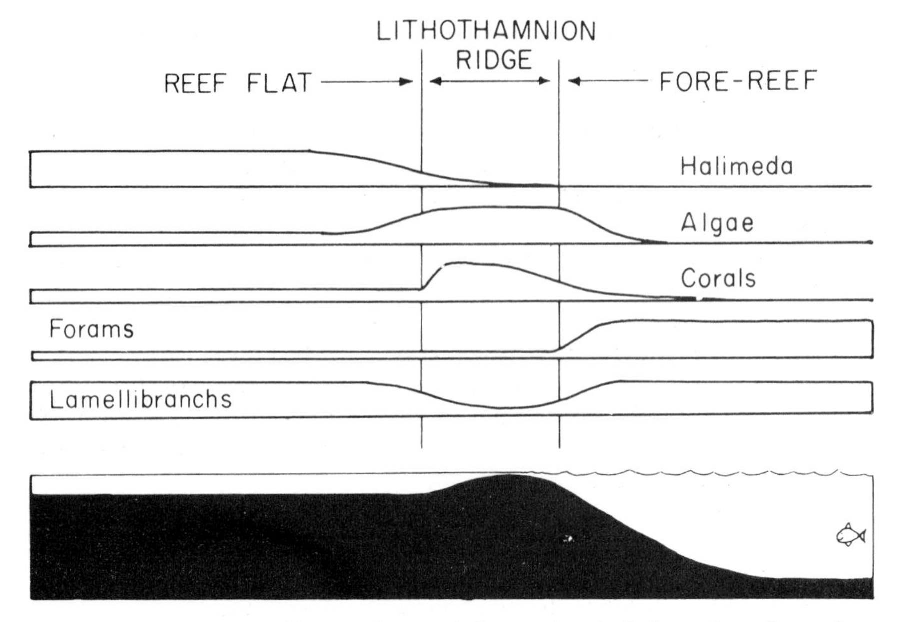

Fig. 151. Cross-section to illustrate the morphology and ecological zonation of a modern Florida reef. (From Ginsburg, 1956.)

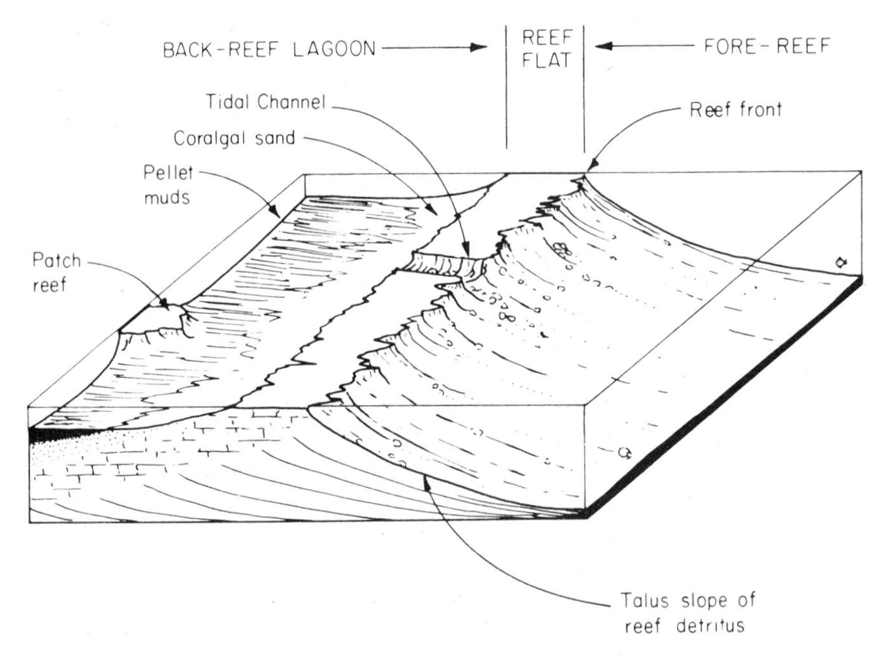

Fig. 152. Physiography and facies of a modern organic reef.

together. Progradation of the scree means that seaward-dipping depositional bedding may be preserved.

The reef itself is composed largely of sessile colonial calcareous frame-builders. The reef front of modern reefs is characteristically encrusted by highly wave-resistant *Lithothamnion* algae. Other organisms grow behind this in ecological zones. The reef itself is flat, because its biota cannot withstand prolonged subaerial exposure; and is generally emergent at low tide. Modern reefs have a high primary porosity present between the skeletal framework. This gradually diminishes through time due to infiltration by lime mud and to cementation (see p. 134).

Behind the reef flat is a third zone, the back-reef. Immediately behind the growing reef there is an area of smashed-up reef debris washed over by storms. The grain size of the back-reef deposits grades away from the reef into micrite and faecal pellet muds of a lagoon. Back-reef lagoons sometimes contain small patch reefs.

The threefold zonation of fore-reef, reef flat and back-reef is locally interrupted by tidal channels which traverse the reef flat.

It is important to remember that reefs do not grow in isolation, but form an integral part of shelf and shelf margin deposits (Fig. 153).

Considered in plan, three main types of modern reefs have long formed part of the geologists' folk-lore: atolls, barrier reefs and fringing reefs. Present-day reefs show complex cross-sectional geometries (where they can be studied), reflecting Pleistocene sea level changes. Likewise, their diagenetic histories have been complex. It is evident, more from the study of ancient reefs than modern ones, that reefs may prograde seawards over their own fore-reef talus, that they may grow vertically, or that they may migrate landward in response to a marine transgression (Fig. 154).

(ii) Ancient "reefs"

Lenses of carbonate contained in other sediments are found in many parts of the world in Phanerozoic rocks. Some of these have been identified as ancient reefs by careful study of their fossils, lithology and facies relationships. Others have been labelled as reefal without the benefit of detailed study. A reef origin is impossible to prove for many ancient carbonate lenses if a rigid definition is used. For example "reefs are bodies of rock composed of the skeletons of organisms which had the ecologic potential to build wave-resistant structures" (Lowenstam, 1950). To avoid such precise and often untenable interpretations, geologists have approached the problem under a blizzard of names (build-up, biostrome, bioherm, mound, organic reef, stratigraphic reef, ecologic reef and so on).

There are two fundamental reasons for the problems of defining a viable ancient-reef model. First, this is because of the confusion of facies and environment; second because many of these rock masses are so extensively recrystallized that their original nature is undetectable.

In no other group of ancient sediments are environments and facies nomenclature so confused (Braithwaite, 1973). Terms like "reef limestone", "intrareef detritus" and "fore-reef facies" are widely used. Many

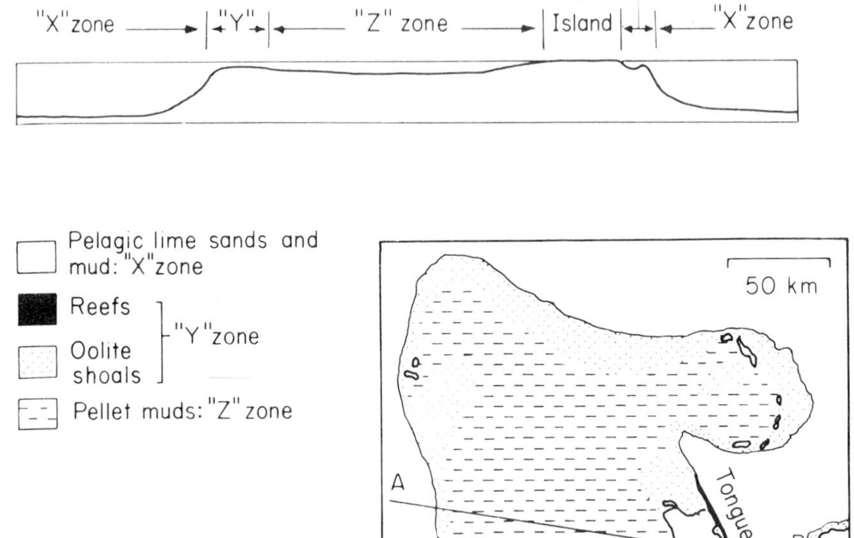

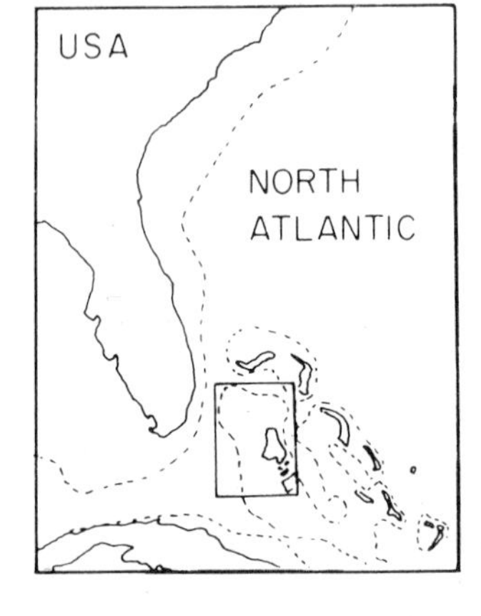

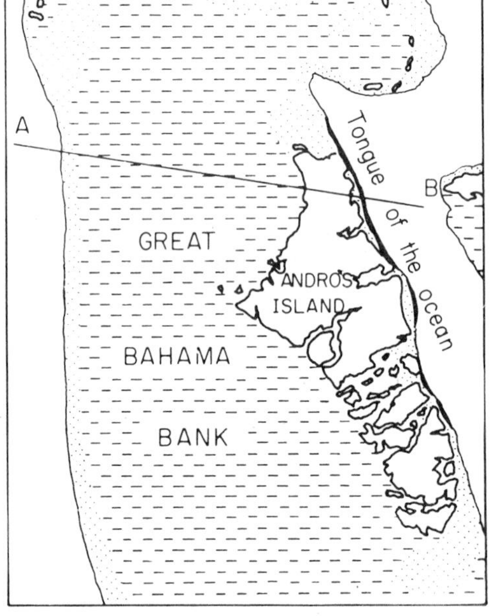

Fig. 153. Cross-section and map of part of the Great Bahama Bank, showing relationship between physiography and facies. (After Purdy, 1963.) Coralgal reefs occur along the east side of Andros Island, forming an integral part of an X, Y, Z zone facies suite.

problems could be avoided by describing ancient carbonates in lithofacies terms only. In particular the term "carbonate build-up" though ugly, has much to commend it for describing a lenticular organic carbonate rock unit of uncertain origin (Stanton, 1967).

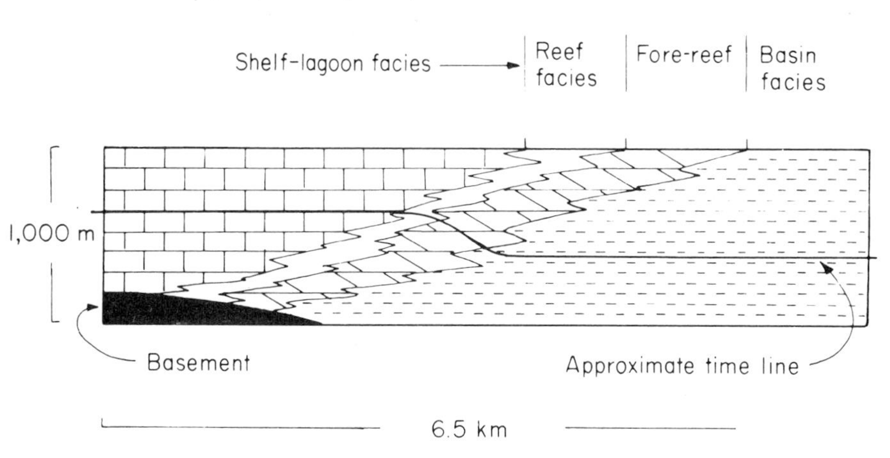

Fig. 154. Diagrammatic cross-section of the Oscar Range reef complex (Devonian) of Western Australia. This shows the basinward progradation of the reef facies produced by a seaward regression of environments. (From Playford, 1969.)

The second factor which has complicated the study of ancient reefs and banks is their extensive diagenesis. This is due to a variety of factors, including their high primary porosity, their unstable (largely aragonitic) composition and their frequent invasion by hypersaline fluids when sea level drops and lagoons evaporate (p. 156). Because of these factors, reefs and banks are frequently recrystallized and dolomitized, so that their primary fabric is undetectable. Many of the Devonian "reefs" of Canada are of this type (p. 160). Because of this diagenesis, it may be impossible to prove whether a lenticular carbonate build-up was actually once a reef in the ecologic sense.

There is a second important point related to the diagenesis of carbonate build-ups. A drop in sea level exposes the crests not only of orthodox reefs, but also of shell banks and carbonate sand shoals. These may all be cemented by early diagenesis. When sea level rises again, all of these features—regardless of their genesis—now form wave-resistant masses complete with talus slopes and lagoons. Repetition of this process can build up wide shelves with margins formed of early cemented carbonates of diverse facies and with aprons of slope deposits which grade down into a basin. The Capitan Formation of West Texas was once interpreted as a classic example of a barrier reef, separating the Delaware basin from the north-western shelf (Newell *et al.*, 1953). The tendency now is to interpret this as a feature which had a wave-resistant scarp, but that this was composed of organic shoals and banks which gained their rigidity, not from the syndepositional activity of encrusting algae, but from cementation during marine regressions (Achauer, 1969; Dunham, 1969; Kendall, 1969).

It has been suggested that it may be helpful to recognize and distinguish between stratigraphic reefs and ecologic reefs (Dunham, 1970). A stratigraphic reef is a carbonate build-up which is inorganically (e.g. sparite) cemented. An ecologic reef is one which is actually composed of *in situ* frame-building skeletons bound by organic means, such as encrusting algae. In the words of Dunham (ibid. p. 1931):

> The stratigraphic reef is an objective concept concerning three dimensional geometric masses. The ecologic reef is a subjective concept concerning inferred interactions of organisms and topographic evolution. One involves geometry, the other topography.

In conclusion an organic reef sedimentary model is definable based on the study of modern tropical coral reefs. Application of this model to ancient sediments is difficult (Elloy, 1972). Lenticular carbonate masses are common, but it is hard to prove that these were always wave-resistant topographic features composed of *in situ* organic skeletons. This is especially hard because many build-ups have lost their primary depositional features by obliterative diagenesis. There may be sufficient data, however, to distinguish whether the build-up originated as a soft sediment bank

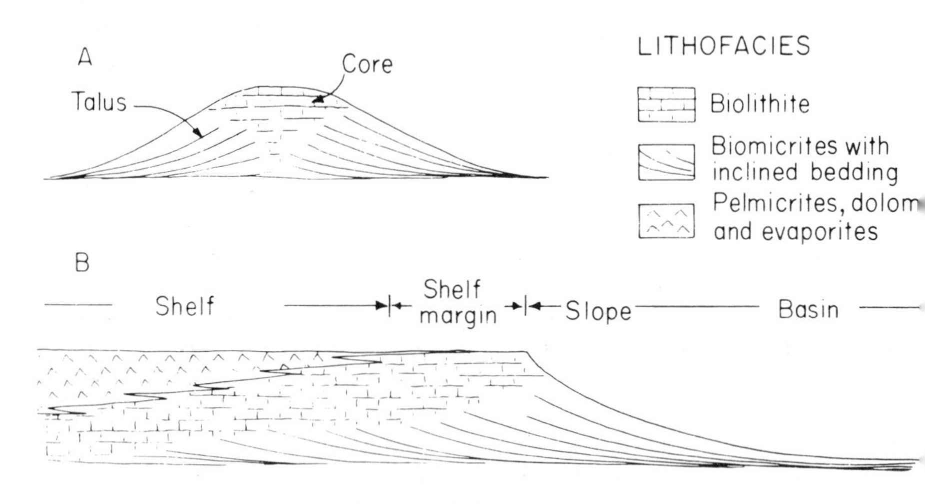

Fig. 155. Non-emotive terminology for (A) carbonate build-ups, and (B) shelf/basin transition facies models. This uses lithologic and physiographic terms; avoiding genetic terms with reefal connotations.

of shells or ooliths, or whether it was an ecologic or a stratigraphic reef.

Though it may be impossible to be specific about the origin of a build-up, or of a shelf margin, yet models for these may be defined based on their lithology and gross morphology (Fig. 155).

9. Turbidites

The processes and deposits of turbidity flows were described in Chapter 6, where it was emphasized that these occur in a wide range of environments.

Nevertheless, there is a particular type of sedimentary environment dominated by turbidite sedimentation. This can be observed on the earth's surface today and its products can be seen in the geologic record. This will be referred to as the turbidite sedimentary model, with the proviso that the turbidite process also occurs elsewhere. This model generates a characteristic sedimentary facies widely termed flysch (defined for example by Dzulinski and Walton, 1965, pp. 1–12).

The essential feature of this model is a basin margin dissected by submarine channel systems. Sediment transported down these by sliding, slumping and grain flow emerges from the channel mouths as true turbidity flows. These build out submarine fans of turbidites which pass distally into the pelagic muds of the basin floor (Shepard, 1971; Gorsline, 1970) (Fig. 156).

Present-day examples of this sedimentary model occur in two settings. They are found on continental margins where the slopes are cut in bedrock, and they are found at the edges of many prograding deltas. The latter arrangement has already been described as an integral part of the high-slope delta model and modern and ancient examples have been cited on p. 292.

Some of the best known modern continental margin turbidite systems occur off the Californian coast (Fig. 157). The submarine valley and fan systems of Monterey and La Jolla are particularly well documented (Gorsline and Emery, 1959; Hand and Emery, 1964; Shepard *et al.*, 1969; Wolf, 1970; Horn *et al.*, 1971).

It appears that the seaward progradation of this type of submarine slope will produce the following sequence: at the top is a facies of sands, tending to be well sorted, often glauconite and with a fraction of skeletal sand. These sediments are cross-bedded, cross-laminated and often burrowed. This facies is deposited by traction currents on the shelf. It passes down, generally abruptly, into the second facies composed of varying amounts of two distinct subfacies.

One consists of laminated clays and silts which were deposited out of suspension in the quieter deeper water of the slope zone. These slope deposits tend to slump and slide because of their high water saturation and unstable situation. Klein *et al.* (1972) have described a typical example of this kind of facies from the front of a Cretaceous delta in the Reconcava basin of Brazil. Many others are known.

The slope shales are cut into by varying amounts of the second slope subfacies. These are the submarine canyon or valley-fill deposits. Reference has already been made to the studies of the La Jolla, Monterey and associated channels of the Californian continental margin. Stanley and Unrug (1972) have synthesized many data on modern submarine channel deposits and used them to identify ancient analogues in Tertiary flysch basins of the Alps and of the Carpathians. Submarine channel deposits consist largely of sands with varying amounts of conglomerate and traces of shale. The sands occur in thick units with erosional bases. They are seldom conspicuously graded, often structureless or slump-bedded, and contain scattered clasts and slump blocks. These sands are attributed to deposition by grain flow; slumping and sliding down the valleys approaching the state of a true water-saturated turbid flow. They are sometimes referred to as fluxoturbidites (see p. 201).

This facies passes downwards into the true turbidite deposits of the

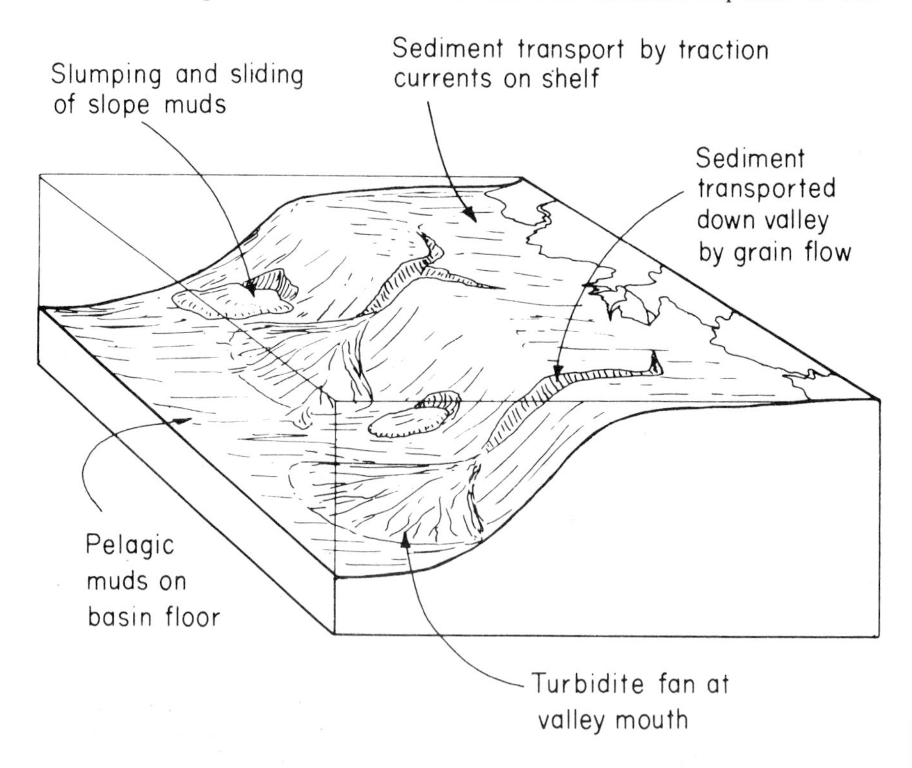

Fig. 156. Sedimentary model for turbidite/slope association.

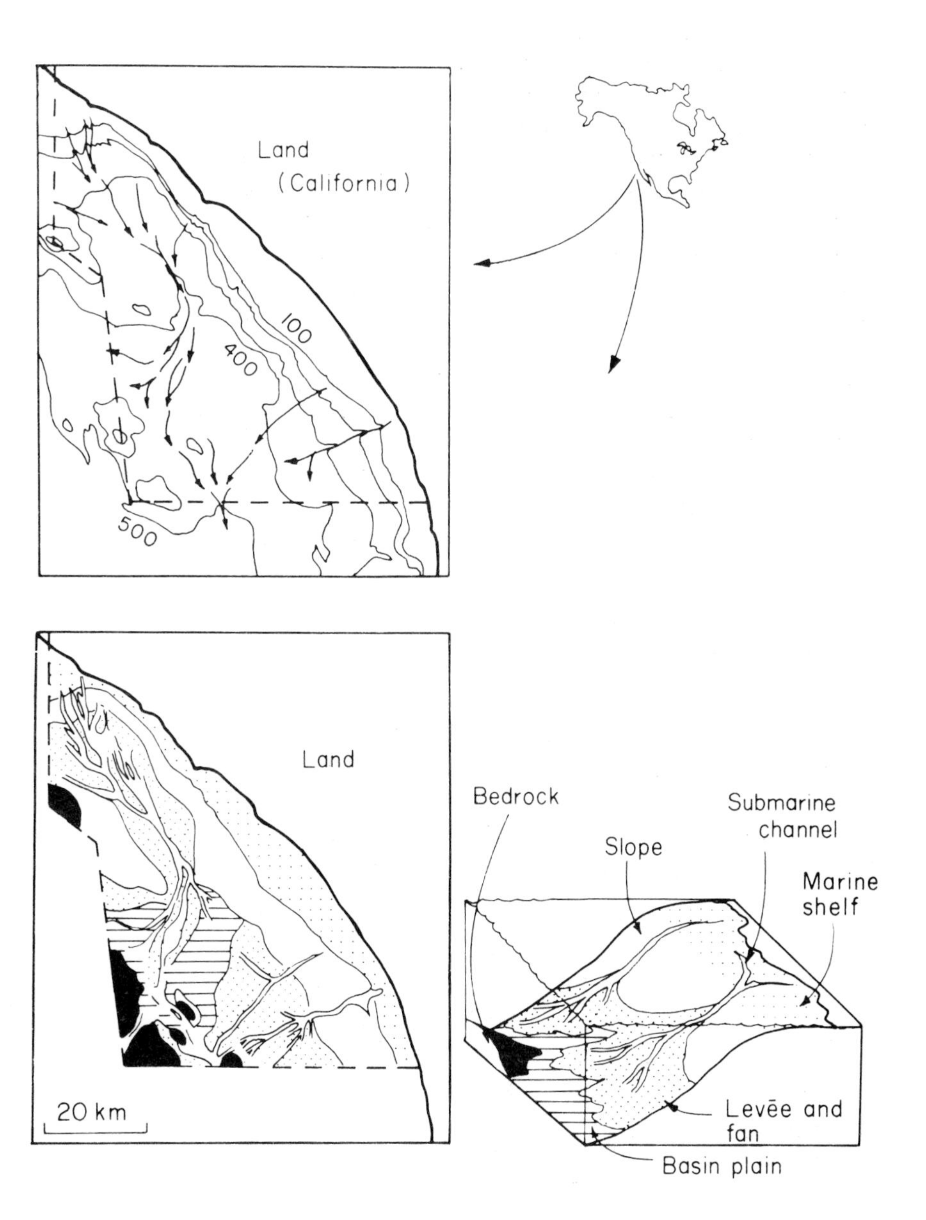

Fig. 157. Physiography of the continental margin off California, from Hand and Emery (1964). Upper: bathymetric chart showing main routes of sediment transport; contours in fathoms. Lower: physiographic units showing the mutual relationship between the slope channels and their associated submarine levées and basin floor fans.

submarine fans at the feet of the channels. This transition is shown by a decline in grain size and bed thickness and by an increase in graded bedding and intervening shale units; simultaneously channels become shallower and wider. Examples of this transition down from channel to fan deposits have been described from the Appennine flysch of Italy by Mutti and Lucchi (1972) and from various delta front deposits cited on p. 292. The sands and shales of the fan facies grade down into thinner-bedded distal turbidites with increasing amounts of fine-grained pelagic muds which settled out in the basin away from the prograding slope.

A characteristic feature of these sediments is that the shales may contain a deep-water (*Nereites*) trace fossils suite, radiolaria and pelagic forams; while the turbidite sands may contain an abraded shallow water fossil assemblage (Fig. 158).

Though there are many documented examples of at least one of the facies transitions of the turbidite slope sedimentary model, it is unusual to be able to find the whole suite present in any one vertical section. An example of this has been described, however, from Eocene sediments of the Santa Ynez mountains, California (Stauffer, 1967; Van de Kamp *et al.*, 1974). In this particular case the slope deposits are represented by extensive grain-flow

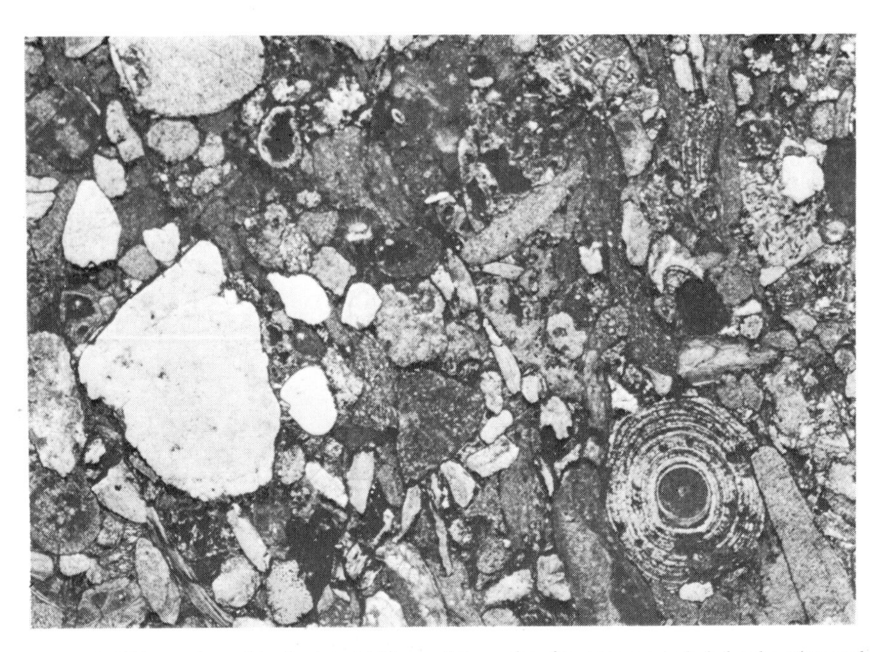

Fig. 158. Thin section of basinal turbidite sandstone showing transported skeletal grains and oolith. Conway Castle Grits (Lower Palaeozoic) North Wales (×20).

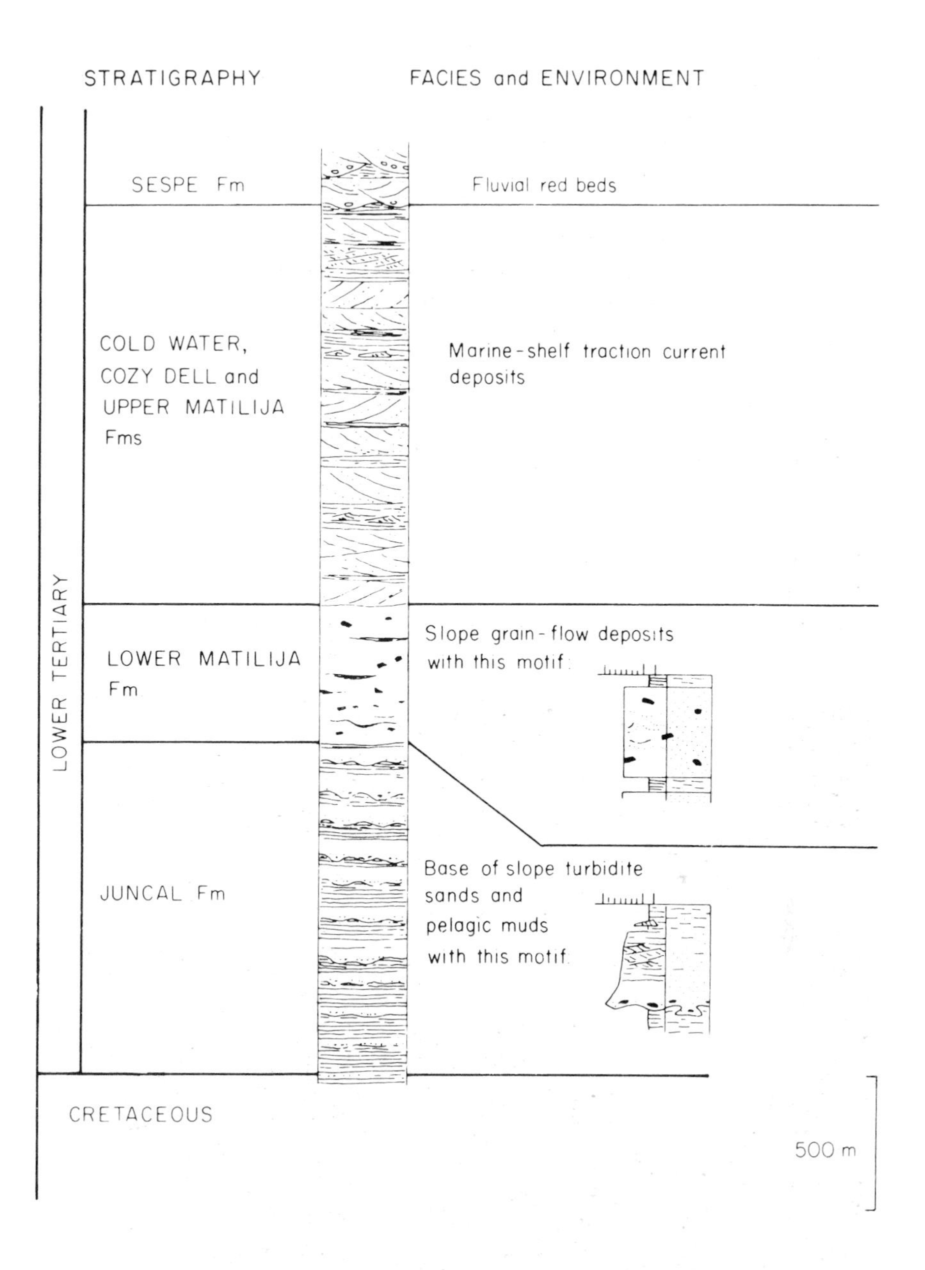

Fig. 159. Stratigraphic section of turbidite slope association in Tertiary sediments in the Santa Ynez mountains, California. (From Stauffer, 1967.) This sequence reflects the basinward progradation of pelagic muds, slope base turbidite sands, slope grain flow sands, marine shelf sands and alluvium.

sandstones. Slope muds and silts are rare and channels are not extensively observed (Fig. 159).

To conclude; a sedimentary model can be defined for the turbidites and associated deposits of slope margin. Examples of this pattern can be seen on modern continental margins and delta fronts. It can be recognized in ancient sediments in prograding sequences where pelagic deposits pass up through turbidites into fluxoturbidite channel and slumped shale facies of the slope environment. Where seaward progradation is sufficiently complete this sequence is overlain by traction current deposits of the overlying platform.

10. Pelagic

Pelagic sediments are the last major facies to consider.

It was pointed out that it is hard to detect the absolute depth of ancient sediments; for this reason a classification into neritic, bathyal and abyssal deposits was avoided (p. 258). The term pelagic is applied to those sediments which were deposited in the sea away from terrestrial influence (Jenkyns, 1978). This term has no depth connotation and pelagic deposits may be found on the distal extremes of broad continental shelves as well as on the abyssal plains. Terrigenous relict sediments of modern shelves and abyssal turbidite wedges are not deemed pelagic.

Modern oceanic sediments have been studied from the Challenger cruise (1872–6) to the more recent Deep Sea Drilling Project (DSDP) and its progeny the International Project for Ocean Drilling (IPOD) (Arrhenius, 1963; Riedel, 1963; Mero, 1965; Kukal, 1971; Hsu and Jenkyns, 1974).

The main sediment types of modern ocean floors are the oozes and the clays. Oozes are defined as pelagic sediments with over 30% organic skeletal detritus. There are three major types of ooze. Pteropod ooze is composed largely of the tests of the microscopic gastropod of that name. Globigerina ooze is made up largely of the foram of that name, together with other species of foram and coccolithic nannoplankton. Radiolarian ooze is dominantly composed of the siliceous tests of radiolaria.

Inorganic detritus also occurs in pelagic sediments. This includes distal turbidites from continental slopes, silt and clay transported out to sea by desert sandstorms, cosmic dust and volcanic ash. When the inorganic component of pelagic sediment is dominant the deposit is referred to as Red Clay.

Not all ocean floors are environments of deposition. In some areas currents scour the sea bed maintaining exposed lithified surfaces (“hard-grounds”, see p. 140) and areas encrusted with manganese nodules.

The distribution of the various modern deep sea sediments is controlled by several factors. Distance from land controls the terrigenous fraction and proximity to volcanoes controls the volcanogenic source. Both of these factors are also related to prevailing wind direction. The presence or absence of ocean floor currents affects sedimentation rates, the distribution of hardgrounds and areas of manganese encrustation. The organic oozes are closely related to depth. The CO_2 concentration of modern oceans increases with depth, and this therefore lowers the pH. The solubility of calcium carbonate thus increases with depth. The aragonitic pteropod oozes are seldom found below 3000 m, while the calcitic Globigerina oozes die out at about 5000 m, below which depth radiolarian ooze or Red Clay prevail. Attempts to apply the concept of the carbonate compensation depth (referred to often just as CCD) to ancient rocks are fraught with difficulty because the CCD is not depth dependant so much as temperature related. Thus assumptions must be made of ancient oceanic temperatures before trying to use the CCD concept to diagnose the depth of ancient pelagic sediments (Van Andel, 1975).

A depositional model for pelagic sediments has been defined by combining observations of modern ocean floors with the romance of plate tectonics (see p. 350). The mid-ocean ridges are composed largely of volcanic bedrock. This is often starved of pelagic sediment, and intensely altered with iron and manganese encrustations. Moving away from the mid-ocean ridge this volcanic layer is progressively blanketed by carbonate ooze. Traced further away the sea floor slopes below the CCD so the carbonate ooze is replaced by siliceous ooze or red clay (Davies and Forsline, 1976). The clays may become gradually replaced by terrigenous sediment as the continental rise is approached or may be digested within a zone of subduction (Fig. 161).

There is a distinct sedimentary facies which may be attributed to a pelagic environment with some confidence. It consists of interbedded laminated radiolarian cherts, micrites and red manganiferous shales. It overlies pillow lavas of the old ocean floor and in turn is overlain by flysch facies. Characteristically this facies is generally tectonically deformed. Ancient pelagic deposits have been described by Aubouin (1965), Garrison and Fischer (1969) and Wilson (1969), to mention but a few.

Specific examples of this facies assemblage include the late Mesozoic "ammonitico rosso" of the Alps, the red cephalopod limestones of the Variscan geosyncline in northern Europe and the Franciscan cherts of California (Fig. 160). This type of facies assemblage is directly comparable to modern carbonate and siliceous oozes and to red manganiferous clays. Though the ancient facies can be interpreted as pelagic their actual depositional depth is a matter for debate.

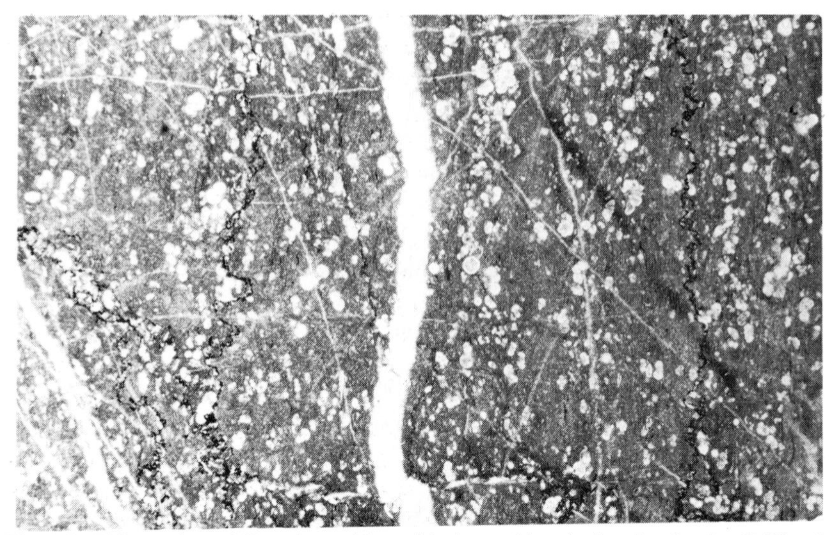

Fig. 160. Thin section of Franciscan (Jurassic) chert with pelagic microfossils, California. This is believed to be a deep water deposit which underwent later intense structural deformation as shown be extensive fracturing. Extensive stylolites with black insoluble residue can also be seen (× 30).

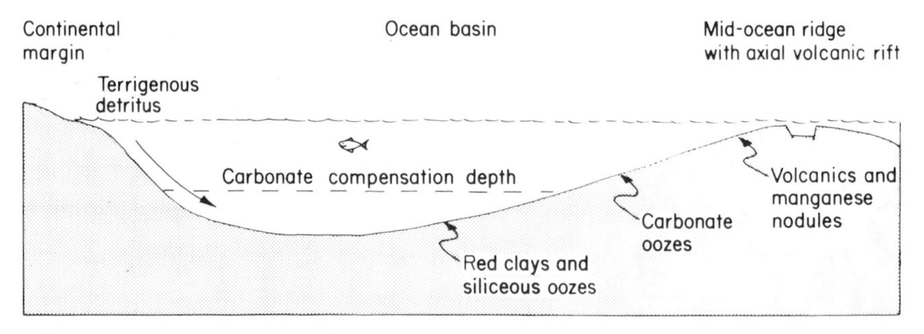

Fig. 161. Generalized cross-section, through an ocean basin showing distribution of pelagic sediments.

IV. SEDIMENTARY MODELS, INCREMENTS AND CYCLES

The preceding pages have described the sedimentary models which can be recognized as major facies generators. At this point it is germane to review the principles behind this approach before examining these sedimentary models in the broader context of tectonic setting.

The fundamental principle behind this chapter is that there are a finite number of sedimentary environments that generate a finite number of distinctive facies. This concept is a good servant, but a bad master. It is useful as a means of communicating ideas, so long as one geologist's submarine valley/fan model is similar to his colleagues. It is dangerous if one man's reef is another's bioherm.

Sedimentary models are useful for demonstrating the relationship between physiography, process and resultant facies. As a student gains experience of real rocks, he must accommodate the model to fit the rocks, not vice versa. He must not be afraid to discard the model altogether if it proves untenable in the face of hard data.

One very important principle can be learnt from the sedimentary models described in this chapter. Example after example showed a pattern of environments which prograded sideways to deposit a series of facies arranged in a predictable vertical sequence. This concept is the key to modern dynamic stratigraphic analysis. It will now be discussed in some detail.

A. Walther's Law

Walther (1893, 1894) continued the development of the facies concept begun by Prevost and Gressley (Middleton, 1973). He recognized that facies are seldom randomly arranged. Environmental analysis shows that vertical sections of strata originated in sequences of environments that are seen side by side on the earth's surface today. Walther coined the term "*facies-bezirk*" (facies tract) for a conformable vertical sequence of genetically related facies. Walther distinguished environments of erosion, equilibrium and deposition and included intraformational erosional intervals as an integral part of a facies tract. Walther realized the significance of this arrangement, writing (1894, p. 979):

> The various deposits of the same facies area and similarly, the sum of the rocks of different facies areas, were formed beside each other in space, but in a crustal profile we see them lying on top of each other . . . it is a basic statement of far-reaching significance that only those facies and facies areas can be superimposed, primarily, that can be observed beside each other at the present time
>
> (translation from Blatt *et al.*, 1972, p. 187).

This principle is termed Walther's law, and may be succinctly stated as: "a conformable vertical sequence of facies was generated by a lateral sequence of environments".

This principle is a vital one in environmental analysis. One of the ways of identifying the environment of a facies is by analysing the environments of the facies above and below to come up with the most logical sequence of events.

B. Genetic Increments of Sedimentation

Walther's concept of "*faciesbezirk*" has been widely used in facies analysis. This idea can be usefully extended to consider not just a vertical sequence, but a whole body of rock; what has been termed a genetic increment of strata (Busch, 1971).

A genetic increment of strata is a mass of sedimentary rock in which the facies or subfacies are genetically related to one another. A typical genetic increment of strata would consist of a single prograding delta sequence containing delta platform, delta front and pro-delta deposits (Fig. 162).

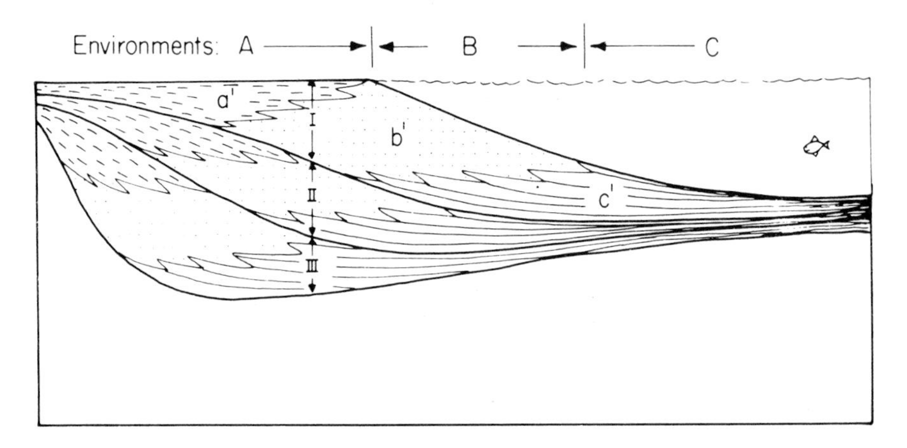

Fig. 162. Cross-section to illustrate the relationship between environments, facies and increments of sedimentation (from Busch, 1971). Environments A, B and C deposited facies a′, b′ and c′ respectively in the increment of sedimentation designated I. The increments I, II and III together constitute a genetic sequence of strata.

Genetic sequences of strata include more than one increment of the same genetic type.

These terms were developed to aid the subsurface mapping of deltaic deposits. The isopach mapping of genetic increments of strata define individual delta lobes. Isopach maps of genetic sequences of strata define larger morphologic units of shelf, hinge line and basin.

It is apparent though, from the data in this chapter, that increments of sedimentation are generated by many sedimentary models. Therefore, incremental mapping is actually feasible with many sedimentary facies. The concept of incremental mapping, though seldom the formal terminology, is widely applied in facies analysis. It forms an integral part of environmental and palaeogeographic analyses to locate favourable hydrocarbon reservoir facies.

C. Sequences and Cycles

Cyclicity, the repetition of genetic increments of strata, is a common feature in many stratigraphic sections. This is a topic which has been widely studied and written about (e.g. Merriam, 1965; Duff *et al.*, 1967).

There has been a tendency to distinguish between symmetric motifs, or cycles, which go ABCBA, and asymmetric motifs, or rhythms, which go ABCABC. Figure 163 shows how these two patterns may be laterally related, with no genetic distinction. The term "cycle" is now generally used for both asymmetric and symmetric motifs.

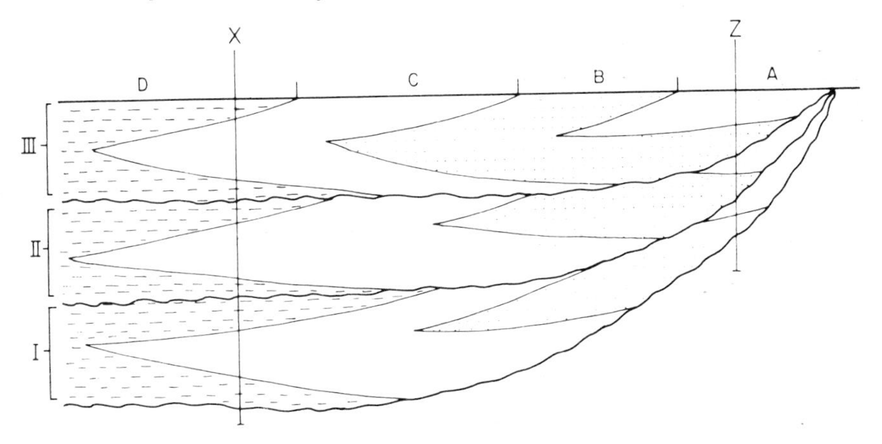

Fig. 163. Illustrating lateral variation of cycle type within genetic increments of strata. Borehole X penetrates increments with symmetric cycle motifs: DCD, DCD, DCD. Borehole Z penetrates the same three increments with asymmetric rhythmic motifs: AB, AB, AB.

It has been shown already, and discussed at length, how many sedimentary models produce predictable sequences of facies. It is important, however, to distinguish sequences from cycles. A genetic increment of strata may contain a sequence (ABC) but a genetic sequence of strata may be cyclic (ABC, ABC, etc.).

The formation of sequences is an obvious and integral part of sedimentation. The origin of cyclicity, the phenomenon of repeated sequences, has aroused lively debate. Beerbower (1964, p. 32) has made an important distinction between *autocyclic mechanisms*, which are an integral part of the sedimentary model, and *allocyclic mechanisms*, which originate outside the model.

The to and fro migration of a river channel and the switching of a delta distributary can generate cyclic sedimentation in a steadily subsiding sedimentary basin. These are examples of autocyclicity.

Not all sedimentary models have this ability, however, and external mechanisms must be invoked to explain cyclicity where it is present.

Mechanisms capable of causing allocyclic sedimentation can be seen operating in the recent past. The world-wide distribution of raised beach deposits, and relict terrestrial sediments on continental shelves, provide evidence of sea level changes correlateable with the waxing and waning of the ice caps. Individual raised beaches are not always of constant elevation when traced along the coast. These testify to tectonic tilting of the earth's crust, sometimes in isostatic response to the vanishing load of ice caps. Cyclic climatic variations are known both in historic times and, from pollen analyses etc. from prehistoric time.

These facts show that transgressions and regressions capable of generating sedimentary cycles may occur due to climatic, eustatic and tectonic changes.

It is not always possible, however, to detect which of these mechanisms caused any specific instance of cyclicity. Furthermore, as at the present time, several cyclic mechanisms may operate simultaneously.

Many studies of cyclic sequences have distinguishes hierarchical systems of "megacyclothems", "cyclothems" and "microcyclothems". Not all of these are imaginary. There are a number of mathematical techniques for statistically testing a section for cyclicity, and for determining cycle wavelength (Krumbein, 1967; Merriam, 1967).

In deltaic deposits there may well be local autocyclic motifs due to delta switching, on which may be superimposed more extensive cycles due to allocyclic mechanisms.

In conclusion it is commonplace to find predictable sequences within genetic increments of strata. Cyclicity is the repetition of these sequences. Autocyclic sedimentation is an integral part of several sedimentary models. Allocyclic sedimentation occurs from time to time in almost all deposits irrespective of environment; it is caused by a variety of mechanisms.

More than one cyclic process may control the deposition of a sedimentary section.

V. REFERENCES

Achauer, C. W. (1969). Origin of Capitan Formation, Guadalupe Mountains, New Mexico and Texas. *Bull. Am. Ass. Petrol. Geol.* **53**, 2314–2323.

Allen, J. R. L. (1964). Studies in fluviatile sedimentation: six cyclothems from the Lower Old Red Sandstone, Anglo-Welsh Basin. *Sedimentology* **3**, 163–198.

Allen, J. R. L. (1965). A review of the origin and characteristics of Recent Alluvial sediments. *Sedimentology* **5**, No. 2. special issue. 191pp.
Allen, J. R. L. and Friend, P. F. (1968). Deposition of the Catskill facies, Appalachian region: with notes on some other Old Red Sandstone basins. *Spec. Pap. geol. Soc. Am.* **206**, 21–74.
Armstrong-Price, W. (1963). Patterns of flow and channelling in tidal inlets. *J. sedim. Petrol.* **33**, 279–290.
Arrhenius, G. (1963). Pelagic sediments. *In* "The Sea" (M. N. Hill, Ed.). Vol. III. 655–727. Interscience, New York.
Asquith, D. O. (1970). Depositional topography and major marine environments, late Cretaceous, Wyoming. *Bull. Am. Ass. Petrol. Geol.* **54**, 1184–1224.
Aubouin, J. (1965). "Geosynclines" Elsevier, Amsterdam. 335pp.
Ball, M. M. (1967). Carbonate sand bodies of Florida and the Bahamas. *J. sedim. Petrol.* **37**, 556–591.
Banner, F. T., Dollins, M. B. and Massie, K. S. (Eds) (1979). The North-West European Shelf Seas: the sea bed and the sea in motion I. "Geology and Sedimentology" Elsevier, Amsterdam. 300pp.
Beerbower, J. R. (1964). Cyclothems and cyclic depositional mechanisms in alluvial plain sedimentation. *Bull. Kans. Univ. geol. Surv.* **169**, 35–42.
Belderson, R. H., Kenyon, N. H. and Stride A. H. (1971). Holocene sediments on the continental shelf west of the British Isles. *In* "The Geology of the East Atlantic Continental Margin" Vol. 2, 70/14, 160–170. Rept. Eur. Inst. Geol. Sci.
Bennacef, A., Beuf, S., Biju-Duval, B., de Charpal, O., Gariel, O. and Rognon, P. (1971). Example of cratonic sedimentation: Lower Paleozoic of Algerian Sahara. *Bull. Am. Ass. Petrol. Geol.* **55**, 2225–2245.
Berg, R. R. and Davis, D. K. (1968). Origin of Lower Cretaceous Muddy sandstone at Bell Creek Field, Montana. *Bull. Am. Ass. Petrol. Geol.* **52**, 1888–1898.
Bernard, H. A., LeBlanc, R. J. and Major, C. F. (1962). Recent and Pleistocene geology of southwest Texas. *In* "Geology of Gulf Coast and Central Texas" 175–224. Houston Geol. Soc.
Bigarella, J. J. (1972). Eolian environments—their characteristics, recognition, and importance. Recognition of Ancient Sedimentary Environments (J. K. Rigby and W. K. Hamblin, Eds). *Spec. Publs Soc. econ. Palaeont. Miner., Tulsa*, **16**, 12–62.
Bigarella, J. J. (1979). Botucatu and Sambaiba Sandstones of South America (Jurassic and Cretaceous). *In* "A Study of Global Sand Seas" (E. D. McKee, Ed.), 233–236. U.S. Geol Surv. Prof. Pap. 1052.
Bigarella, J. J. and Salamuni, R. (1961). Early Mesozoic wind patterns as suggested by dune-bedding in the Botucatu Sandstone of Brazil and Uruguay. *Bull. geol. Soc. Am.* **72**, 1089–1106.
Blackwelder, E. (1928). Mudflow as a geologic agent in semi-arid mountains. *Bull. geol. Soc. Am.* **39**, 465–483.
Blatt, H., Middleton, G. and Murray, R. (1978). "Origin of Sedimentary Rocks" (2nd Edition). Prentice-Hall, New Jersey. 782pp.
Blissenbach, E. (1954). Geology of alluvial fan in Southern Nevada. *Bull. geol. Soc. Am.* **65**, 175–190.
Bluck, B. J. (1964). Sedimentation of an alluvial fan in Southern Nevada. *J. sedim. Petrol.* **34**, 395–400.
Bluck, B. J. (1967). Deposition of some Upper Old Red Sandstone conglomerates in the Clyde area: a study in the significance of bedding. *Scott. J. geol.* **3**(2), 139–167.

Boyd, D. R. and Dyer, B. F. (1966). Frio Barrier Bar system of S. Texas. *Bull. Am. Ass. Petrol. Geol.* **50**, 170–178.

Bradley, W. H. (1948). Limnology and the Eocene lakes of the Rocky Mountain region. *Bull. geol. Soc. Am.* **59**, 635–648.

Braithwaite, C. J. R. (1973). Reefs: just a problem of semantics? *Bull. Am. Ass. Petrol. Geol.* **57**, 1100–1116.

Broussard, M. L. (Ed.) (1975). "Deltas, Models for Exploration" Houston Geol. Soc. Houston. 555pp.

Burke, K. (1972). Longshore drift, submarine canyons, and submarine fans in development of Niger delta. *Bull. Am. Ass. Petrol. Geol.* **56**, 1975–1983.

Burke, R. A. (1958). Summary of oil occurrence in Anahuac and Frio Formations of Texas and Louisiana. *Bull. Am. Ass. Petrol. Geol.* **42**, 2935–2950.

Busch, D. A. (1971). Genetic units in delta prospecting. *Bull. Am. Ass. Petrol. Geol.* **55**, 1137–1154.

Chorley, R. J. (Ed.) (1969). "Introduction to Fluvial Processes" Methuen, London. 218pp.

Coleman, J. M. and Gagliano, S. M. (1965). Sedimentary structure: Mississippi River deltaic plain. Primary Sedimentary Structures and their Hydrodynamic Interpretation. *Spec. Publs Soc. econ. Palaeont. Miner., Tulsa* **12**, 133–148.

Coleman, J. M., Gagliano, S. M. and Smith, W. C. (1970). Sedimentation in a Malaysian high tide tropical delta. Deltaic Sedimentation Modern and Ancient (J. P. Morgan and R. H. Shaver, Eds) *Spec. Publs Soc. econ. Palaeont. Miner., Tulsa* **15**, 185–197.

Collinson, J. D. (1978). Alluvial sediments. *In* "Sedimentary Environments and Facies" (H. G. Reading, Ed.), 15–60. Blackwell Scientific, Oxford.

Crosby, E. J. (1972). Classification of sedimentary environments. Recognition of Ancient Sedimentary Environments. *Spec. Publs Soc. econ. Palaeont. Miner., Tulsa* **16**, 1–11.

Curray, J. R. (1965). Late Quaternary history, continental shelves of the United States. *In* "Quaternary of the United States" (H. E. Write and D. C. Frey, Eds). Princeton. 922pp.

Curtis, D. M. (Ed.) (1978). Environmental Models in Ancient Sediments. *Soc. Econ. Pal. Min.* Reprint Series **6**, 240pp.

Davies, D. K., Ethridge, F. G. and Berg, R. R. (1971). Recognition of barrier environments. *Bull. Am. Ass. Petrol. Geol.* **55**, 550–565.

Davies, J. L. (1973). "Geographical Variation in Coastal Development" Oliver and Boyd, Edinburgh. 435pp.

Davies, T. A. and Gorsline, D. S. (1976). Oceanic sediments and sedimentary processes. *In* "Chemical Oceanography" Vol. 5 (2nd Edition) (J. P. Riley and G. Skirrow, Eds), 1–80. Academic Press, London and New York.

De Graaff, F. R. van (1972). Fluvial-deltaic facies of the Castlegate sandstone (Cretaceous), East-Central Utah. *J. sedim. Petrol.* **42**, 558–571.

Denny, C. S. (1965). Alluvial fans in the Death Valley Region, California and Nevada. *Prof. Pap. U.S. geol. Surv.* **466**. 62pp.

De Raaf, J. F. M. and Boersma, J. R . (1971). Tidal deposits and their sedimentary structures. *Geologie Mijnb.* **50**, 479–504.

De Raaf, J. F. M., Reading, H. G. and Walker, R. G. (1964). Cyclic sedimentation in the Lower Westphalian of North Devon. *Sedimentology* **74**, 373–420.

Doeglas, D. J. (1962). The structure of sedimentary deposits of braided rivers. *Sedimentology* **1**, 167–190.

Dott, R. H. (1966). Eocene deltaic sedimentation at Coos Bay, Oregon. *J. Geol.* **74**, 373–420.
Duff, P. McL. D., Hallam, A. and Walton, E. K. (1967). "Cyclic Sedimentation" Elsevier, Amsterdam. 280pp.
Dunbar, C. O. and Rodgers, J. (1957). "Principles of Stratigraphy" John Wiley, New York. 356pp.
Dunham, R. J. (1969). Vadose pisolite in the Capitan Reef (Permian), New Mexico and Texas. Depositional Environments of Carbonate Rocks. *Spec. Publs Soc. econ. Palaeont. Miner., Tulsa* **14**, 182–191.
Dunham, R. J. (1970). Stratigraphic reefs versus ecologic reefs. *Bull. Am. Ass. Petrol. Geol.* **54**, 1931–1932.
Dzulinski, S. and Walton, E. K. (1965). "Sedimentary Features of Flysch and Greywackes" Elsevier, Amsterdam. 274pp.
Elloy, R. (1972). Réflexions sur quelques environments récifaux de Paleozoique. *Bull. Cent. Rech. Pau.* **6**, 1–106.
Emery, K. O. (1956). Sediments and water of Persian Gulf. *Bull. Am. Ass. Petrol. Geol.* **40**, 2354–2383.
Emery, K. O. (1968). Relict structures on continental shelves of world. *Bull. Am. Ass. Petrol. Geol.* **52**, 445–464.
Erben, H. K. (1964). Facies developments in the Marine Devonian of the old world. *Proc. Ussher Soc.* **1**, pt. 3, 92–118.
Eskola, P. (1915). Om sambandet mellan kemisk och mineralogisk sammansattning hos Orijarvitraktens metamorfa bergarter. *Bull. Commn géol. Finl.* **44**, 1–145.
Eugster, H. P. and Surdam, R. C. (1973). Depositional environment of the Green River Formation of Wyoming. A preliminary report. *Bull. geol. Soc. Am.* **84**, 1115–1120.
Evans, G. (1965). Intertidal flat sediments and their environments of deposition in the Wash. *Q. Jl geol. Soc. Lond.* **121**, 209–245.
Evans, G. (1970). Coastal and nearshore sedimentation: a comparison of clastic and carbonate deposition. *Proc. geol. Ass.* **81**, 493–508.
Evans, G., Schmidt, V., Bush, P. and Nelson, H. (1969). Stratigraphy and geologic history of the sabkha, Abu Dhabi, Persian Gulf. *Sedimentology* **12**, 145–159.
Ferm, J. C. (1970). Allegheny deltaic deposits. Deltaic Sedimentation Modern and Ancient (J. P. Morgan and R. H. Shaver, Eds). *Spec. Publs Soc. econ. Paleont. Miner., Tulsa* **15**, 246–255.
Fischer, A. G. and Garrison, R. E. (1967). Carbonate lithification on the sea floor. *J. Geol.* **75**, 488–496.
Fisher, W. L., Brown, L. F., Scott, A. J. and McGowen, J. H. (1969). Delta systems in the exploration for oil and gas. Bur. Econ. Geol. Univ. Tex. 78pp.
Fisk, H. N. (1955). Sand facies of Recent Mississippi delta deposits. Proc. 4th Wld Petrol. Cong. Rome. Section 1, 377–398.
Freeman, W. E. and Visher, G. S. (1975). Stratigraphic analysis of the Navajo Sandstone. *J. sediment. Petrol.* **45**, 651–8.
Friend, P. F. (1965). Fluviatile sedimentary structures in the Wood Bay Series (Devonian) of Spitsbergen. *Sedimentology* **5**, 39–68.
Friend, P. F. and Moody-Stuart, M. (1972). Sedimentation of the Wood Bay Formation (Devonian) of Spitsbergen: regional analysis of a late orogenic basin. *Norsk. Polarinstitutt. Skrift.* **157**, 77pp.
Garrison, R. E. and Fischer, A. G. (1969). Deepwater limestones and radiolarites of the Alpine Jurassic. *In* "The Depositional Environment of Carbonate Rocks" (G. M. Friedman, Ed.). *Spec. Publs Soc. econ. Palaeont. Miner., Tulsa* **14**, 20–56.

Ginsburg, R. N. (1956). Environmental relationships of grain size and constituent particles in some S. Florida sediments. *Bull. Am. Ass. Petrol. Geol.* **40**, 2384–2427.

Glennie, K. W. (1970). "Desert Sedimentary Environments" Elsevier, Amsterdam. 222pp.

Glennie, K. W. (1972). Permian Rotliegendes of northwest Europe interpreted in light of modern sediment studies. *Bull. Am. Ass. Petrol. Geol.* **56**, 1048–1071.

Gorsline, D. S. (1970). Submarine canyons, an introduction. *Mar. Geol.* **8**, 183–186.

Gorsline, D. S. and Emery, K. O. (1959). Turbidity current deposits in San Pedro and Santa Monica basins off Southern California. *Bull. geol. Soc. Am.* **70**, 279–290.

Gould, H. R. (1970). The Mississippi delta complex. Deltaic Sedimentation Modern and Ancient (J. P. Morgan, and R. H. Shaver, Eds). *Spec. Publs Soc. econ. Palaeont. Miner., Tulsa* **15**, 3–30.

Greensmith, J. T. (1968). Paleogeography and rhythmic deposition in the Scottish Oil-Shale group. Proc. U.N. Symp. Dev. Util. Oil Shale Res. Tallin. 16pp.

Gregory, K. J. (Ed.) (1977). "River Channel Changes" John Wiley, New York. 448pp.

Gressly, A. (1838). Observations geologiques sur le Jura Soleurois. *Neue Denkschr. allg. schweiz. Ges. ges. Naturw.* **2**, 1–112.

Guilcher, A. (1970). Symposium on the evolution of shorelines and continental shelves in their mutual relations during the Quaternary. *Quaternaria* **12**, 229pp.

Hallam, A. (1967). Editorial comment. *In* "Depth indicators in marine sedimentary environments". Mar. Geol. Sp. Issue **5**, No. 5/6. pp. 329–332.

Hand, B. M. and Emery, K. O. (1964). Turbidites and topography of north end of San Diego Trough, California. *J. Geol.* **72**, 526–552.

Heckel, P. H. (1972). Recognition of Ancient shallow marine environments. Recognition of Ancient Sedimentary Environments (J. K. Rigby and W. K. Hamblin, Eds). *Spec. Publs Soc. econ. Palaeont. Miner., Tulsa* **16**, 226–286.

Hollenshead, C. T. and Pritchard, R. L. (1961). Geometry of producing Mesaverde Sandstones, San Juan basin. *In* "Geometry of Sandstone Bodies" (J. A. Peterson, and J. C. Osmond, Eds), 98–118. Am. Ass. Petrol. Geol.

Horn, D. (1965). Zur geologischen Entwicklung der sudlichen Schleimundung im Holozan. *Meyniana* **15**, 42–58.

Horn, D. R., Ewing, M., Delach, M. N. and Horn, B. M. (1971). Turbidites of the northeast Pacific. *Sedimentology* **16**, 55–69.

Horowitz, D. H. (1966). Evidence for deltaic origin of an Upper Ordovician sequence in the Central Appalachians. *In* "Deltas in their Geologic Framework" (M. L. Shirley and J. A. Ragsdale, Eds), 159–169. Houston Geol. Soc.

Houbolt, J. J. H. C. (1968). Recent sediments in the southern bight of the North Sea. *Geologie Mijnb.* **47**, 245–273.

Hoyt, J. H. (1967). Barrier island formation. *Bull. geol. Soc. Am.* **78**, 1125–1136.

Hoyt, J. H. and Henry, V. J. J. (1967). Influence of island migration on barrier island sedimentation. *Bull. geol. Soc. Am.* **78**, 77–86.

Hoyt, J. H., Weimer, R. J. and Vernon, J. H. (1964). Late Pleistocene and Recent sedimentation, central Georgia coast. *In* "Deltaic and Shallow Marine Deposits" (L. M. J. U. Van Straaten, Ed.), 170–176. Elsevier, Amsterdam.

Hsu, K. J. and Jenkyns, H. C. (Eds) (1974). Pelagic sediments: on Land and under the Sea. *Spec. Pub. Int. Ass. Sediment.* **1**, 447pp.

Hulsemann, J. (1955). Grossrippeln und Schragschichtungs-Gefuge im Nordsee-Watt und in der Molasse. *Senckenberg. Leth.* Band **36**, 359–388.

Imbrie, J. and Buchanan, H. (1965). Sedimentary structures in modern carbonate sands of the Bahamas. Primary Sedimentary Structures and Their Hydrodynamic Interpretation". *Spec. Publs Soc. econ. Palaeont. Miner., Tulsa* **12**, 149–172.

Irwin, M. L. (1965). General theory of epeiric clear water sedimentation. *Bull. Am. Ass. Petrol. Geol.* **49**, 445–459.

Jefferies, R. (1963). The stratigraphy of the *Actinocamax plenus* subzone (Turonian) in the Anglo-Paris basin. *Proc. geol. Ass.* **74**, 1–34.

Jenkyns, H. C. (1978). Pelagic Environments. *In* "Sedimentary Environments and Facies" (H. G. Reading, Ed.), 314–371. Blackwell Scientific, Oxford.

Johnson, H. D. (1978). Shallow Siliclastic Seas. *In* "Sedimentary Environments and Facies" (H. G. Reading, Ed.), 207–258. Blackwell Scientific, Oxford.

Jordan, G. F. (1962). Large submarine sand waves. *Science, N. Y.* **136**, 839–847.

Jordan, W. M. (1969). The enigma of Colorado Plateau eolian sandstones (abs.) *Bull. Am. Ass. Petrol. Geol.* **53**, 725.

Kazmi, A. H. (1964). Report on the geology and ground water investigations in Rechna Doab, West Pakistan. *Rec. geol. Surv. Pakist.* **10**, 26pp.

Keller, G. H. and Richards, A. F. (1967). Sediments of the Malacca Strait, Southeast Asia. *J. sedim. Petrol.* **37**, 102–127.

Kendall, C. G. St. C. (1969). An environmental reinterpretation of the Permian evaporite/carbonate shelf sediments of the Guadalupe Mountains. *Bull. geol. Soc. Am.* **80**, 2503–2525.

Kenyon, N. H. and Stride, A. H. (1970). The tide-swept continental shelf sediments between the Shetland Isles and France. *Sedimentology* **14**, 159–173.

King, C. A. M. (1972). "Beaches and Coasts" (2nd Edition). Edward Arnold, London. 570pp.

King, L. C. (1962). "Morphology of the Earth" Oliver and Boyd, Edinburgh, 699pp.

Klein, G. de Vries (1962). Triassic sedimenation, Maritime Provinces, Canada. *Bull. geol. Soc. Am.* **73**, 1127–1146.

Klein, G. de Vries (1971). A sedimentary model for determining paleotidal range. *Bull. geol. Soc. Am.* **82**, 2585–2592.

Klein, G. de V. (1977). "Clastic Tidal Facies" CEPCO, Champaign, Illinois. 149pp.

Klein, G. de V. (1980). "Sandstone Depositional Models for Exploration for Fossil Fuels" CEPCO, Burgess, Minneapolis. 149pp.

Klein, G. de Vries, de Melo, U. and Favera, J. C. D. (1972). Subaqueous gravity processes on the front of Cretaceous deltas, Reconcavo basin, Brazil. *Bull. geol. Soc. Am.* **83**, 1469–1491.

Kolb, C. R. and Van Lopik, J. R. (1966). Depositional environments of the Mississippi River deltaic plain—southeastern Louisiana. *In* "Deltas in their Geologic Framework" (M. L. Shirley and J. A. Ragsdale, Eds), 17–62. Houston Geol. Soc.

Krumbein, W. C. (1967). Fortran IV computer programs for Markov chain experiments in geology. Comput. Contri. Kans. Univ. geol. Surv. No. **13**.

Krumbein, W. C. and Sloss, L. L. (1959). "Stratigraphy and Sedimentation" W. H. Freeman, San Francisco. 497pp.

Kukal, Z. (1971). "Geology of Recent Sediments" Academic Press, London and New York. 490pp.

Laming, D. J. C. (1966). Imbrication, paleocurrents and other sedimentary features in the Lower New Red Sandstone, Devon, England. *J. sedim. Petrol.* **36**, 940-957.
Laporte, L. F. (1979). "Ancient Environments" (2nd Edition). Prentice-Hall, Englewood Cliffs, New Jersey. 163pp.
Lawson, A. C. (1913). The petrographic designation of alluvial-fan formations. *Univ. Cal. Publs geol. Sci.* **7**, 325-334.
Leopold, L. B., Solman, M. G. and Miller, J. P. (1964). "Fluvial Processes in Geomorphology" W. H. Freeman, San Francisco. 522pp.
Lowenstam, H. A. (1950). Niagaran reefs of the Great Lakes area. *J. Geol.* **58**, 430-487.
Lyell, C. (1865). "Elements of Geology" John Murray, London. 794pp.
McKee, E. D. (Ed.) (1979). A study of global sand seas. *U.S. Geol. Surv. Prof. Pap.* **1052**, 431pp.
Magleby, D. C. and Klein, I. E. (1965). Ground water conditions and potential pumping resources above the Corcoran Clay. U.S. Bureau Reclamation Open File Rept 21pp.
Maksimova, S. V. (1972). Coral reefs in the Arctic and their paleogeographic interpretation. *Int. Geol. Rev.* **14**, 764-769.
Matter, A. and Tucker, M. E. (Eds.) (1978). "Modern and Ancient Lake Sediments" Sp. Publ. Int. Assn. Sedol. Blackwell Scientific, Oxford. 289pp.
Mero, J. L. (1965). "The Mineral Resources of the Sea" Elsevier, Amsterdam. 312pp.
Merriam, D. F. (Ed.) (1965). Symposium on cyclic sedimentation. *Bull. Kans. Univ. geol. Surv.* **169**, 636pp.
Merriam, D. F. (Ed.) (1967). Computer applications in the Earth Sciences. Computer Contr. Kans. geol. Surv. No. 18.
Miall, A. D. (Ed.) (1978). Fluvial Sedimentology. Can. Soc. Pet. Geol. Sp. Pub. No. 5, 859pp.
Miall, A. D. (1977). A review of the braided river depositional environment. *Earth Sci. Rev.* **13**, 1-62.
Middleton G. V. (1973). Johannes Walther's law of correlation of facies. *Bull. geol. Soc. Am.* **84**, 979-988.
Mojsisovics, M. E. von (1879). "Die Dolomit-Riffe Von Sud Tirol und Venetien" A. Holder, Vienna. 552pp.
Moore, D. (1959). Role of deltas in the formation of some British Lower Carboniferous cyclothems. *J. Geol.* **67**, 522-539.
Moore, G. T. and Asquith, D. O. (1971). Delta: term and concept. *Bull. geol. Soc. Am.* **82**, 2563-2568.
Moore, R. C. (1949). Meaning of facies. *In* "Sedimentary Facies in Geologic History" 1—34. Geol. Soc. Am. Mem. No. 39.
Morgan, J. P. (1970). Depositional processes and products in the deltaic environment. Deltaic Sedimentation Ancient and Modern (J. P. Morgan and R. H. Shaver, Eds). *Spec. Publs Soc. econ. Palaeont. Miner., Tulsa* **15**, 31-47.
Morgan, J. P. and Shaver, R. H. (1970). Deltaic Sedimentation Modern and Ancient. *Spec. Publs Soc. econ. Palaeont. Miner., Tulsa* **15**, 312pp.
Mutti, E. and Lucchi, F. R. (1972). Le torbiditii dell'-Appennino settentrionale: introduzione all'-analisi di facies. *Memorie Soc. geol. ital.* **11**, 161-199.
Narayan, J. (1970). Sedimentary structures in the Lower Greensand of the Weald, England, and Bas-Boulonnais, France. *Sedimentary Geol.* **6**, 73-109.

Newell, N. D. and Rigby, J. K. (1957). Geological studies of the Great Bahama bank. Regional Aspects of Carbonate Deposition. *Spec. Publs Soc. econ. Palaeont. Miner., Tulsa* **5**, 15–72.

Newell, N. D., Rigby, J. K., Fischer, A. G., Whiteman, A. J., Hickox, J. E. and Bradbury, J. S. (1953). *In* "The Permian Reef Complex of the Guadalupe Mountains Region, Texas and New Mexico". Freeman, San Francisco. 236pp.

Opdyke, N. D. (1961). The Paleoclimatological significance of desert sandstone. *In* "Descriptive Paleoclimatology" (A. E. M. Nairn, Ed.) 390–405. Interscience, New York.

Pettijohn, F. J. (1956). "Sedimentary Rocks" Harper Bros, New York. 718pp.

Pettijohn, F. J., Potter, P. E. and Siever, R. (1972). "Sand and Sandstone" Springer-Verlag, Berlin. 618pp.

Pezzetta, J. M. (1973). The St. Clair River delta: sedimentary characteristics and depositional environments. *J. sedim. Petrol.* **43**, 168–187.

Picard, M. D. (1967). Paleocurrents and shoreline orientations in Green River Formation (Eocene), Raven Ridge and Red Wash areas, north-eastern Unita basin, Utah. *Bull. Am. Ass. Petrol. Geol.* **5**, 383–392.

Picard, M. D. and High, L. R. (1968). Sedimentary cycles in the Green River Formation (Eocene) Uinta basin, Utah. *J. sedim. Petrol.* **38**, 378–383.

Picard, M. D. and High, L. R. Jr (1972a). Criteria for recognizing lacustrine rocks. Recognition of Ancient Sedimentary Environments (J. K. Rigby and W. K. Hamblin, Eds). *Spec. Publs Soc. econ. Palaeont. Miner., Tulsa* **16**, 108–145.

Picard, M. D. and High, L. R. Jr. (1972b). Paleoenvironmental reconstructions in an area of rapid facies change, Parachute Creek member of Green River Formation (Eocene), Uinta basin, Utah. *Bull. geol. Soc. Am.* **83**, 2689–2708.

Picard, M. D. and High, L. P. (1979). Lacustrine Stratigraphic Relations. *In* "Some Sedimentary Basins and Associated Ore Deposits of South Africa" Geol. Soc. S. Af. Sp. Pub. No. 6. 1–22.

Playford, P. E. and Marathon Oil Company (1969). Devonian carbonate complexes of Alberta and Western Australia: a comparative study. Geol. Surv. Western Australia Report 1, 43pp.

Poole, F. G. (1964). Paleowinds in Western U.S.A. *In* "Problems of Paleoclimatology" (A. E. M. Nairn, Ed.), 390–405. Interscience, New York.

Potter, P. E. (1959). Facies model conference. *Science, N.Y.* **129**, No. 3558, 1292–1294.

Potter, P. E. (1962). Late Mississippian sandstones of Illinois. *Illinois S. geol. Surv.* Cic. **340**, 36pp.

Potter, P. E. (1967). Sand bodies and sedimentary environments: a review. *Bull. Am. Ass. Petrol. Geol.* **51**, 337–365.

Potter, P. E. and Pettijohn, F. J. (1977). "Paleocurrents and Basin Analysis" (2nd Edition) Springer-Verlag, Berlin, Gottingen and Heidelberg. 425pp.

Prevost, C. (1838). *Bull. Soc. geol. Fr.* **9**, 90–95.

Pryor, W. A. (1971). Petrology of the Permian Yellow Sands of northeast England and their North Sea basin equivalents. *Sedimentary geol.* **6**, 221–254.

Purdy, E. G. (1961). Bahamian oolite shoals. *In* "The Geometry of Sandstone Bodies" A symposium, 53–62. Am. Ass. Petrol. Geol.

Purdy, E. G. (1963). Recent calcium carbonate facies of the Great Bahama bank. *J. Geol.* **71**, 334–355, 472–497.

Purser, B. H. (Ed.) (1973). "The Persian Gulf" Springer-Verlag, Berlin. 471pp.

Reading, H. G. (Ed.) (1978). "Sedimentary Environments and Facies" Blackwell Scientific, Oxford. 557pp.

Reineck, H. E. (1963). Sedimentgefuge im Bereich der sudlichen Nordsee. *Abh. senckenb. naturforsch. Ges.* **505**, 138pp.

Reineck, H. E. (1971). Marine sandkörper, rezent und fossil (Marine sandbodies recent and fossil). *Geol. Rdsch.* **60**, 302–321.

Riedel, W. R. (1963). The preserved record:paleontology of Pelagic sediments. *In* "The Sea" Vol. 3 (M. N. Hill, Ed.) 866–887. Interscience, New York.

Rust, B. R. (1972). Structure and process in a braided river. *Sedimentology* **18**, 221–246.

Rust, B. R. (1978). A classification of alluvial channel systems. *In* "Fluvial Sedimentology" (A. D. Miall, Ed.). Can. Soc. Pet. Geol. Mem. No. 5. Calgary. 859pp.

Sabins, F. F. (1963). Anatomy of stratigraphic traps, Bisti field, New Mexico. *Bull. Am. Ass. Petrol. Geol.* **47**, 193–228.

Schumm, S. A. (1977). "The Fluvial System" John Wiley, New York. 338pp.

Schwarz, M. L. (1971). The multiple causality of barrier islands. *J. Geol.* **79**, 91–94.

Selley, R. C. (1978). "Ancient Sedimentary Environments" (2nd Edition) Chapman and Hall, London. 287pp.

Selley, R. C. (1972). Diagnosis of marine and non-marine environments from the Cambro-Ordovician sandstones of Jordan. *Q. Jl geol. Soc. Lond.* **128**, 135–150.

Shantser, E. V. (1951). Alluvium of river plains in a temperate zone and its significance for understanding the laws governing the structure and formation of alluvial suites. *Akad. Nauk. S.S.S.R. geol. Ser.* **135**, 1–271.

Sharma, G. D. (1972). Graded sedimentation on Bering shelf. Rep. 24th int. Geol. Cong. Montreal Section 8, 262–271.

Shaw, A. B. (1964). "Time in Stratigraphy" McGraw-Hill, New York. 365pp.

Shawe, D. R., Simmons, G. C. and Archbold, N. L. (1968). Stratigraphy of Slick Rock district & vicinity, San Miguel and Dolores counties, Colorado. *Prof. pap. U.S. geol. Surv.* **576**—A.

Shelton, J. W. (1965). Trend and genesis of Lowermost sandstone unit of Eagle Sandstone at Billings, Montana, *Bull. Am. Ass. Petrol. Geol.* **49**, 1385–1397.

Shelton, J. W. (1967). Stratigraphic models and general criteria for recognition of alluvial, barrier bar and turbidity current sand deposits. *Bull. Am. Ass. Petrol. Geol.* **51**, 2441–2460.

Shepard, F. P. (Ed.) (1960). "Recent Sediments, North-western Gulf of Mexico" Am. Ass. Petrol. Geol. Mem. 394pp.

Shepard, F. P. (1963). Submarine canyons. *In* "The Sea" (M. N. Hill, Ed.) Vol. 3, 480–506. John Wiley, New York.

Shepard, F. P. (1971). "Submarine Canyons and Other Sea Valleys" John Wiley, New York. 381pp.

Shepard, F. P., Dill, R. F. and Von Rad, U. (1969). Physiography and sedimentary processes of La Jolla submarine fan and Fan-Valley, California. *Bull. Am. Ass. Petrol. Geol.* **53**, 390–420.

Shirley, M. L. and Ragsdale, J. A. (Eds) (1966). "Deltas in Their Geologic Framework" Houston Geol. Soc. 251pp.

Shotton, F. W. (1937). The Lower Bunter Sandstones of North Worcestershire and East Shropshire. *Geol. Mag.* **74**, 534–553.

Stanley, D. J. and Unrug, R. (1972). Submarine channel deposits, fluxoturbidites and other indicators of slope and base of slope environments in modern and ancient marine basins. Recognition of Ancient Sedimentary Environments (J. K. Rigby and W. K. Hamblin, Eds). *Spec. Publs Soc. econ. Palaeont. Miner., Tulsa* **16**, 287–340.

Stanley, K. O., Jordan, W. M. and Dott, R. H. (1971). New hypothesis of early Jurassic paleogeography and sediment dispersal for Western United States. *Bull. Am. Ass. Petrol. Geol.* **55**, 10–19.

Stanley, D. J. and Swift, D. J. P. (1976). "Marine Sediment Transport and Environment Management" John Wiley, New York. 602pp.

Stanton, R. J. (1967). Factors controlling shape and interval facies distribution of organic build-ups. *Bull. Am. Ass. Petrol. Geol.* **51**, 2462–2467.

Stauffer, P. H. (1967). Grainflow deposits and their implications, Santa Ynez Mountains, California. *J. sedim. Petrol.* **37**, 487–508.

Steers, J. A. (1971). "Introduction to Coastline Development" Macmillan, London. 229pp.

Stille, H. (1924). "Grundfragen der Vergleichenden Tektonik" Gebr. Borntraeger, Berlin. 443pp.

Stride, A. H. (1963). Current-swept sea floors near the southern half of Great Britain. *Q. Jl geol. Soc. Lond.* **119**, 175–200.

Stride, A. H. (1970). Shape and size trends for sandwaves in a depositional zone of the North Sea. *Geol. Mag.* **107**, 469–478.

Sundborg, A. (1956). The River Klaralven: a study of fluvial processes. *Geogr. Annlr.* **38**, 127–316.

Swett, K., Klein, G. de Vries and Smit, D. E. (1971). A Cambrian tidal sand body—the Eriboll Sandstone of Northwest Scotland: an ancient: recent analogue. *J. Geol.* **79**, 400–415.

Swift, D. J. P. (1969). Evolution of the shelf surface, and the relevance of the modern shelf studies to the rock record. *In* "The NEW Concepts of Continental Margin Sedimentation" (D. J. Stanley, Eds). AGI, Washington.

Swift, D. J. P., Duane, D. B. and Orrin, H. P. (1973). "Shelf Sediment Transport: Process and Pattern" John Wiley, Chichester. 670pp.

Swift, D. J. P., Duane, D. B. and O. H. Pilkey (1972). "Shelf Sediment Transport: Process and Pattern" Dowden, Hutchinson and Ross, Stroudsberg. 405pp.

Swift, D. J. B. and Palmer, H. D. (Eds) (1978). "Coastal Sedimentation." Benchmark Papers in Geology, Vol. 42. Dowden, Hutchinson and Ross, Stroudsberg. 339pp.

Tanner, W. F. (1968). Tertiary sea-level fluctuations. *Paleogeog. Paleoclimatol. Paleoecol.* Sp. issue, 178pp.

Teichert, C. (1958). Cold and deep-water coral banks. *Bull. Am. Ass. Petrol. Geol.* **42**, 1064–1082.

Teichert, C. (1958). Concept of facies. *Bull. Am. Ass. Petrol. Geol.* **42**, 2718–2744.

Thompson, D. B. (1969). Dome-shaped aeolian dunes in the Frodsham Member of the so-called "Keuper" sandstone formation (Scythian; ?Anisian: Triassic) at Frodsham, Cheshire (England). *Sedimentary Geol.* **3**, 263–289.

Till, R. (1978). Arid Shorelines and Evaporites. *In* "Sedimentary Environments and Facies" (H. G. Reading, Ed.) 178–206. Blackwell Scientific, Oxford.

Trowbridge, A. C. (1911). The terrestrial deposits of Owens Valley, California. *J. Geol.* **19**, 707–747.

Turnbull, W. J., Krinitsky, E. S. and Johnson, L. J. (1950). Sedimentary geology of the Alluvial valley of the Mississippi River and its bearing on foundation problems. *In* "Applied Sedimentation" (P. D. Trask, Ed.), 210–226. John Wiley, New York.

Twenhofel, W. H. (1926). "Treatise on Sedimentation" Dover Publishing, New York. 926pp.

Van Andel, Tj., H. (1975). Mesozoic/Cenozoic calcite compensation depth and the global distribution of calcareous sediments. *Earth Plan. Sci. Lett.* **26**, 187–194.

Van Houten, F. B. (1964). Cyclic lacustrine sedimentation, Upper Triassic Lockatong Formation, central New Jersey and adjacent Pennsylvania. *Bull. Kans. Univ. geol. Surv.* **169**, 497–531.

Van Houten, F. B. (Ed.) (1977). "Ancient Continental Deposits." Benchmark Papers in Geology, Vol. 43. Dowden, Hutchinson and Ross, Stroudsberg. 367pp.

Van Straaten, L. M. J. U. (1965). Coastal barrier deposits in south and north Holland in particular in the areas around Scheveningen and Ijmuiden. *Meded. geol. Sticht.* **17**, 41–75.

Visher, G. S. (1965). Use of vertical profile in environmental reconstruction. *Bull. Am. Ass. Petrol. Geol.* **49**, 41–61.

Visher, G. S. (1971). Depositional processes and the Navajo sandstone. *Bull. geol. Soc. Am.* **82**, 1421–1424.

Walker, R. G. (1966). Shale grit and Grindslow shales: transition from Turbidite to shallow water sediments in the Upper Carboniferous of Northern England. *J. sedim. Petrol.* **36**, 90–114.

Walker, R. G. (1970). Review of the Geometry and facies organization of turbidites and turbidite-bearing basins. Flysch Sedimentology in North America (J. Lajoie, Ed.). *Spec. Pap. geol. Ass. Can.* 219–251.

Walker, R. G. (Ed.) (1979). "Facies Models." Geoscience Canada Reprint Series. No. 1. Toronto. 211pp.

Walther, J. (1893). "Einleithung in die Geologie als Historische Wissenschaft" Bd. I. Beobachtungen uber die Bildung der Gesteine und ihter organischen Einschlusse. G. Fischer, Jena. 196pp.

Walther, J. (1894). "Einleitung in die Geologie als Historische Wissenschaft" Bd. 3. Lithogenesis der Gegenwart, 535–1055. G. Fischer, Jena.

Wanless, H. R., Baroffio, J. R., Gamble, J. C., Horne, J. C., Orlopp, D. R., Rocha-Campos, A., Souter, J. E., Trescott, P. C., Vail, R. S. and Wright, C. R. (1970). Late Paleozoic deltas in the central and eastern United States. Deltaic Sedimentation Modern and Ancient (J. R. Morgan and R. H. Shaver, Eds). *Spec. Publs Soc. econ. Palaeont. Miner., Tulsa* **15**, 215–245.

Weber, K. J. (1971). Sedimentological aspects of oil fields in the Niger Delta. *Geologie Mijnb.* **50**, 559–576.

Weimer, R. J. (1961). Spatial dimensions of Upper Cretaceous Sandstones, Rocky Mountain area. *In* "Geometry of Sandstone Bodies" (J. A. Peterson and J. C. Osmond, Eds), 82–97. Am. Ass. Petrol. Geol.

Weimer, R. J. (1970). Rates of deltaic sedimentation and intrabasin deformation, Upper Cretaceous of Rocky Mountain region. Deltaic Sedimentation Modern and Ancient (J. P. Morgan and R. H. Shaver, Eds). *Spec. Publs Soc. econ. Palaeont. Miner., Tulsa* **15**, 270–293.

Williams, G. E. (1969). Characteristics and origin of a PreCambrian pediment. *J. Geol.* **77**, 183–207.

Williams, G. E. (1971). Flood deposits of the sandbed emphemeral streams of Central Australia. *Sedimentology* **17**, 1–40.

Wilson, J. L. (1975). "Carbonate Facies in Geologic History" Springer-Verlag, Berlin. 471pp.

Wilson, H. H. (1969). Late Cretaceous and eugeosynclinal sedimentation, gravity tectonics and ophiolite emplacement in Oman Mountains, southeast Arabia. *Bull. Am. Ass. Petrol. Geol.* **53**, 626–671.

Withrow, P. C. (1968). Deposition environment of Pennsylvanian Red Fork Sandstone in northeastern Anadarko basin, Oklahoma. *Bull. Am. Ass. Petrol. Geol.* **52**, 1638–1654.

Wolf, S. C. (1970). Coastal currents and mass transport of surface sediments over the shelf regions of Monterey Bay, California. *Mar. Geol.* **8**, 321–336.

Wood, G. V. and Wolfe, M. J. (1969). Sabkha cycles in the Arab/Darb Formation off the Trucial Coast of Arabia. *Sedimentology* **12**, 165–191.

Wright, L. D. and Coleman, J. M. (1973). Variations in morphology of major river deltas as functions of ocean wave and river discharge regimes. *Bull. Am. Ass. Petrol. Geol.* **57**, 370–417.

9 Sedimentary Basins

I. ENVIRONMENTS, BASE LEVELS AND TECTONISM

The preceding chapter discussed sedimentary environments and facies. This chapter attempts to put sedimentation in its broader perspective. Why should sedimentation occur in one place at a particular time? What is the spatial organization of large volumes of sediment and what are the factors which control their facies? These are the questions which this chapter attempts to answer.

Thick sedimentary sequences are essentially attributable to tectonic subsidence for which there are three causes (Stoneley, 1969).

They may occur where subcrustal displacement of the mantle leads to down-dragging and compressional warping of the crust. This occurs principally at what are called zones of subduction, linear features which are the site of geosynclinal sedimentation. This topic is discussed in more detail later in this chapter.

Sedimentation may also occur on a large scale where changes in the mantle cause foundering and subsidence of the crust. This process is responsible for intracratonic basins. Conversely these changes can cause the crust to dome. Thick volcanic and sedimentary sequences may form in crestal rift basins.

Finally, thick sedimentary sequences may form where the weight of the sediment itself causes isostatic depression of the crust. This process obviously requires an outside mechanism to create an initial crustal void, since it poses the old problem of which came first, the hen or the egg? The most likely place for such a process is the continental margin, where a whole ocean basin waits to be infilled. Sedimentation at the foot of the continental slope may cause isostatic depression of the crust. Some geophysical data support this (e.g. Drake *et al.*, 1968, Fig. 2).

For any of these three situations to occur, a hole in the ground is a necessary prerequisite: sedimentation needs subsidence. Considerable attention has been paid to the actual mechanics of basin subsidence, a problem requiring geophysics and structural analysis for its elucidation (e.g. Bott and Johnson, 1967).

Directly relevant to sedimentation studies is the structural level of the faults which permit an area to subside. Two types of basement/sediment interface can be defined.

In one variety deposition occurs on platforms or gently subsiding basins which are undefaced by syndepositional surface faults. Faulting may be taking place, however, at deeper crustal levels to accommodate basin subsidence.

The second type of sediment/basement interface is actually disrupted by faults which have penetrated to the surface. These may be horst and graben fractures within the basement. The Sirte basin of Libya, for example, shows this type of basin floor. Alternatively, faults may be generated within the sediment cover, independent of the tectonic style of the basement. Growth faults, as these are termed, are due to sediment compaction. They are characteristic of areas of rapid deposition and are, therefore, especially common in deltas such as the Mississippi and the Niger (e.g. Shelton 1968; Carver, 1968; Weber, 1971; Evamy *et al.*, 1978).

This distinction between smooth basement/sediment interfaces and faulted ones is important. Not only does shallow faulting cause rapid lateral facies changes, but its presence may actually be responsible for the development of additional facies. This point will be elaborated shortly.

Having analysed what may be termed the tectonic base level, it is relevant to consider other base levels which control the nature and distribution of sedimentary facies. These include the depositional surface, the water level and wave base.

The depositional surface is the boundary which separates sediment from the overlying fluid, air or water. This surface may be horizontal or inclined at angles of up to about 30°. This angle is the depositional dip.

The water level is a horizontal surface. Mean sea level is obviously particularly critical in differentiating sedimentary facies.

Wave base level is a rather more difficult parameter to define. Wave base is generally understood to mean the water depth which separates high-energy traction deposits from low-energy suspension deposits (including turbidites). This boundary depends not just on waves, however, but also on the power of tidal currents and on the amount of suspended sediment load. Effective wave base does not, therefore, parallel water level at a constant depth. For example, on the European continental shelf it ranges from 30 to 70 m in the North Sea and Celtic Sea respectively (McCave, 1971). Nevertheless, despite its variability, wave base is one of the most important parameters of environmental (and therefore of facies) control. Rich (1951) classified marine environments into three types: shelf, slope and basin floor (or clinothem, undothem and fundothem in his terminology). He tabulated the characteristics of the facies generated in each setting. His clinothem/

undothem boundary largely corresponds to wave base. This may occur at a shelf/basin hinge line, whose position is controlled generally by a fault, either at the surface or at depth. Alternatively the position of wave base may be unrelated to tectonics, as in the shelf sea model (p. 303).

The point of this tedious analysis of tectonics, depositional surface, water level and wave base becomes apparent when they are integrated.

Three main types of sedimentary pattern can be defined according to whether the sediment/basement interface is faulted, whether it has a smooth platform, or a basin shape. These three types are further subdivided into a total of nine assemblages, depending on whether the depositional surface is above, at, or below water level (Fig. 164). Table XXXII cites examples of these nine assemblages.

This scheme incorporates all the facies models defined in the previous chapter and shows that the dominant factors which control their genesis and geometry are tectonic setting, and elevation of the depositional surface with respect to sea level and wave base.

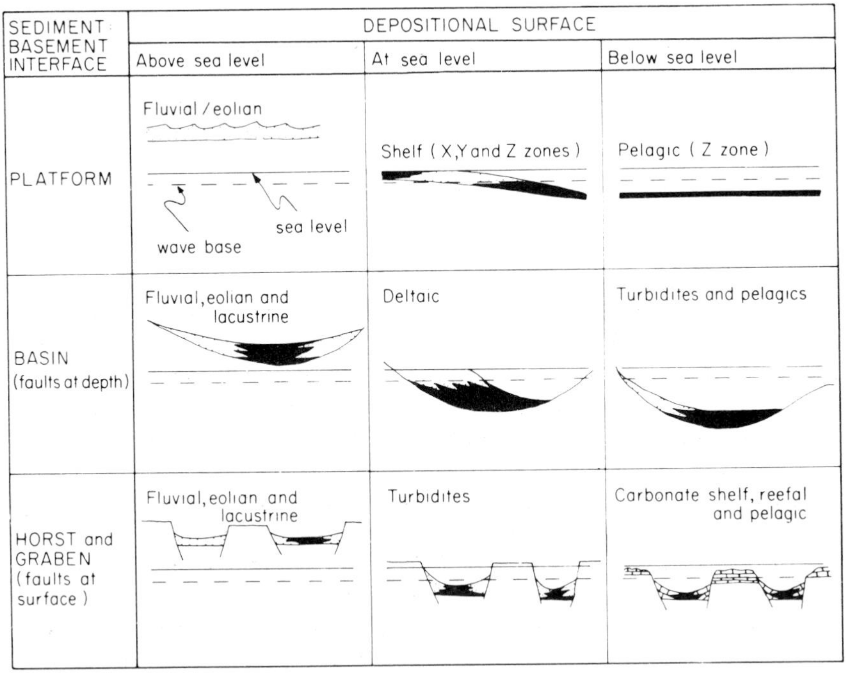

SEDIMENT: BASEMENT INTERFACE	DEPOSITIONAL SURFACE		
	Above sea level	At sea level	Below sea level
PLATFORM	Fluvial / eolian wave base sea level	Shelf (X,Y and Z zones)	Pelagic (Z zone)
BASIN (faults at depth)	Fluvial, eolian and lacustrine	Deltaic	Turbidites and pelagics
HORST and GRABEN (faults at surface)	Fluvial, eolian and lacustrine	Turbidites	Carbonate shelf, reefal and pelagic

Fig. 164. Diagram to show how environments relate to the tectonic geometry of the sediment/basement interface, coupled with sea level and wave base.

Table XXXII
Examples of the nine sedimentary patterns illustrated in Fig. 164

Sediment/ basement interface	Depositional surface		
	Above sea level	At sea level	Below wave base
Platform	Saharan "Nubian" Sandstones	Arabian Gulf Mesozoic	Sunda shelf
Basin	Alpine Molasse	Illinois basin, Pennsylvanian	Modern ocean basins and early phase of geosynclinal cycle
Block-faulted	Rocky Mountain range and basin, Tertiary	Ventura basin (California)	English L. Carboniferous

This scheme must not be made an article of faith. Its main use is that it links the concepts of environments and facies with the broader sedimentologic aspects of regional basin analysis which is the main theme of this chapter.

II. SEDIMENTARY BASINS

A. Concepts and Classifications

The preceding chapter discussed the depositional environments of sedimentary facies. The preceding section discussed facies in their local tectonic setting with respect to the crustal level of faulting and of aqueous and sedimentary base levels. The wider geographic distribution and organization of sediments will now be described; a topic which may be loosely referred to as basin analysis.

Sedimentary rocks cover a large part of the earth's surface, including some 75% of the land areas. Yet sedimentary rocks make up only 5% of the lithosphere (data from Pettijohn, 1957, p. 7). From this it follows that sedimentary rocks cover the earth only as a thin and superficial veneer. This cover is not evenly distributed. Thick sedimentary deposits were laid down in localized areas; loosely termed sedimentary basins. Large areas of the continents lack a thick sedimentary cover; Pre-Cambrian igneous and metamorphic rocks occur at or near the surface. These stable continental cores are termed cratons.

Sedimentary basins are of many types, ranging from small alluvial intermontane valleys to vast mountain ranges of contorted sediment, kilometres thick. Before proceeding to define and detail these various basins, some terms and concepts will first be defined.

Basins are of three types: topographic, structural and sedimentary. Topographic basins are low-lying areas of the earth's surface naturally surrounded by higher areas. Topographic basins are both subaerial and subaqueous. Subaerial topographic basins range from bolsons (intermontane plains), to transcontinental alluvial valleys such as the Amazon basin. Subaqueous topographic basins range from periglacial pingo ponds to oceans. The existence of a topographic basin is necessary for the genesis of a sedimentary basin.

A sedimentary basin is an area of gently folded centripetally dipping strata. It is a matter of great importance to distinguish tectonic, or post-depositional basins from synsedimentary basins. In tectonic basins facies trends and palaeocurrents are unrelated to the basin architecture, indicating that subsidence took place after deposition of the deformed strata. In syndepositional basins, by contrast, the facies trends, palaeocurrents and depositional thinning of strata towards the basin margin all indicate contemporaneous movement (Fig. 165).

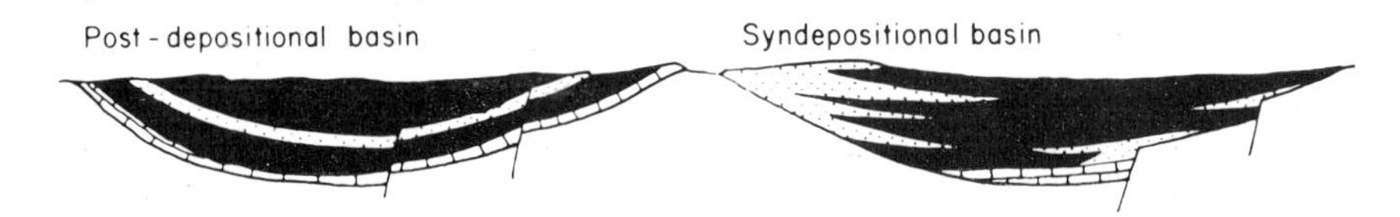

Fig. 165. Section to illustrate the differences between post-depositional basins and syndepositional basins. The sedimentary facies of syndepositional basins reflect the position of the basement margin and the movement of faults. In post-depositional basins stratigraphy is discordant with structure.

Sedimentary basins generally cover tens of thousands of square kilometres, but their sizes are not diagnostic. Distinction is made, however, between basins (*sensu stricto*), embayments and troughs (Fig. 166). Basins (*sensu stricto*) are saucer-shaped and subcircular in plan.

Embayments are areas which are not completely closed structurally, but which open out into a deeper area. Troughs are linear basins.

These three basin types are essentially unmetamorphosed and undeformed by tectonism. These features distinguish them from the linear metamorphosed tectonized sedimentary troughs which are termed "geosynclines".

The thinning of the sedimentary cover towards the basin margin may be erosional or non-depositional, shown by intraformational thinning. These margin types differentiate post-depositional from syndepositional tectonism. The axis of a basin is a line connecting the lowest structural points of the basin, as in a synclinal axis. Similarly the axes of troughs may plunge. The depocentre is the part of the basin with the thickest sedimentary fill.

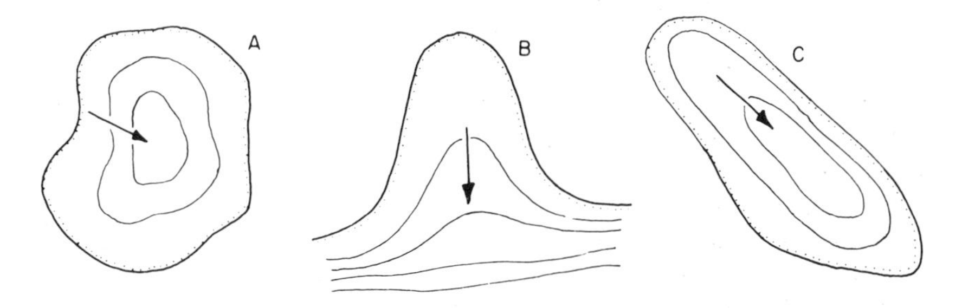

Fig. 166. Diagrams to illustrate the nomenclature of basins. Lines indicate structure contours, arrows indicate depositional palaeoslope. A: a basin *sensu stricto*, subcircular in plan and structurally closed. B: an embayment, opening out at one side to a structurally lower area. C: a trough, structurally closed and elongated.

It is very important to note that the depocentre and basin axis need not be coincident, neither need they coincide with the topographic axis of the basin (Fig. 167). This is particularly true of asymmetric basins with large amounts of terrigenous sedimentation on the limb of maximum uplift. In gentle basins, with pelagic fine-grained and turbidite fill, depocentre, axis and topographic nadir may coincide. These points should be considered when deciding whether a regional isopach map gives a valid picture of syn-depositional basin architecture.

It is a common feature of many basins that the depocentre moves across

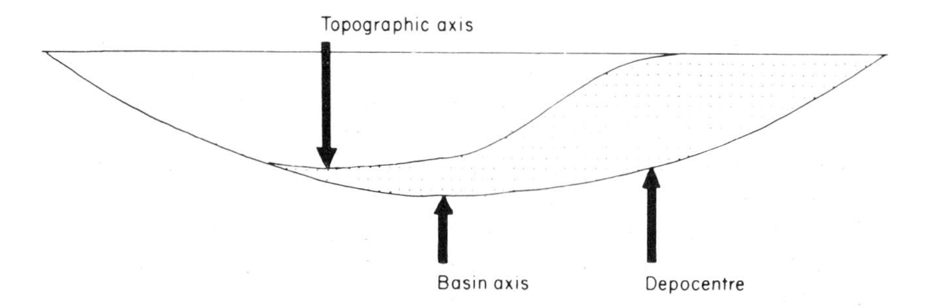

Fig. 167. Cross-section to illustrate that topographic and structural axes need neither be coincident with each other, nor with the site of maximum sedimentation (the depocentre). Isopach maps are not necessarily, therefore, reliable indicators of palaeotopography.

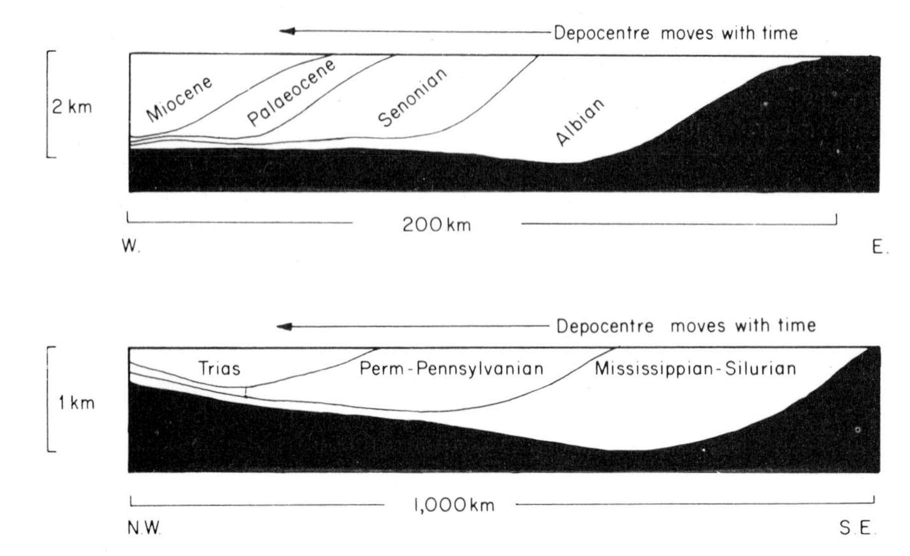

Fig. 168. Cross-sections of two sedimentary basins showing the lateral migration of depocentres with time. Upper: the Gabon basin of W. Africa, an ocean margin basin. (Data from Belmonte *et al.*, 1965.) Lower: the Maranhao basin, of Brazil, an intracratonic basin. (Data from Mesner and Wooldridge, 1964.)

Table XXXIII
A classification of sedimentary basins

Class	Type	Suite
I. Basins (*sensu stricto*)	Intracratonic Epicratonic	Cratonic suite — associated with crustal stability
II. Troughs	Miogeosyncline Eugeosyncline Molasse	Geosynclinal suite — associated with zones of crustal subduction
III. Rifts	Intramontane (post-orogenic)	(Geosynclinal suite)
	Intracratonic Intercratonic	Rift: drift suite — associated with zones of crustal spreading
IV. Ocean margin basins		(Rift: drift suite)

Examples and explanations of this scheme are given in the text.

the basin in time (Fig. 168). This may reflect a migration of the topographic axis of the basin, or merely a lateral progradation of the main locus of deposition across an essentially stable basin floor.

Basins are separated one from another by raised linear areas where the sediment cover is thin or absent. These are variously termed arches, palaeo-highs, schwelle, axes of uplift or positive areas. Similarly, major basins are commonly divisible into sub-basins, troughs and embayments by smaller

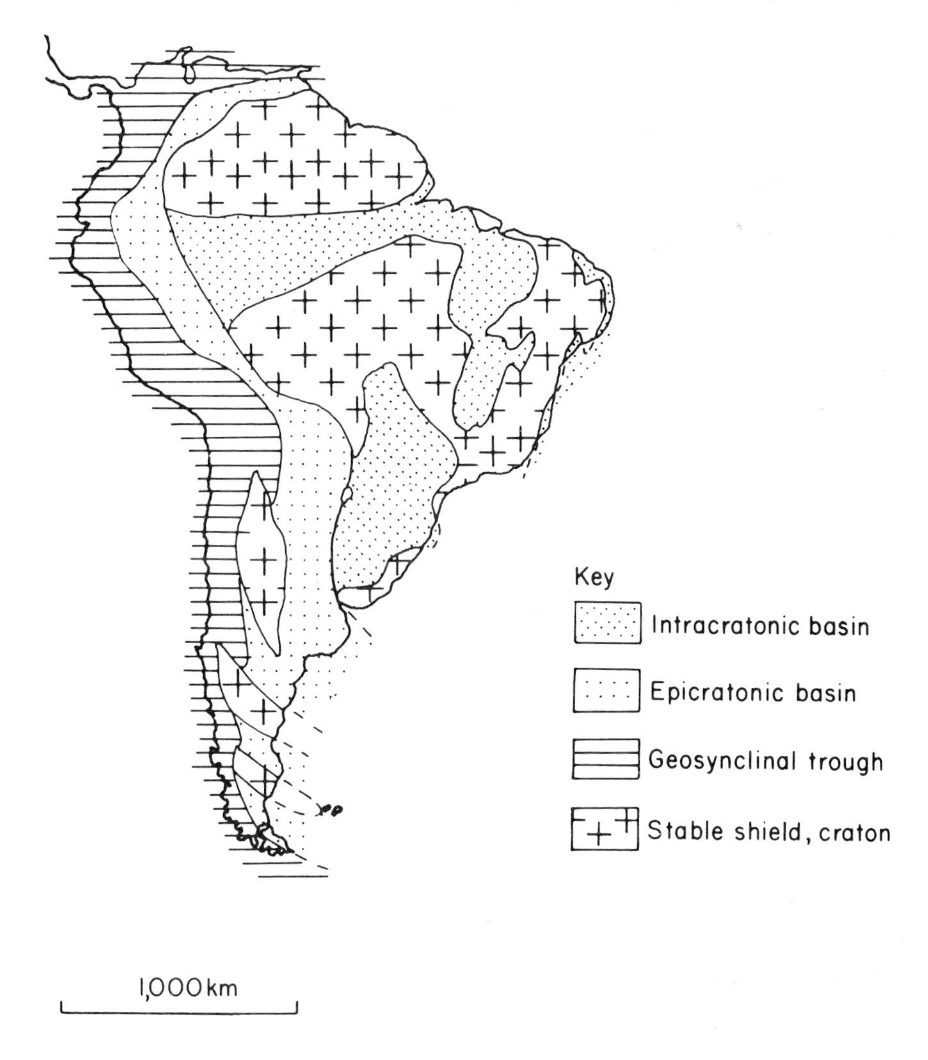

Fig. 169. Map of South America to show the arrangement of different types of sedimentary basins. (After Harrington, 1962.)

positive features. Figure 169 illustrates the distribution of different basin types on the South American continent.

Attempts to classify the various types of sedimentary basins have been made by many geologists, notably Weeks (1958), Halbouty *et al.* (1970), Perrodon (1971) and Klemme (1980). These classifications vary according to the defining parameters which have been chosen, and according to the purpose for which a scheme was drawn up. As with most geological phenomena, basins can be broadly grouped into several fairly well-established families, whose limits are ill-defined. The attempt at basin classification given in Table XXXIII owes much to those already cited. This scheme will be used as a framework for the following description and discussion of sedimentary basins.

B. Basins

A sedimentary basin, in the restricted sense of the term, in an essentially saucer-shaped area of sedimentary rocks. It is, therefore, sub-rounded in plan view. Strata dip and thicken centripetally towards the centre of the basin. It is an interesting exercise, however, to draw a cross-section of a basin to scale, allowing for the curvature of the earth (Dallmus, 1958). This reveals that basins are in fact veneers of sediment on the earth's surface, which are convex to the heavens.

Simple basins of this type are divisible into two groups. Intracratonic basins lie within the continental crust. Epicratonic basins lie on continental crust but are partially open to an ocean basin. These two types often occur adjacent to one another with little fundamental difference in genesis or fill (Fig. 170). Descriptions now follow.

1. Intracratonic basins

Intracratonic basins are the classic type of sedimentary basin. Notable examples include the Williston, Michigan and Illinois basins of North America, the Maranhao basin of Brazil, and the Murzuk and Kufra basins of the Sahara (Darling and Wood, 1958; Smith *et al.*, 1958; Cohee and Landes, 1958; Swann and Bell, 1952; Mesner and Woolridge 1964; Conant and Goudarzi, 1967; Klitzsch, 1970). Figures 171–174 illustrate the shapes of these basins and shows that they have a common geometry, intracratonic setting and scale.

The Williston basin is the classic example of an intracratonic basin. It contains some 3 km of rocks of all periods from Cambrian to Tertiary, with notable gaps only in the Permian and Triassic. Sedimentation spanned a range of environments including fluvial and marine sands, reefal carbonates, evaporites and subwave-base pelagic muds. Deep-sea, turbidite and deltaic facies, igneous activity and shallow syndepositional faulting are all absent.

The lengthy history, diverse facies and structural simplicity of the Williston basin are also found in the other basins cited as examples of

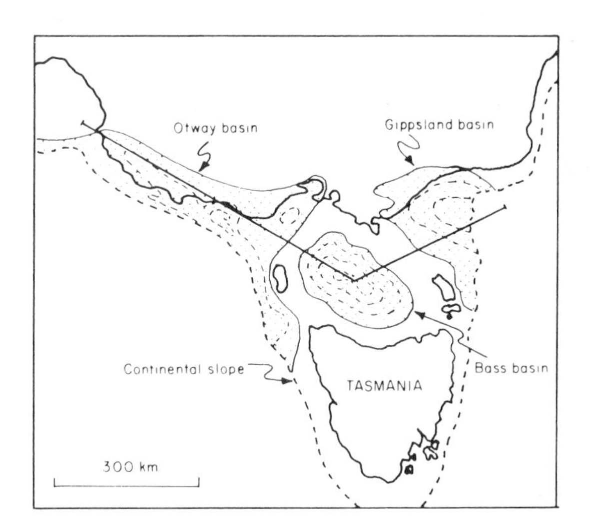

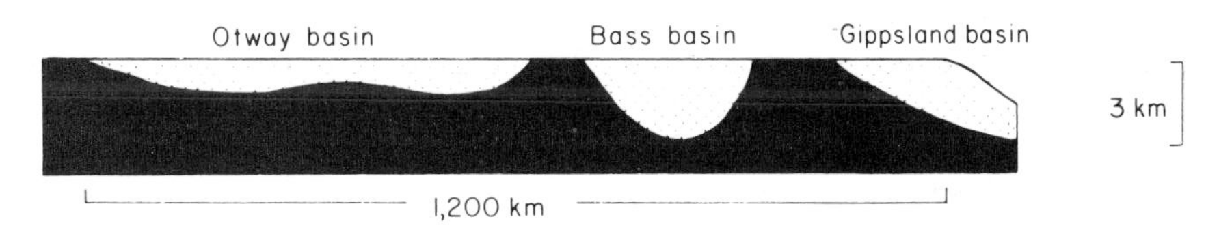

Fig. 170. Map and section of three Bass Strait basins, Australia, showing their mutual relationships. Dashed lines are isopachs at 1000 m intervals. The Otway and Gippsland basins are, strictly speaking, embayments open to the ocean. Only the Bass basin is a true closed basin. (From Weeks and Hopkins, 1967.)

intracratonic basins. It is important to note that the sedimentary facies, though diverse in lithology and environment, are seldom indicative of deep water or abrupt subsidence of the basin floor. Deposition took place close to sea level. Subsidence was thus a gradual, if erratic event, with sedimentation being sufficiently rapid to keep the basin nearly filled at any point in time.

Basins of this type are found well within the present limit of the continental margins, but it is obvious that they were frequently connected to the sea. This is shown by their intermittent phases of marine carbonate and evaporite sedimentation. During these periods they might, therefore, be more truly termed embayments rather than basins.

Modern analogues include the Hudson Bay and the Baltic Sea, which lie on the Canadian and Scandinavian shields respectively. Intermittent uplift

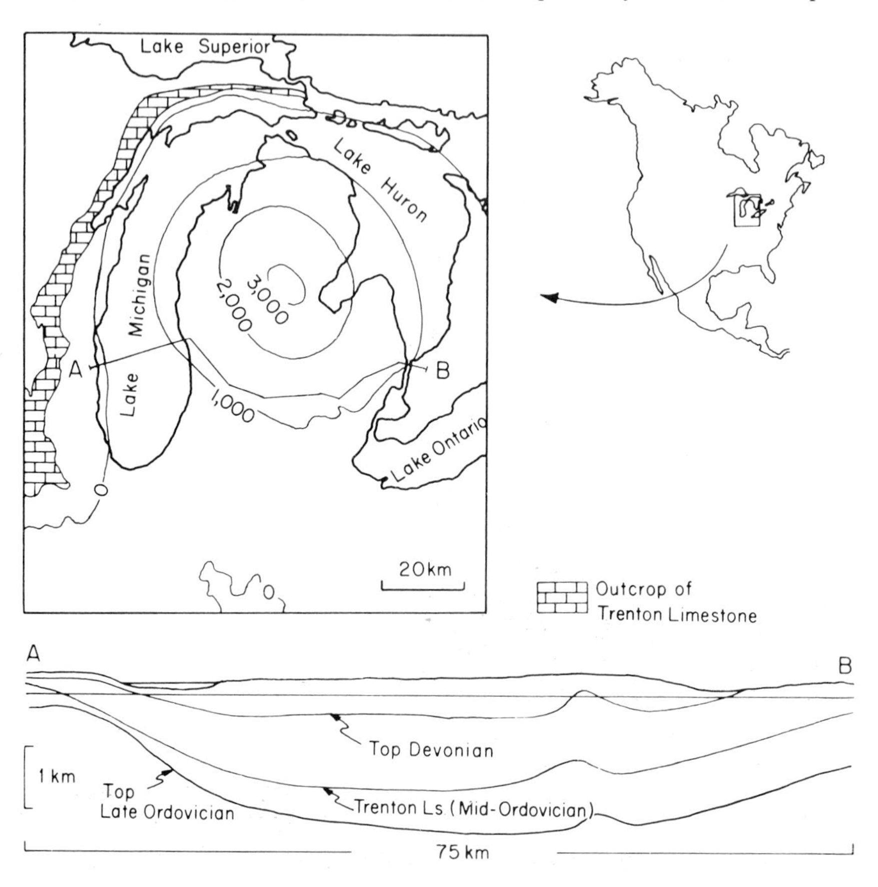

Fig. 171. Map and cross-section of the Michigan basin of the Great Lakes, North America. The subcircular basin shape is shown by a structure contour map of the top of the Trenton Limestone, Middle Ordovician. (From Cohee and Landes, 1952.)

of the open rim of the embayment closes the basin off from oceanic influence. Evaporite or continental sedimentation follows. An interesting example of this is provided by the Murzuk basin of Libya. The northern rim of this basin was separated from the Tethyan Ocean by the Gargaf arch, a tectonic feature which controlled deposition through much of Palaeozoic and Mesozoic time. Nevertheless, palaeocurrent analysis of fluvial

Cambro-Ordovician and Mesozoic strata indicate a northerly palaeoslope over the Gargaf arch (Fig. 174). These data show that the Murzuk basin alternately fluctuated from embayment to basin until attaining its present structurally closed intracratonic basin shape.

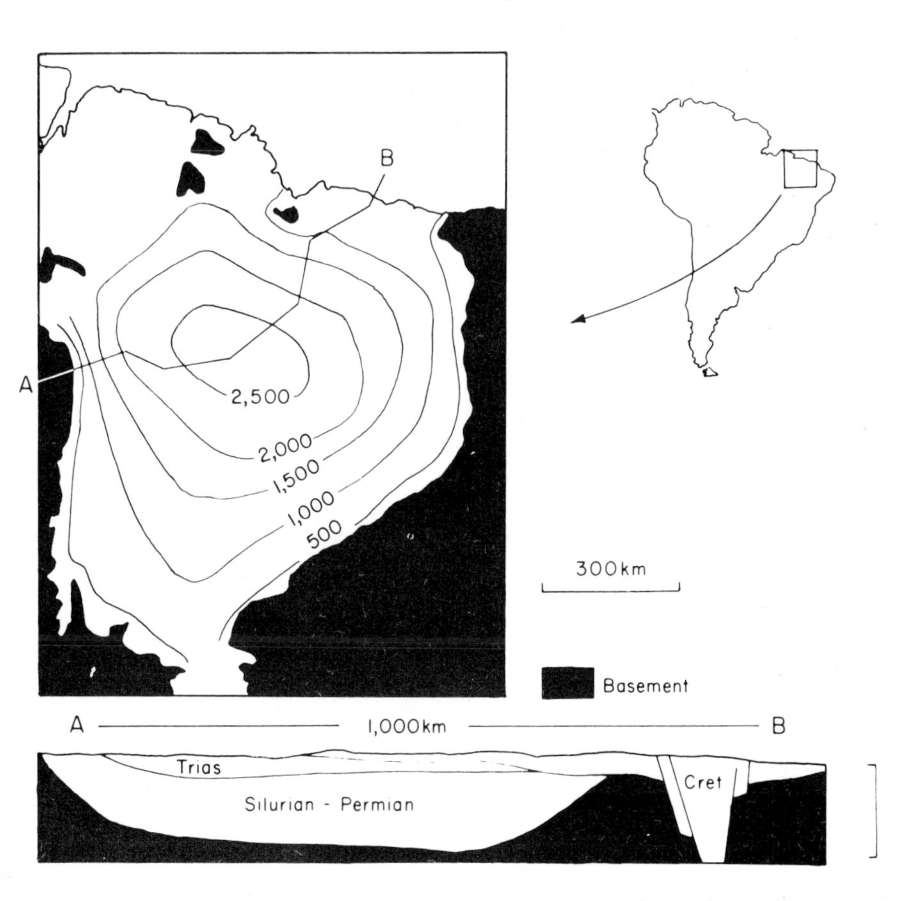

Fig. 172. Upper: isopach map showing the total sedimentary thickness of the Maranhao basin, Brazil, demonstrating its subcircular form. Lower: cross-section. (Data from Mesner and Wooldridge, 1964.)

2. *Epicratonic basins*

Epicratonic basins are those which lie on the edge of continental crust. Though subcircular in plan, by definition, they tend to be embayed and open towards the adjacent ocean basin. The axis of an epicratonic basin may plunge to the floor of the ocean or be interrupted by a sill-like feature at the rim of the continental margin. Examples of epicratonic basins include those of the Mississippi Gulf coast, the Niger delta basin and the Sirte basin of North Africa (Wilhelm and Ewing, 1972; Walcott, 1972; Weber, 1971; Burke, 1972; Sanford, 1970; Evamy *et al.*, 1978). These are summarily

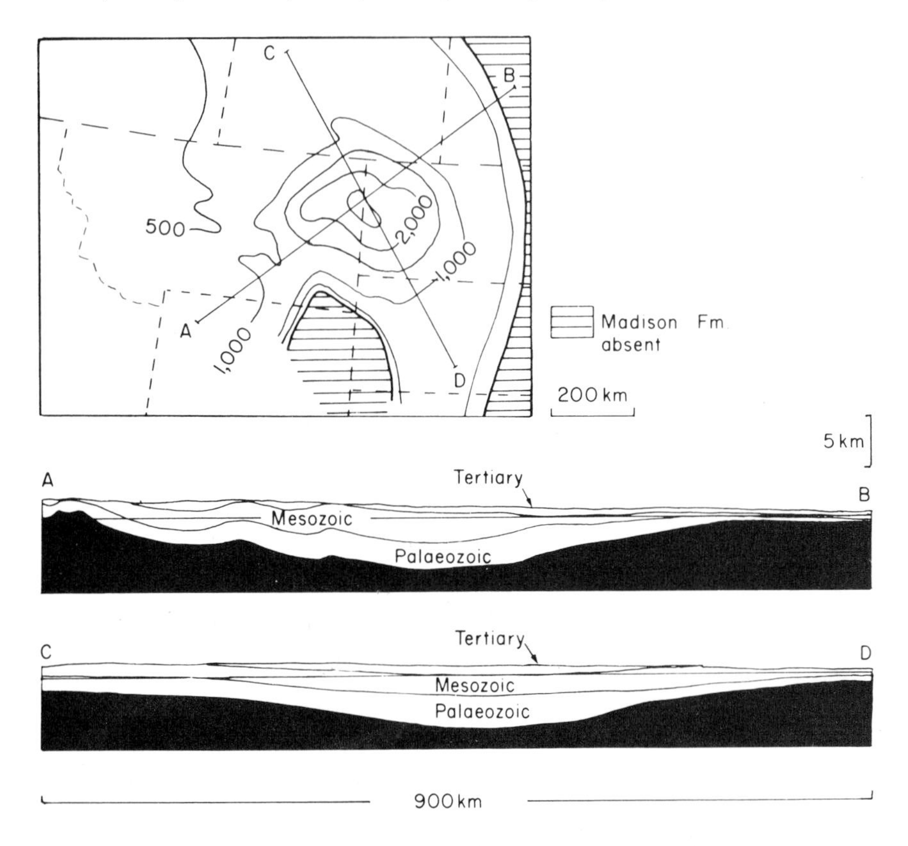

Fig. 173. Structure contour map on the top of the Madison Limeston (Missippian) showing the subcircular shape of the Williston basin. Cross-sections show the many ages of the rocks which infill it. (From Dallmus, 1958.)

illustrated in Figs 175 and 176. The Mississippi and Niger delta basins are very similar. Both originated towards the end of the Mesozoic and continued to be sites of active sedimentation until the present day.

Both basins contain a basal layer of salt, which diapirically intrudes younger sedimentary rocks. The Louann salt of the Gulf basin is of Jurassic age. The basal salt of the Niger delta basin is believed to be of Albian–Aptian age (Mascle *et al.*, 1973).

The basins are infilled by prisms of terrigenous clastics which were deposited in a range of environments. On the landward sides of the

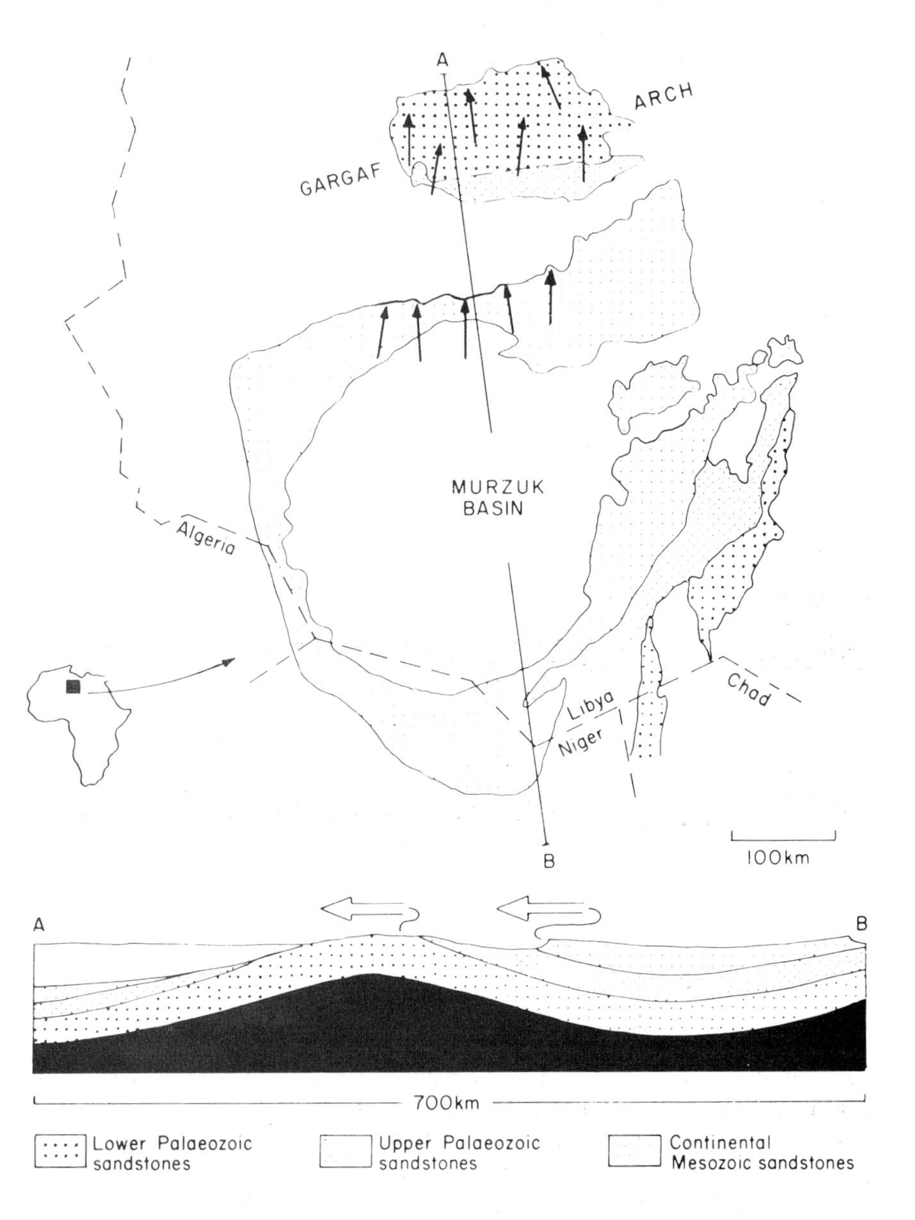

Fig. 174. The Murzuk basin of Libya is subcircular in shape, lying between the Sahara shield in the south and Tethys in the North. Palaeocurrent analysis in fluvial Cambro-Ordovician and Mesozoic sediments show that for much of its existence it was an embayment open to the north. (Data from McKee, 1965 and Burollet and Byramjee, 1969.)

embayments alluvial deposits predominate. These pass basinward into diverse shoreline facies, which include both barrier and deltaic deposits. These thin and grade seawards into marine slope muds, with some development of turbidite sand facies at the base of the slope. Geophysical data suggest a gradual seaward thinning of the continental crust beneath both the Mississippi Gulf coast and Niger delta basins (Walcott, 1972 and Burke, 1972, respectively).

The Sirte basin of North Africa shares many features with the Gulf coast and Niger basins. It too is essentially an embayment which opens out to an oceanic basin (the Mediterranean). The Sirte basin originated in the end of the Cretaceous and was infilled more or less continuously throughout the Tertiary.

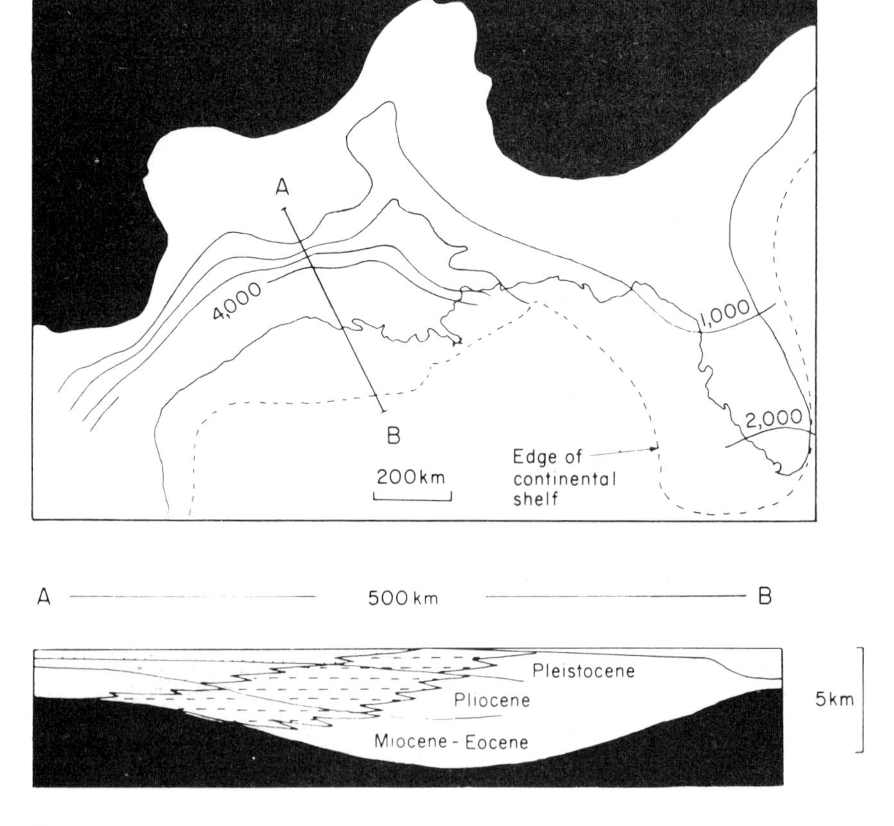

Fig. 175. Upper: isopach map of the Tertiary-Recent sedimentary fill of the Mississippi embayment. (From Murray, 1960.) Lower: cross-section of the embayment showing prograding sedimentary motif stippled which are alluvial sands. Dashed: deltaic sands, shales and coals. Blank: prodelta and shelf shales.

By contrast with the previous two examples, the Sirte basin was predominantly a site of carbonate sedimentation. Basal sands and thin evaporites are overlain by deep-water Upper Cretaceous and Palaeocene shales. These are thickest in intrabasinal troughs, while reefal carbonates were deposited on adjacent horsts. Throughout the Eocene the Sirte embayment was infilled by nearly a kilometre of interbedded carbonates and evaporites. The final phase of basin infilling during the Oligocene and Miocene involved terrigenous and carbonate sedimentation in both marine and continental environments. The active history of the basin was concluded by a bout of basaltic volcanism in the Pleistocene.

These brief reviews of three epicratonic basins show how they differ from the previously described intracratonic examples. Epicratonic basins tend to

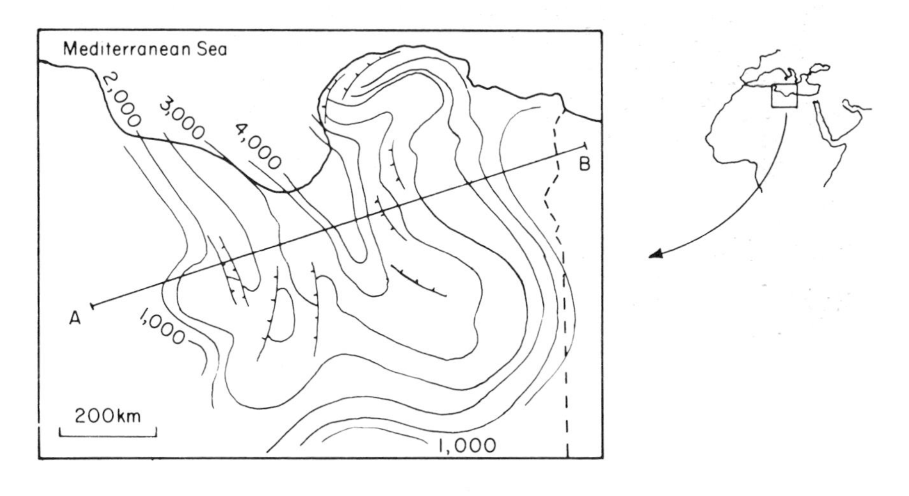

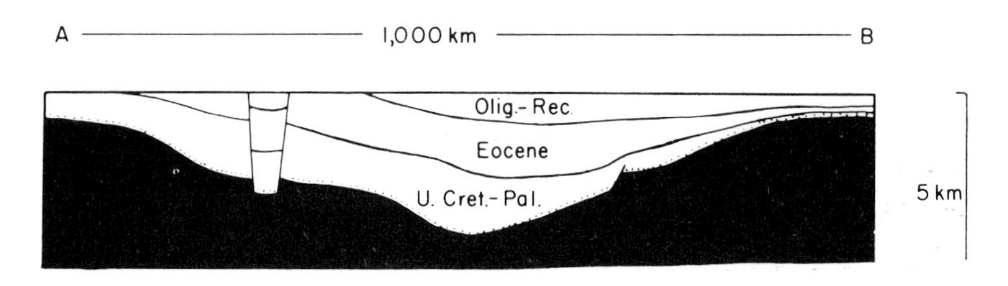

Fig. 176. Upper: structure contour map on mid-Cretaceous unconformity, Sirte embayment, from Sanford (1970). Lower: cross-section of the Sirte embayment. By contrast with the Mississippi embayment the Sirte is infilled largely by carbonates, evaporites and shales, with only minor amounts of sand in the initial and terminal phases.

be much less stable than intracratonic ones, due to their situation at continent margins. Initial basin subsidence can be rapid, resulting in an early phase of deep-water sedimentation. The floor of the Sirte basin was extensively faulted and there was some igneous activity. Like intracratonic basins, however, epicratonic embayments can be infilled by both carbonates and terrigenous sediment. This differentiation is a function of the degree of uplift of the adjacent crust.

C. Troughs

As previously defined, a trough is an elongated sedimentary basin. Three main types of trough may be recognized. The most dramatic is the geosyncline. This is a trough of highly tectonized sediment which forms mountain belts kilometres deep and hundreds of kilometres long. The two other types of trough both develop adjacent to a geosynclinal furrow in the fore deep. The first of these occurs synchronously with geosynclinal sedimentation. The second develops as a later response to uplift of the mountain chain.

These three types of trough, together with the intramontane rifts described in the next section, may be referred to as the geosynclinal suite of sedimentary basins (Table XXXIII). To understand basins in general, and the geosynclinal suite of basins in particular, it is necessary to know something of the larger-scale morphology and mechanics of the earth.

1. Plate tectonics: a necessary digression

Once upon a time geology students were taught that the continents were composed largely of silica and alumina (sial) and floated isostatically on denser oceanic crust composed largely of silica and magnesia (sima). Mountain chains occurred where continents bumped together pushing up folded belts of the sediments which had been deposited in the troughs between the continents. The compression of the deposits of the Tethys Ocean to form the Alpine mountain chain was the classic example. The motive power in this case being the convergence of the European and African shields.

Mountain chains such as the Appalachians and the Andes were hard to fit into so simple a scheme, as half of the vice was absent. One explanation offered was that a continent had foundered on the oceanic side of such mountain chains, apparently in defiance of the principles of isostasy. An alternative proposal was that continents could drift horizontally across the face of the earth. The close geographic and geologic fit of the *circum*-Atlantic continents was the keystone of this thesis (Wegener, 1924; du Toit, 1937).

These ideas have been rejuvenated by recent advances in oceanography and geophysics; advances which have lead to the formulation of a refined model variously referred to as "plate tectonics" or "the new global tectonics".

There is a vast literature on this topic. Source books include Wyllie (1971), Tarling and Runcorn (1973), Seyfert and Sirkin (1973), Fischer and Judson (1975) and Davis and Runcorn (1980). The following brief account of these concepts is given to illustrate their relevance to basin analysis in general and geosynclines in particular.

The new global tectonics is based on the concept that the surface of the earth is made up of a mosaic of rigid plates (Fig. 177). These rigid plates, termed the lithosphere, consist of a continuous layer of basaltic rocks above which is a discontinuous layer of continental crust. The upper mantle layer corresponds essentially to "sima" and forms the floor of ocean basins. The crust forms the granitoid continental areas, corresponding to "sial".

The evidence now assembled suggests to many workers, not that the crust

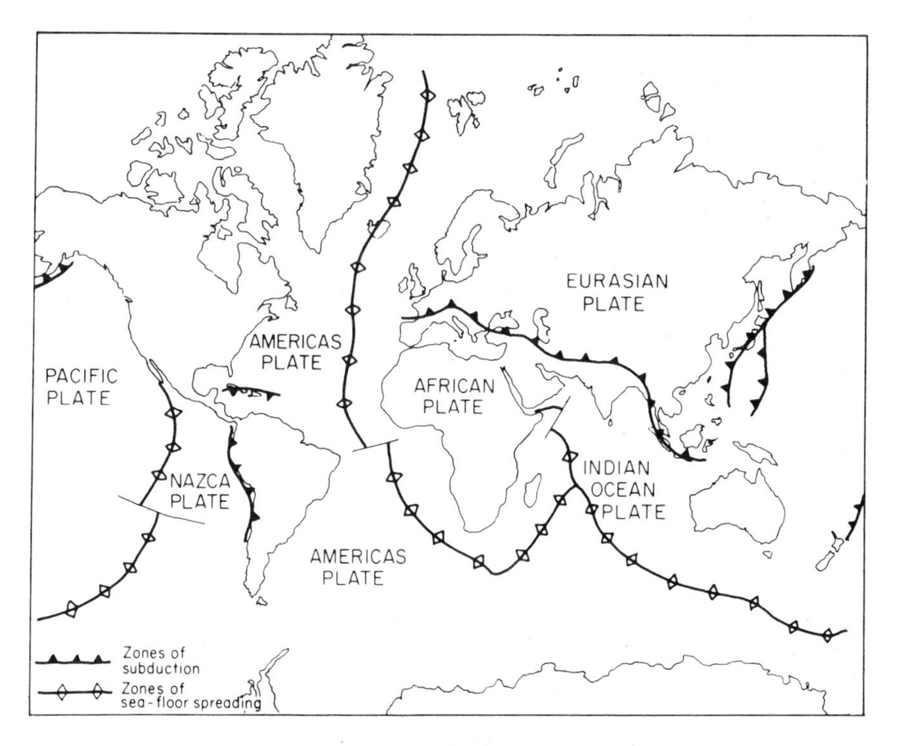

Fig. 177. World map showing the major zones of sea-floor spreading and zones of subduction. New crust is generated at the former by upwelling along axial volcanic rifts, and drawn down into the mantle at the latter (simplified from various sources).

moves over the upper mantle, but that the lithosphere (crust and upper mantle together) may ride over the deeper asthenosphere. Present geophysical data suggest that the boundary between rigid lithosphere and plastic asthenosphere lies between 100–150 km beneath the earth's surface (Fig. 178).

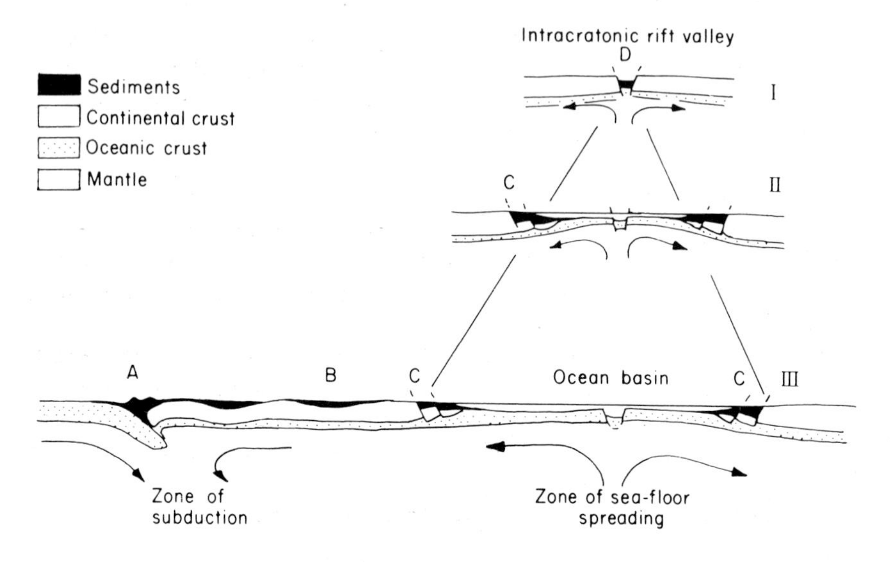

Fig. 178. Cross-sections illustrating the basic concepts of plate tectonics. I: an axis of sea-floor spreading develops, in this instance beneath continental crust. Updoming occurs and a volcanic rift valley is formed (D). The East African rifts are a modern example. II: with continued upwelling the two new plates are torn apart. New oceanic crust is formed by extensive vulcanicity between two flanking half-graben basins (C). The Red Sea is now at this stage. III: New crust continues to be formed at the axial zone of sea-floor spreading. The two continental shields, now far apart, have ocean margin basins on their opposing shores (C). At far left is a zone of subduction down which crust is drawn. This is the site of a geosynclinal trough (A). Intracratonic basins occur on the continental crust (B). A cross-section from the Andes to Africa is similar to this sketch.

New lithospheric material is added to each plate at zones of sea-floor spreading. These are the mid-oceanic ridges. These ridges are seismically and volcanically active, they have a high rate of heat flow, and are topographically and palaeomagnetically symmetrical on either side of axial rift valleys. The ridges are composed of young basalts which are overlain by progressively older pelagic sediment away from the ridge.

The new lithosphere generated at these zones of sea-floor spreading is carried across the plate, and it ultimately drawn down into the earth along zones of subduction on the far side of the plate. Some zones of subduction occur at the junction of continental and oceanic crustal plates, as for example under Central America, and off the central and southern Andes. Island arcs are formed where two plates of oceanic crust meet in a zone of subduction. Continents drift across the plates as if on a conveyor belt.

The concept of plate tectonics is far more complex than the preceding few sentences suggest, and it is not universally accepted (see for example Meyerhoff and Meyerhoff, 1972). Nevertheless, it is one of the most stimulating concepts of the present time, impinging upon all branches of geology. Plate tectonics invites a reappraisal of concepts of basin analysis especially in the case of the tectonized linear troughs termed geosynclines (see, for example, Reading and Mitchell, 1978, and Crostalla, 1977).

2. Geosynclines

The concept of the geosyncline has been one of the most stimulating ideas in geology, because it has brought together workers from many branches including structural, igneous, metamorphic, stratigraphic and geophysical approaches.

The concept of the geosyncline was born from the work of Hall (1850) and Dana (1873) in the Appalachians. Here a mountain range composed of a vast thickness of shallow-water sediment, pointed to the continued subsidence of a linear trough over a long time-span, followed by tectonism and uplift.

The geosynclinal theory was introduced to Europe and applied to the Alps by Haug (1900).

Subsequently the term geosyncline became so widely used and ill-defined that it embraced all types of sedimentary basin. Significant papers on geosynclines and their classification were published by Schuchert (1923), Stille (1936), Kay (1944, 1947) and Glaessner and Teichert (1947). Aubouin's definitive monograph (1965) expounded the geosynclinal concept as commonly understood by most geologists, and documented his model with several case histories.

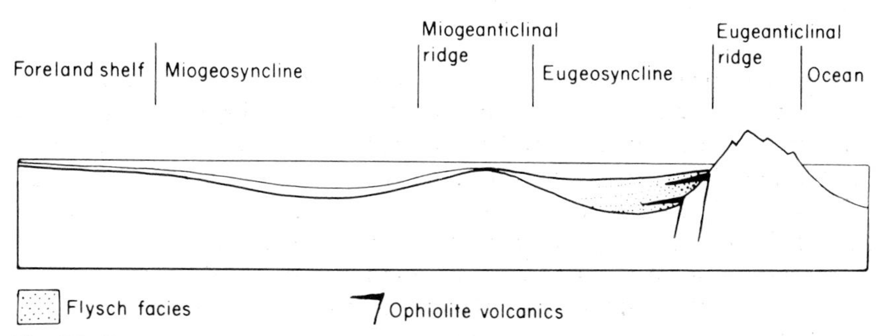

Fig. 179. Sketch cross-section of a geosyncline, showing the various tectonic elements. (From Aubouin, 1965.)

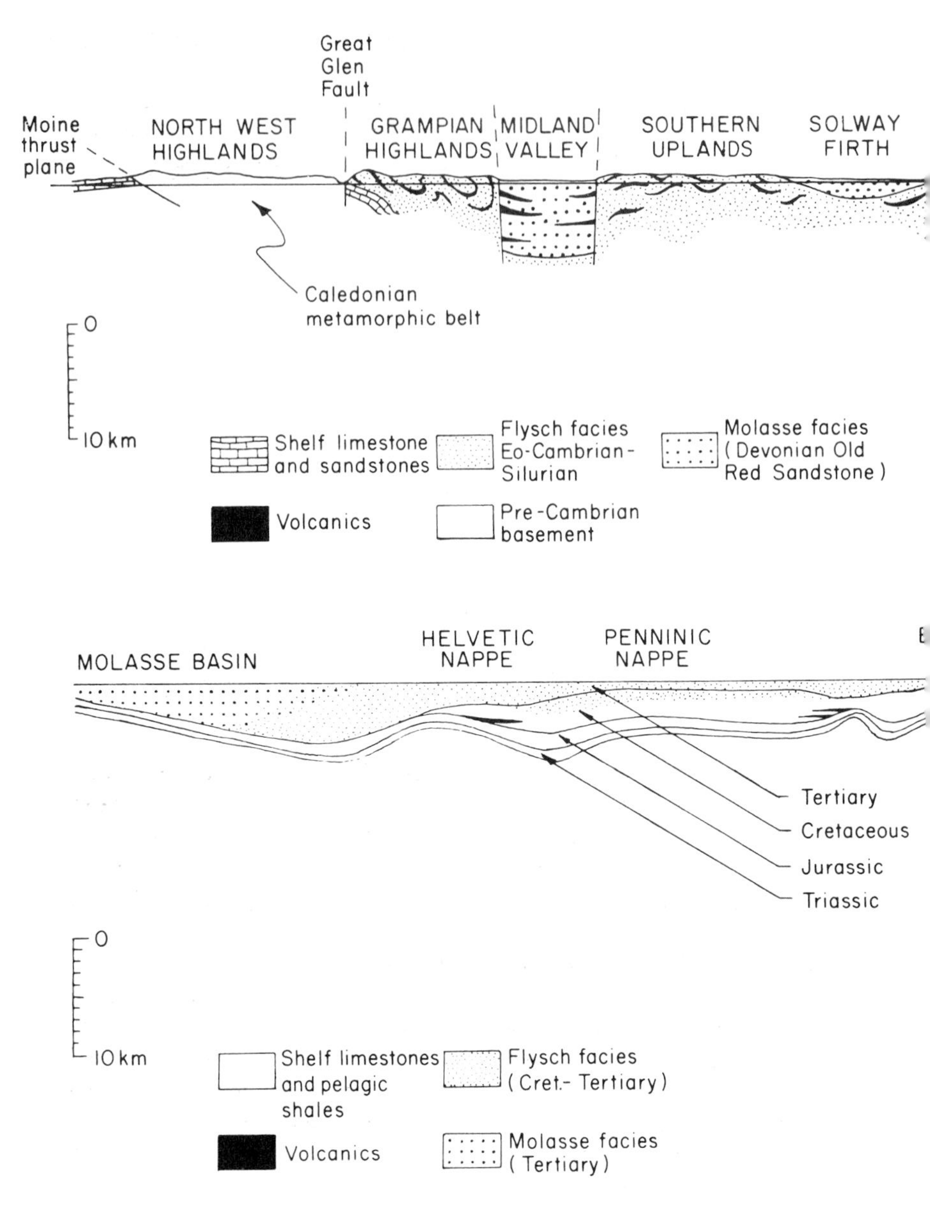

Fig. 180. Upper: cross-section of the Caledonian geosyncline of the British Isles, based on Stoneley (1969). Length of section is 800 km. The flysch facies, largely greywacke turbidites, range in age from Eo-Cambrian-Silurian. They are extensively regionally metamorphosed in the Scottish eugeosyncline, but only gently folded in the Welsh miogeosyncline. The Devonian Old Red Sandstone is the post-orogenic molasse of the Caledonian orogeny. Lower: Palinspastic cross-section of the Alps showing axial zone of thick Mesozoic and Tertiary sediments flanked by younger post-orogenic molasse basins. (Based on Bernoulli *et al.*, 1974.) Length of section is 600 km. The effects of the Alpine nappe-forming orogenesis have been removed in this figure.

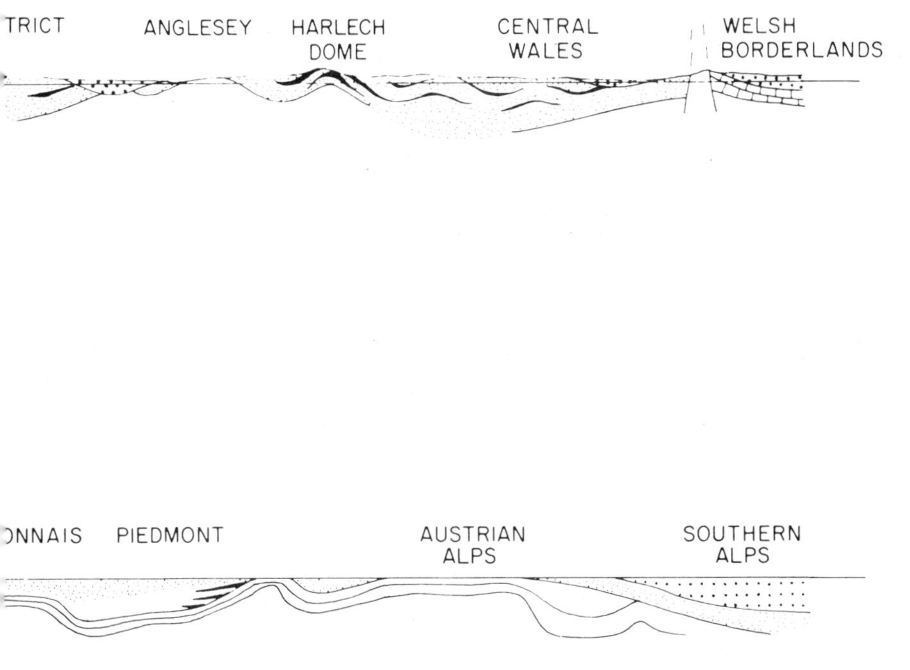

The present custom is to restrict the term to linear tectonized sedimentary troughs. More specifically geosynclinal troughs are believed to show a complex, yet characteristic, morphology, and a regular sequence of structural evolution, sedimentary facies and igneous activity. Fig. 179 illustrates the geosynclinal model defined by Aubouin. Figure 180 illustrates two examples.

Aubouin's studies, largely based on the Hellenide fold belt of the eastern Mediterranean, divide geosynclines into several tectonomorphic zones. A stable foreland on the craton passes laterally into a shallow trough termed the miogeosynclinal furrow (fore-deep of some writers). This is separated by a positive axis, the miogeanticline, from the eugeosynclinal furrow which is the main active and unstable trough. The eugeosyncline is separated from the open oceanic basin by the eugeanticline. This second positive feature is an island arc of rising volcanics.

The evolution of a geosyncline commences with the development of a tensional sag on the site of the future eugeosyncline. Pelagic limestones and radiolarian cherts develop in the basin in synchronization with shelf carbonates and shelf sands on the adjacent stable platforms.

The sag deepens to form the eugeosynclinal furrow. Subaqueous igneous activity causes the cherts to be interbedded with pillow lavas, ophiolites and spilites (e.g. Hynes *et al.*, 1972). This is the Steinmann Trinity discussed on

p. 90. Simultaneously, the miogeosynclinal furrow forms, separated from the eugeosyncline by the miogeanticline.

Sedimentation in the miogeosyncline is still largely of shelf and shallow marine facies because subsidence is generally sufficiently gradual for deposition to keep pace. The sediments of the miogeosyncline differ from those of the foreland in thickness rather than facies. Miogeosynclines are in many ways similar to epicratonic basins.

Continued tectonic activity leads to uplift of the eugeanticline. This becomes a major source of terrigenous sediment, immature in texture and mineralogy and often volcaniclastic. These deposits are laid down within the eugeosyncline largely as turbidites to form flysch facies.

Flysch is a term which has had a lengthy and much abused history in the literature of geology. Like the geosynclinal concept, however, it seems to have a high preservation potential. Detailed accounts of flysch facies have been given by Dzulinski and Walton (1965) and Lajoie (1970). Hsu (1970) has discussed flysch semantics and Reading (1972) has reviewed flysch facies in the light of the new global tectonics. Flysch is generally regarded as the synorogenic deposit of geosynclinal troughs. Sedimentologically most flysch show the features of turbidites. Flysch is petrographically diverse but in older Palaeozoic and Proterozoic geosynclines it is characteristically of greywacke type. Carbonate flysch is known however, and, in the Hellenide geosyncline, troughs of detrital carbonate turbidites were presumably supplied from the adjacent miogeanticline.

The main orogenic phase of the geosyncline is diverse and beyond the scope of this text. It may range from the high-angle thrust faulting and tight synclinal folding of the Rockies, to the recumbent folds and low-angle thrusts of the Alps. Orogenesis also involves regional metamorphism and plutonic activity within the eugeosyncline. During orogenesis the axis of the rising eugeanticline migrates towards the craton. This forces the depositional axis of the flysch facies in the same direction spilling sediment over the miogeanticline into the fore deep. During this migration the deposits change from flysch to molasse facies. The term molasse, like flysch, has had a chequered career since it was first used in the French Alps by Bertrand in 1897. Molasse deposits are generally taken to be "late orogenic clastic wedges deposited in the linear fore-deep on the flank of a craton" (Van Houten, 1973). Molasse facies are largely coarse terrigenous clastics with abundant conglomerates, few shales and negligible limestones. These beds are generally laid down in non-marine fluvial and fanglomerate environments. Grain size decreases and beds thin away from the mountain front. In the type area of the Alps, however, some molasse is marine.

The final phase of geosynclinal evolution is regional epeirogenic uplift. This is typically accompanied by block faulting within the new mountain chain. Intramontane troughs are infilled with thick non-marine clastic sequences, often accompanied by basalt effusion along the faults. The range and basin province of the Rocky Mountains is of this type. The Devonian rocks of the North Atlantic environs also provide good examples of post-orogenic sedimentary basins. These continental red beds include both post-orogenic molasse deposited in fore deeps, such as the Pocono and Catskill facies of the Appalachians, and intramontane rift basins such as the Midland Valley of Scotland (Friend, 1969).

In summary, this brief review of geosynclinal sedimentation shows that the evolution of a geosyncline may involve the genesis of four different types of basin, termed the geosynclinal suite (Table XXXIII). Each basin has a characteristic morphology, tectonic setting, time of development and type of fill. The eugeosynclinal basin is composed of pre-orogenic pelagic sediments overlain by synorogenic flysch facies. These are generally tectonically deformed and metamorphosed. The main flysch trough is often host to minor rifted troughs filled with post-orogenic continental clastics. Two basin types develop in the fore-deep between the main geosynclinal axis and the craton. An earlier epicratonic marine basin is often partially obliterated by a post-orogenic molasse trough filled with terrigenous continental deposits shed from the rising mountain chain.

The geosynclinal model outlined above has been reviewed in recent years in the light of the new global tectonics. The concept has been criticized by Ahmad (1968) and Coney (1970) largely on the grounds that the time connotations of the geosynclinal cycle with a beginning, a middle and an end, are hard to reconcile with the continuous processes of subduction and sea-floor spreading postulated by the new plate theory.

More constructive attempts to integrate geosynclines and plate tectonics have been made by Mitchell and Reading (1969, 1978), Schwab (1971) and Roberts (1972). Mitchell and Reading have shown that most geosynclines, ancient and modern, fall into four main categories.

Atlantic-type geosynclines occur at the edges of continents which are not zones of subduction (as around the present Atlantic). Sedimentation takes place here essentially as shelf carbonates, and as sands and ocean-rise flysch which Schwab (1971) compares with miogeosynclinal and eugeosynclinal types. These sediments are laid down in a non-orogenic setting. The flysch facies may be due, not so much to turbidite processes, as to oceanic-bottom traction currents (see p. 188), and Stanley (1970) has suggested that such Atlantic margin deposits should be termed "flyschoid" to differentiate them from synorogenic turbidite flysch.

Andean-type geosynclines occur at zones of subduction where oceanic crust descends beneath a submarine trench adjacent to a continental mountain arc.

Island arc-type geosynclines occur where a zone of subduction involves only oceanic crust.

The fourth type, Japan sea-type geosynclines, occur where small ocean basins lie between a continent and an island arc.

This fourfold scheme successfully integrates the geosynclinal and plate theories and helps to explain many types of ancient geosynclinal troughs.

D. Rifts

Rift basins are long fault-bounded troughs which occur in various tectonic settings and show a corresponding diversity of sediment fill. They are of considerable economic importance as sources of hydrocarbons, evaporites and metals. Furthermore, rift basins contain many clues to the understanding of the new global tectonics. Important publications on this topic include works by Unesco (1965), Illies and Mueller (1970) and Tarling and Runcorn (1973, pp. 731–788).

Essentially there are four types of rift basin, each characterized by a particular type of tectonic setting and sedimentary fill.

The post-orogenic intramontane rifts of geosynclinal belts were described in the previous section. They are genetically distinct from all other rift basins save in their tensional origin.

The three other types of rift basin all develop at zones of upwelling of the mantle which may ultimately become belts of sea-floor spreading.

More of less continuous lines of rift valleys are found along the mid-ocean ridge systems of the world and can be traced into the continental crust of Africa and Europe. The sedimentary basins of the mid-ocean ridge rifts are relatively little known. Their fill appears to be largely of pelagic suspended sediment interbedded with the volcanic assemblage of new oceanic crust which is generated in these zones of intense igneous and seismic activity.

Rifts are initiated by an updoming of the crust. The fracture system which results consists of a triradiate rift system whose node is referred to as a triple-rift (or colloquially triple-R) junction. When a zone of sea floor spreading develops beneath this point the crust separates and the floors of the rifts subside to be infilled by a regular sequence of sedimentary facies (Schneider, 1972). Initially the rifts lie above sea level, so they are infilled with coarse fluvial and lacustrine sediments, often associated with volcanics. When the rift floor reaches sea level it provides a shallow restricted trough which (given the right climate) favours evaporite development, followed by restricted marine sedimentation as the rift subsides permanently, below the sea. This favours oil source rock formation. With continued subsidence and separation the rift becomes infilled with open marine sediments, clastic or carbonate. As the two sides of the new zone of sea-floor spreading move apart two of the rift arms develop into incipient oceans, each flanked with half-graben sedimentary basins. The third branch of the triple-R junction will not separate, though it may be infilled by sediments similar in facies and thickness to those in the other two rifts. This is referred to as an aborted rift or failed arm (Fig. 181).

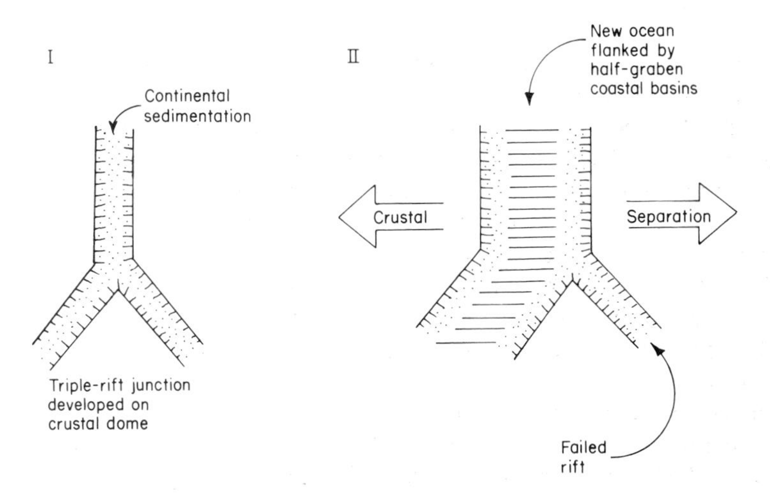

Fig. 181. Diagrams to illustrate the formation of a triple-rift junction in continental crust (left), followed by crustal separation and the development of an aborted rift (right).

The North Sea rift is a fine example of a failed arm (Woodland, 1975; Illing and Hobson, 1981).

Thus rift basins form a continuous spectrum related to the progressive breaking up and lateral drift of continental margins. The various basin types will now be described in more detail.

1. Intracontinental rift basins

The first phase of rift development is seen within cratonic areas of continental crust. Localized up-doming generates crestal tensional rifts. Triradial rift valley systems commonly diverge from the culmination of the dome. The rifts become infilled with continental fluviolacustrine deposits shed from the faulted basin margins. Igneous activity contributes lavas and volcaniclastic detritus to the basin fill.

Best known of these intracontinental rift systems are those of Baikal in Siberia, of the German Rhine Valley, and of East Africa. The Baikal rift developed in a basement of Palaeozoic and Pre-Cambrian rocks. It is some

400 km in length and 50 km wide. Down-warping began in the Miocene; active rifting in the Pliocene. The Baikal rift basin now has a fill of some 5 km of non-marine clastics (Salop, 1967).

The Rhine graben transects the Rhine shield with the Vosges mountains and the Black mountains on its west and east flank respectively (Fig. 182). Rifting appears to have begun in the Middle Eocene. The Rhine graben contains over three kilometres of diverse non-marine facies and is still seismically active and presumably subsiding. Where best developed the basin is about 40 km wide and some 250 km long (Illies, 1970). Enthusiasts, however, can demonstrate that the Rhine graben is actually part of a line of rift basins traceable from the North Sea, via the Rhine, to the Rhone Valley, and thence across the Mediterranean to the Hon graben of Libya.

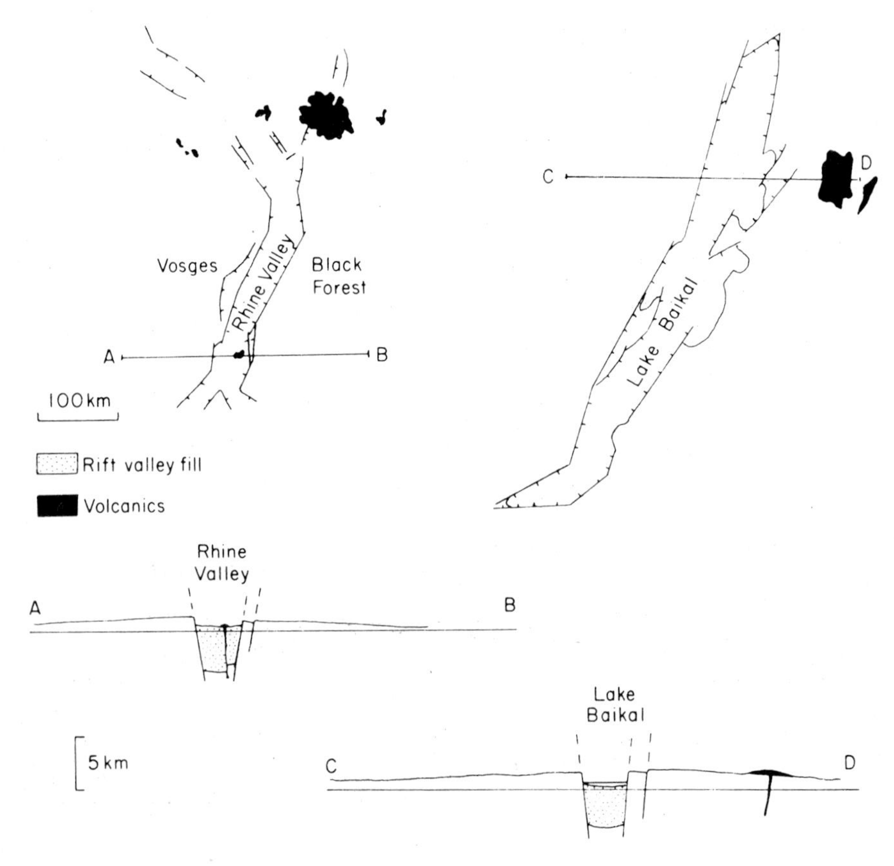

Fig. 182. Maps and cross-sections of the Rhine Valley (Germany) and Baikal (USSR) rift basins drawn to the same scale. These are largely infilled with Tertiary non-marine clastic sediments and volcanics.

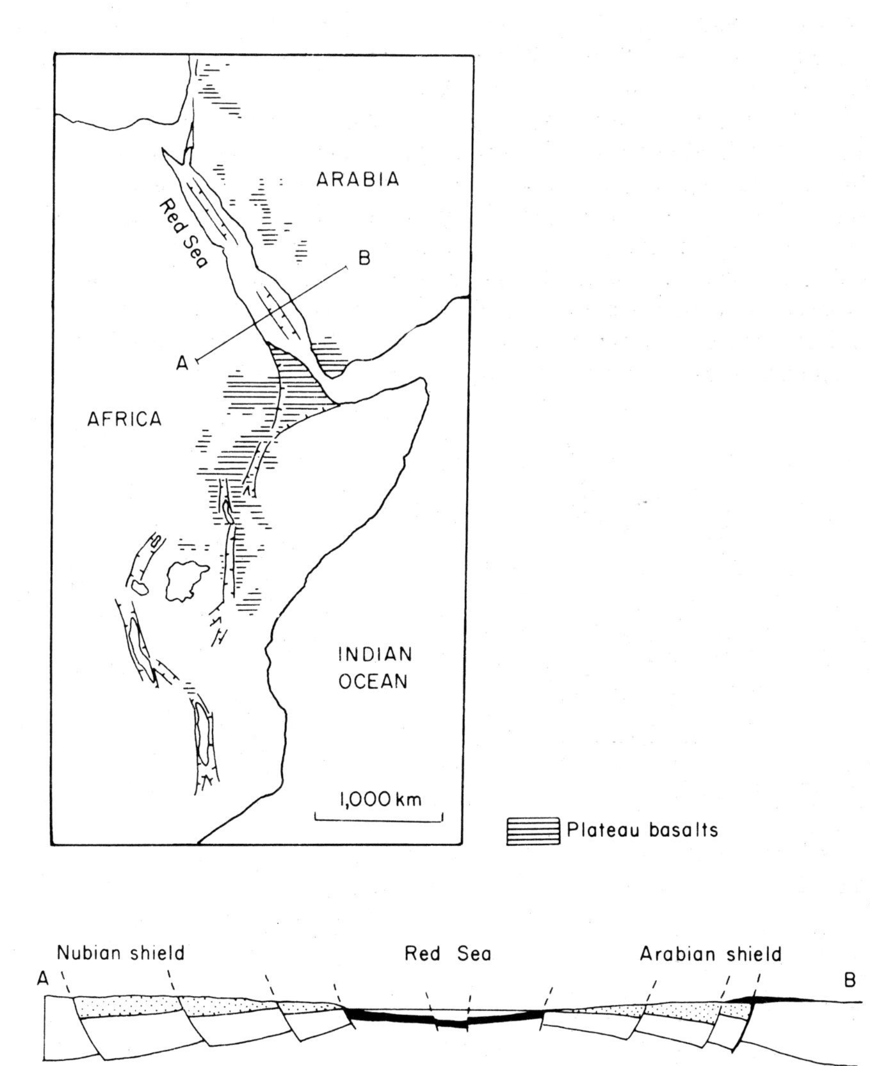

Fig. 183. Upper: map showing the extent of the East African rift valley basins and their continuation into the Red Sea. Lower: cross-section of the Red Sea showing the coastal sedimentary basins and the axial rift where volcanic activity forms new oceanic crust. (From Lowell and Genik, 1972.)

A second line of rifts subparallel to the first runs from Suez down the Red Sea and up to the great lakes of East Africa (Fig. 183). These East African rift valleys are intracontinental basins analogous to the Rhine graben. They are at present seismically and volcanically active. Let down into the African Pre-Cambrian metamorphic shield, they are infilled by Late Tertiary and Recent volcanics and fluviolacustrine deposits (Darrcott *et al.*, 1973; Baker *et al.*, 1972).

2. Intercratonic rifts

Further development of tension and thinning of continental crust beneath a rift, depresses its floor to sea level.

Sea water floods the rift basin from time to time, but intermittent seismic uplift during net subsidence can trap saline water within the trough. Thus the continental clastic facies are overlain by evaporites.

The Suez graben is an example of this next stage of rift basin evolution. Pre-rift basement rocks are overlain by some 4 km of Miocene sediments which consist of a lower clastic part, the Gharandal group, overlain by the Evaporite group. These are a diverse assemblage of gypsum–anhydrite, marls, rock salt, dolomites and algal limestones (Heybroek, 1965).

The Suez gulf opens out southwards into the Red Sea. This is a true intercratonic rift basin. It was formed by the crustal rifting of the Arabo-Nubian Pre-Cambrian craton (Fig. 183). The axis of the Red Sea consists of thin marine sediments of pelagic facies, which overlie volcanics. Strange "hot spots" on the sea floor are infilled with brines and are areas where native metals are precipitated on the sea floor (Degens and Ross, 1969). Volcanic rocks lie at shallow depths and magnetic and gravity data suggest that oceanic crust is not far below (Lowell and Genik, 1972, 1975).

The margins of the Red Sea are marked by two parallel sedimentary basins. They dip regionally away from the axis of the Red Sea and have complex horst and graben floors. Essentially these marginal basins of the intercratonic rift are stratigraphically contiguous with the Suez graben. Basement is overlain by red beds which are in turn overlain by thick evaporites in the grabens and reefal carbonates on the horsts (Lowell and Genik, 1972, 1975). These are Miocene in age and are unconformably overstepped by Pliocene and younger sediments, both marine, non-marine, terrigenous and carbonate.

The southern entrance of the Red Sea is continuous with the Carlsberg Ridge of the Indian Ocean. Thus the intracratonic and intercratonic rift basins are genetically related to the oceanic ridges and axial rifts which are the hallmarks of zones of sea-floor spreading.

From the evidence of sea-floor spreading seen in its early stages in the Red Sea and adjacent rifts, it follows that analogous half graben basins with similar facies sequences should be present, for example, along the Atlantic coasts. This is indeed the case in the Southern Atlantic. The coasts of Brazil and of Africa, from Senegal to Gabon, possess a series of tensional horst and graben coastal basins. In each basin basal Cretaceous non-marine clastics are overlain by evaporites, which diapirically intrude thick wedges of younger marine deposits (Fig. 184). These are essentially the geosynclines of Atlantic type defined by Reading and Mitchell (1969).

So we have come full circle, demonstrating the limitations of classifying such complex features as sedimentary basins.

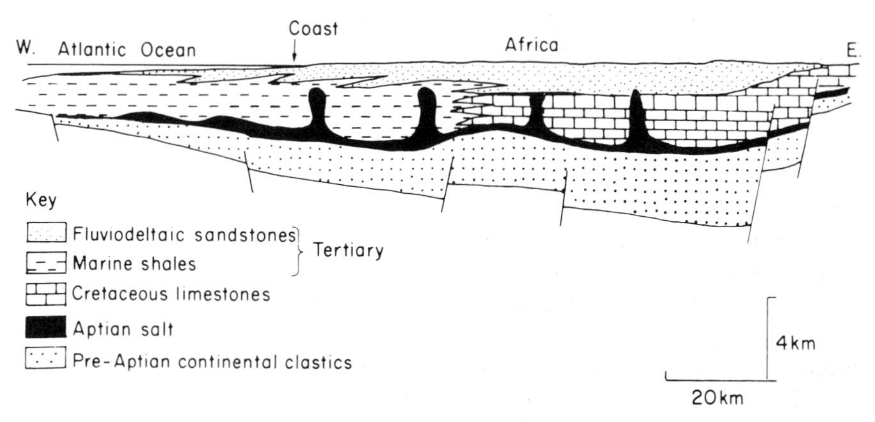

Fig. 184. Cross-section of the Gabon basin from Brink (1974). This is a typical example of an ocean margin basin, the end product of a rift/drift sequence. The characteristic basin fill shows a vertical sequence of continental clastics, evaporites and marine sediments.

III. REFERENCES

Ahmad, F. (1968). Orogeny, geosynclines and continental drift. *Tectonophysics* **5**, 177–189.

Aubouin, J. (1965). "Geosynclines" Developments in Geotectonics, Vol. 1. Elsevier, New York. 335pp.

Baker, B. H., Mohr, P. A. and Williams, L. A. J. (1972). Geology of the eastern rift system of Africa. Spec. Pap. geol. Soc. Am. **136**.

Belmonte, Y., Hirtz, P. and Wenger, R. (1965). The salt basins of the Gabon and Congo (Brazzaville). *In* "Salt Basins Around Africa" 55–78. Inst. Petrol., London.

Bernoulli, D., Laubscher, H. P., Trumpy, R. and Wenk, K. (1974). Central Alps and Jura Mountains. *In* "Mesozoic-Cenozoic Orogenic Belts" 85–108. Geol. Soc. Lond. Sp. Pub. No. 4.

Bertrand, M. (1897). Structure des Alpes Françaises et recurrence de certaines facies sedimentaires. Rep. Int. Geol. Cong. 6th Session. Comptes Rendus, 161–177.

Bott, M. H. P. and Johnson, G. A. L. (1967). The controlling mechanism of Carboniferous cyclic sedimentation. *Q. Jl geol. Soc. Lond.* **122**, 421–441.

Brink, A. H. (1974). Petroleum geology of Gabon basin. *Bull. Am. Ass. Petrol. Geol.* **58**, 216–235.

Burke, K. (1972). Longshore drift, submarine canyons, and submarine fans in development of Niger Delta. *Bull. Am. Ass. Petrol. Geol.* **56**, 1975–1983.

Burollet, P. F. and Byramjee, R. (1969). Sedimentological remarks on Lower Paleozoic Sandstones of South Libya. *In* "Geology, Archaeology and Prehistory of the Southwestern Fezzan, Libya" (W. H. Kanes, Ed.), 91–102. Petrol. Explor. Soc. Libya, Tripoli.

Carver, R. E. (1968). Differential compaction as a cause of regional contemporaneous faults. *Bull. Am. Ass. Petrol. Geol.* **52**, 414–419.

Cohee, G. V. and Landes, K. K. (1958). Oil in the Michigan basin. *In* "The Habitat of Oil" (L. G. Weeks, Ed.), 473–493. Am. Ass. Petrol. Geol. Tulsa.

Conant, L. C. and Goudarzi, G. H. (1967). Stratigraphic and tectonic framework of Libya. *Bull. Am. Ass. Petrol. Geol.* **51**, 719–730.

Coney, P. J. (1970). The geotectonic cycle and the new global tectonics. *Bull. geol. Soc. Am.* **81**, 739–747.

Crostella, A. (1977). Geosynclines and Plate Tectonics in Banda Arcs, Eastern Indonesia. *Bull. Am. Ass. Petrol. Geol.* **61**, 2063–2081.

Dallmus, K. F. (1958). Mechanics of basin evolution and its relation to the habitat of oil. *In* "The Habitat of Oil" (L. G. Weeks, Ed.), 883–931. Am. Ass. Petrol. Geol. Tulsa.

Dana, J. D. (1873). On some results of the earth's contraction from cooling, including a discussion of the origin of mountains and the nature of the earth's interior. *Am. J. Sci.* **5**, 423–443.

Dana, J. D. (1873). On some results of the earth's contraction from cooling, including a discussion of the origin of mountains and the nature of the earth's interior. *Am. J. Sci.* **6**, 6–14, 104–115, 161–172.

Darling, G. B. and Wood, P. W. J. (1958). Habitat of oil in the Canadian portion of the Williston basin. *In* "The Habitat of Oil" (L. G. Weeks, Ed.), 129–148. Am. Ass. Petrol. Geol. Tulsa.

Darcott, B. W., Girdler, R. W., Fairhead, J. D. and Hall, S. A. (1973). The East African rift system. *In* "Implications of Continental Drift to the Earth Sciences" (D. H. Tarling and S. K. Runcorn, Eds), 757–766. Academic Press, London and New York.

Davies, P. A. and Runcorn, S. A. (1980). "Mechanics of Continental Drift and Plate Tectonics" Academic Press, London and New York.

Degens, E. T. and Ross, D. A. (Eds) (1969). "Hot Brines and Recent Heavy Metal Deposits in the Red Sea" Springer-Verlag, Berlin. 600pp.

De Sitter, L. U. (1964). "Structural Geology" McGraw-Hill, London. 551pp.

Drake, C. L., Ewing, J. I. and Stokand, H. (1968). The continental margin of the United States. *Can. Jl Earth Sci.* **5**, 99–110.

Dzulinski, S. and Walton, E. K. (1965). "Sedimentary Features of Flysch and Greywacke" Elsevier, Amsterdam. 300pp.

Evamy, B. D., Haremboure, J., Kamerling, P., Knapp, W. A., Malloy, F. A. and Rowlands, P. H. (1978). Hydrocarbon habitat of Tertiary Niger delta. *Bull. Am. Ass. Petrol. Geol.* **62**, 1–39.

Fischer, A. G. and Judson, S. (Eds.) (1975). "Petroleum and Global Tectonics" Princeton Univ. Press, Princeton. 322pp.

Friend, P. F. (1969). Tectonic features of Old Red Sedimentation in North Atlantic borders. *In* "North Atlantic—Geology and Continental Drift" (M. Kay, Ed.), 703–710. Am. Ass. Petrol. Geol. Mem. No. 12.

Glaessner, M. F. and Teichert, C. (1947). Geosynclines: a fundamental concept in geology. *Am. J. Sci.* **245**, 465–482, 571–591.

Grunau, H. R. (1965). Radiolarian cherts and associated rocks in space and time. *Ecolog. geol. Helv.* **58**, 157–208.

Haile, N. S. (1968). Geosynclinal theory and the organizational pattern of the north-west Borneo geosyncline. *Q. Jl geol. Soc. Lond.* **124**, 171–194.

Halbouty, M. T., Meyerhoff, A. A., King, R. E., Dott, R. H., Klemme, H. D. and Shabad, T. (1970). World's giant oil and gas fields, geologic factors affecting their formation and basin classification. *In* "Geology of Giant Petroleum Fields" (M. T. Halbouty, Ed.), 502–555. Am. Ass. Petrol. Geol. Mem. No. 14.

Hall, A. J. (1859). "Natural History of New York" Vol. 3, 1–96. Paleontology. Appleton Century Crofts, New York.

Harrington, H. J. (1962). Paleogeographic development of South America. *Bull. Am. Ass. Petrol. Geol.* **46**, 1773–1814.

Haug, E. (1900). Les geosynclinaux et les aires continentales. *Geol. Soc. Fr. Bull.* **28**, 617–711.

Heybroek, F. (1965). The Red Sea Miocene Evaporite basin. *In* "The Salt Basins Around Africa" 17–40. Inst. Petrol. London.

Hsu, JK. J. (1970). The meaning of the work flysch, a short historical search. *In* "Flysch Sedimentology in North America" 1–11. Spec. Pap. Geol. Ass. Can. No. 7.

Hynes, A. J., Nisbet, E. G., Smith, A. G., Welland, M. J. P. and Rex, D. P. (1972). Spreading and emplacement age of some ophiolites in the Othris region, eastern central Greece. *Z. pdt. geol. Ges.* **123**, 455–468.

Illies, J. H. (1970). Graben tectonics as related to crust—mantle interaction. *In* "Graben Problems" (J. H. Illies and E. St. Mueller, Eds), 4–27. Schweizerbart'-sche Verlagsbuchhandlung, Stuttgart.

Illies, J. H. and Mueller, E. St. (1970). "Graben Problems" Swcheizerbart'sche Verlagsbuchhandlung, Stuttgart. 316pp.

Illing, L. V. and Hobson, G. D. (eds) (1981). "Petroleum Geology of the Continental Shelf of Northwest Europe" Heyden, London. 521pp.

Kay, M. (1944). Geosynclines in continental development. *Science, N.Y.* **99**, 461–462.

Kay, M. (1947). Geosynclinal nomenclature and the craton. *Bull. Am. Ass. Petrol. Geol.* **31**, 1289–1293.

Klemme, H. D. (1980). Petroleum basins—classification and characteristics. *J. Petrol. Geol.* **3**, 187–207.

Klitzsch, E. (1970). Die Struktargeschichte der Zentralsahara. *Geol. Rdsch.* **59**, 495–527.

Lajoie, J. (Ed.) (1970). "Flysch Sedimentology in North America" Spec. Pap. geol. Soc. Can. No. 7, 272pp.

Lowell, J. D. and Genik, G. J. (1972). Seafloor spreading and structural evolution of the southern Red Sea. *Bull. Am. Ass. Petrol. Geol.* **56**, 247–259.

Lowell, J. D. and Genik, G. J. (1975). Geothermal Gradients, Heat Flow, and Hydrocarbon Recovery. *In* "Petroleum and Global Tectonics" (A. G. Fischer and S. Judson, Eds), 129–156. Princeton Univ. Press, Princeton.

Mascle, J. R., Bornhold, B. D. and Renard, V. (1973). Diapiric structures off Niger delta. *Bull. Am. Ass. Petrol. Geol.* **57**, 1672–1678.

McCave, I. N. (1971). Wave effectiveness at the sea bed and its relationship to bed-forms and deposition of mud. *J. sedimn. Petrol.* **41**, 89–96.

McKee, E. D. (1965). Origin of Nubian and similar sandstones. *Geol. Rdsch.* **52**, 551–587.

Mesner, J. C. and Wooldridge, L. C. P. (1964). Maranhao Paleozoic basin and Cretaceous coastal basins. *Bull. Am. Ass. Petrol. Geol.* **48**, 1475–1512.

Meyerhoff, A. A. and Meyerhoff, H. A. (1972). The new global tectonics: major inconsistencies. *Bull. Am. Ass. Petrol. Geol.* **56**, 269–336.

Mitchell, A. H. and Reading, H. G. (1969). Continental margins, geosynclines, and ocean floor spreading. *J. Geol. 77*, 629–646.

Mitchell, A. H. G. and Reading, H. G. (1978). Sedimentation and Tectonics. *In* "Sedimentary Environments and Facies" (H. G. Reading, Ed.), 439–476. Blackwell Scientific, Oxford.

Murray, G. E. (1960). Geologic framework of Gulf Coastal Province of United States. *In* "Recent Sediments, Northwest Gulf of Mexico" (F. P. Shepard, F. B. Phleger and T. H. Van Andel, Eds), 5–33. Am. Ass. Petrol. Geol., Tulsa.

Perrodon, A. (1971). Classification of sedimentary basins: an essay. *Sci. Terre.* **16**, 193–227.

Pettijohn, F. F. (1957). "Sedimentary Rocks" Harper Bros, New York. 718pp.

Ramsay, J. G. (1963). Stratigraphy, structure and metamorphism in the western Alps. *Proc. geol. Soc. Lond.* 357–392.

Reading, H. G. and Mitchell, A. H. (1969). Continental margins, geosynclines, and ocean floor spreading. *J. Geol.* **77**, 629–646.

Reading, H. G. (1972). Global tectonics and the genesis of flysch successions. Rep. 24th Int. Geol. Cong. Montreal, 1972. Section 6, Stratigraphy and Sedimentology, 59–66.

Rich, J. L. (1951). Three critical environments of deposition and criteria for recognition of rocks deposited in each of them. *Bull. geol. Soc. Am.* **62**, 1–19.

Roberts, R. J. (1972). Evolution of the Cordilleran fold belt. *Bull. geol. Soc. Am.* **83**, 1989–2004.

Salop, L. I. (1967). "Geology of the Baikal Region" Vol. 2. Izd. Nedra, Moscow. (In Russian.)

Sanford, R. M. (1970). Sarir oil field, Libya-desert surprise. *In* "Geology of Giant Petroleum Fields" (M. T. Halbouty, Ed.) 447–476. Am. Ass. Petrol. Geol. Mem. No. 14.

Schneider, E. D. (1972). Sedimentary evolution of rifted continental margins. Mem. geol. Soc. Am. **132**, 109–118.

Schuchert, C. (1923). Sites and nature of the North American geosynclines. *Bull. geol. Soc. Am.* **34**, 151–230.

Schwab, F. L. (1971). Geosynclinal compositions and the new global tectonics. *J. sedim. Petrol.* **41**, 928–938.

Seyfert, C. K. and Sirkin, L. A. (1973). "Earth History and Plate Tectonics" Harper and Row, New York. 544pp.

Shelton, J. W. (1968). Role of contemporaneous faulting during basinal subsidence. *Bull. Am. Ass. Petrol. Geol.* **52**, 399–413.

Smith, G. W., Summers, G. E., Wallington, D. and Lee, J. L. (1952). Mississippian oil reservoirs in Williston basin. *In* "The Habitat of Oil" (L. G. Weeks, Ed.), 149–177. Am. Ass. Petrol. Geol., Tulsa.

Stanley, D. J. (1970). Flyschoid sedimentation on the outer Atlantic margin off northeast North America. Flysch Sedimentology in North America, 179–210. *Spec. Pap. geol. Ass. Can.* **7**.

Stille, H. (1936). Wege unde ergebnisse der geologisch-tektonischen forschung. 25 *Jber. K. Wilhelm Gesellsch. Ford Wissensch.* 84–85.

Stoneley, R. (1969). Sedimentary thicknesses in orogenic belts. *In* "Time and Space in Orogeny" 215–238. Geol. Soc. Lond.

Swann, D. H. and Bell, A. H. (1952). Habitat of oil in the Illinois basin. *In* "The Habitat of Oil" (L. G. Weeks, Ed.), 447–472. Am. Ass. Petrol. Geol., Tulsa.
Tarling, D. H. and Runcorn, S. K. (1973). "Implications of Continental Drift to the Earth Sciences" Vols 1 and 2. Academic Press, London and New York. 1184pp.
du Toit, A. L. (1937). "Our Wandering Continents" Oliver and Boyd, Edinburgh. 361pp.
Unesco (1965). "East African Rift System, Upper Mantle Committee" University College, Nairobi. 91pp.
Van Houten, F. B. (1973). Meaning of molasse. *Bull. geol. Soc. Am.* **84**, 1973–1976.
Walcott, R. I. (1972). Gravity flexure and the growth of sedimentary basins at a continental edge. *Bull. geol. Soc. Am.* **83**, 1845–1848.
Weber, K. J. (1971). Sedimentological aspects of oil fields in the Niger delta. *Geologie Mijnb.* **50**, 559–576.
Weeks, L. G. (1958). Factors of sedimentary basin development that control oil occurrence. *Bull. Am. Ass. Petrol. Geol.* **32**, 1093–1160.
Weeks, L. G. and Hopkins, B. M. (1967). Geology and exploration of three Bass Strait basins, Australia. *Bull. Am. Ass. Petrol. Geol.* **51**, 742–760.
Wegener, A. (1924). "The Origin of the Continents and Oceans" Methuen, London. 212pp.
Wilhelm, O. and Ewing, M. (1972). Geology and history of the Gulf of Mexico. *Bull. geol. Soc. Am.* **83**, 575–600.
Woodland, A. W. (Ed.). "Petroleum and the Continental Shelf of North West Europe" Vol. 1. Applied Science, London. 501pp.
Wyllie, P. J. (1971). "The Dynamic Earth" Wiley-Interscience, New York. 416pp.

10 Applied Sedimentology

I. INTRODUCTION

The preceding chapters have ranged widely across the fields of sedimentology. This book has progressed from describing the characteristics of sand grains to those of sedimentary basins. The purpose of this last chapter is to show how sedimentology can be useful in the world today.

Sedimentology may be studied as a subject in its own right, arcane and academic; an end in itself. On the other hand, sedimentology has a contribution to make to the exploitation of natural resources and to the way in which man manipulates the environment. This book has been written primarily for the reader who is, or intends to be, an industrial geologist. It is not designed for the aspiring academic. It is relevant, therefore, to conclude with a chaper on the applications of sedimentology. This is a topic so vast that it merits a book to itself (Table XXXIV). This final chapter is not a comprehensive review, but seeks only to demonstrate some of sedimentology's uses by discussing certain fields in which it has been extensively applied.

Most of the intellectual and financial stimulus to sedimentology has come from the oil industry, and, to a lesser extent, the mining industry. The applications of sedimentology in these fields will be examined in some detail to indicate the reasons for this fact.

First, however, some of the other uses of sedimentology will be briefly reviewed.

The emphasis throughout this book has been on sedimentology and its relationship to ancient lithified sedimentary rocks. It is important to note, however, that a large part of sedimentology is concerned with modern sediments and depositional processes. This is not just to better interpret ancient sedimentary rocks. These studies are of vital importance in the manipulation of our environment (Knill, 1970). For example, the construction of modern coastal erosion defences, quays, harbours and submarine pipelines all require detailed site investigation. These investigations include the study of the regime of wind, waves and tides and of the

Table XXXIV
Illustrative of some of the applications of sedimentology

		Application	Related fields
I. Environmental		Sea-bed structures Pipelines Coastal erosion defences Quays, jetties and harbours	Oceanography
		Opencast excavations and tunnelling	Engineering geology
		Foundations for motorways Airstrips and tower blocks	Soil mechanics and rock mechanics
II. Extractive	A. Whole rock removed	Sand and gravel aggregates Clays Limestones	Quarrying
		Coal Phosphate Evaporites Sedimentary ores	Mining geology
	B. Pore fluid removed	Water	Hydrology
		Oil Gas	Petroleum geology

physical properties of the bedrock. Such studies also include an analysis of the present path and rate of movement of sediment across the site and the prediction of how these will alter when the construction work is completed. It is well known how the construction of a single pier may act as a trap for longshore drifting sediment, causing coastal erosion on one side and beach accretion on the other.

Turning inland, studies of modern fluvial processes have many important applications. The work of the US Army Corps of Engineers in attempting to prevent the Mississippi from meandering is an example. Studies of fluvial channel stability, flood frequency and flood control are an integral part of any land utilization plan or town development scheme.

Engineering geology is another field in which sedimentology may be applied. In this case, however, most of the applications are concerned with the physical properties of sediments once they have been deposited and their response to drainage or to the stresses of foundations for dams, motorways or large buildings. These topics fall under the disciplines of soil mechanics and rock mechanics.

Thus before proceeding to examine the applications of sedimentology to the study of ancient sedimentary rocks, the previous section demonstrates some of its many applications in environmental problems concerning recent sediments and sedimentary processes.

Most applications of sedimentology to ancient sedimentary rocks are concerned with the extraction of raw materials. These fall into two main groups: the extraction of certain strata of sediment, and the extraction of fluids or gases from pores, leaving the strata intact.

There are many different kinds of sedimentary rock which are of economic value. These include recent unconsolidated sands and gravels which are useful in the construction industry. Their effective and economic exploitation requires accurate definition of their physical properties such as size, shape and sorting, as well as the volume and geometry of individual bodies of potentially valuable sediment. Thus in the extraction of river gravels it is necessary to map the distribution of the body to be worked, be it a palaeochannel or an old terrace, and to locate any ox-bow lake clay plugs which may diminish the calculated reserves of the whole deposit.

Similarly, consolidated sandstones have many uses as aggregate and building stones. Clays have diverse applications and according to their composition, may be used for bricks, pottery, drilling mud and so forth. Limestones are important in the manufacture of cement, fertilizer and as a flux in the smelting of iron. The use of all these sedimentary rocks involves two basic problems. The first is to determine whether or not the rock conforms to the physical and chemical specifications required for a particular purpose. This involves petrography and geochemistry. The second problem is to predict the geometry and hence calculate the bulk reserves of the economic rock body. This involves sedimentology and stratigraphy. Here geology mingles with problems of quarrying, engineering and transportation. Geology is nevertheless of extreme importance. It is no use building a brand new cement works next to a limestone crag if, when quarrying commences, it is discovered that the limestone is not a continuous formation, but a reef of local extent.

Coal is another sedimentary rock of vital importance to the energy budget of most modern industrial countries. Coal technology is itself a major field of study with its own text books (e.g. Williamson, 1967). Like the other economic sedimentary rocks coal mining hinges on two basic geological problems: quality and quantity. The quality of the coal is determined by specialized petrographic and chemical techniques. The quantitative aspect of coal mining involes both problems of structural geology and mining engineering as well as careful facies analysis. Classic examples of ancient coal-bearing deltaic rocks have been documented in the literature (e.g. the Circulars of the Illinois State Geological Survey). These studies have been made possible by a combination of closely spaced core holes and data from modern deltaic sediments.

Using this information, facies analysis can delineate the optimum stratigraphic and geographic extent of coal bearing facies (e.g. Jansa, 1972). Detailed environmental studies can then be used to map the distribution of individual coal seams. Coal can form in various deltaic subenvironments, such as interdistributary bays, within, or on the crests of channel sands, as well as regionally uniform beds. The coals which form in these various subenvironments may be different both in composition and areal geometry (e.g. Dapples and Hopkins, 1969).

Evaporite deposits are another sedimentary rock of great economic importance, forming the basis for chemical industries in many parts of the world. The main evaporite minerals, by bulk, are gypsum ($CaSO_4.2H_2O$) and halite (NaCl). Many other salts are rarer but of equal or greater economic importance. These include carbonates, chlorides and sulphates. The genesis of evaporite deposits has been extensively studied both in nature and experimentally in the laboratory (e.g. Richter-Bernburg, 1972; Kirkland and Evans, 1973). The study of evaporites is important, not just because of the economic value of the minerals, but because of the close relationship between evaporites and petroleum deposits (Buzzalini *et al.*, 1969).

Turning now from the applications of sedimentology in the extraction of rock *en masse*, let us consider those uses where only the pore fluids are sought and removed.

The world shortage of potable water may soon become as important as the present energy crisis. It has been argued that hydrogeology is "merely petroleum geology upside-down". This is a shallow statement, yet it contains much truth. Many sedimentary rocks are excellent aquifers and sedimentology and stratigraphy may be used to locate and exploit these. Hydrogeology, like petroleum geology, is largely concerned with the quest for porosity. It is necessary for an aquifer, like an oil reservoir, to have both the pore space to contain fluid and the permeability to give it up.

Whereas an oil reservoir requires a cap rock to prevent the upward dissipation of oil and gas, an aquifer requires an impermeable seal beneath to prevent the downward flow of water.

Despite these fundamental differences, there are many points in common between the search for water and for hydrocarbons. Both use regional stratigraphic and structural analyses to determine the geometry and attitude of porous beds. Both can use facies studies and environmental analysis to fulfill these objectives. Unlike the search for many minerals, the search for water, oil and gas lack direct sensing tools. The only way of testing a location is by the drill and the production test.

The foregoing brief review demonstrates the wide applications of sedimentology. Space does not allow them all to be explained in full. The following two sections examine the applications of sedimentology to the search for sedimentary ores and to the petroleum industry. These provide examples from the two extractive classes of Table XXXIV and are probably the two fields in which sedimentology has been most extensively applied.

II. OIL AND GAS

There is no doubt that the search for oil and gas has been the major driving force behind the rapid expansion in sedimentology over the last quarter of a century.

Almost all oil and gaseous hydrocarbons are found within sedimentary rocks. The majority of petroleum geologists believe that oil and gaseous hydrocarbons originate from sedimentary deposits (but see the opinion of some Russian workers such as Porfir'ev, 1974). The geology of hydrocarbons is, therefore, closely related to sedimentology. Petroleum geology is beyond the scope of this volume, but this book may be regarded as preliminary reading before embarking on the texts of petroleum geology (Dott and Reynolds, 1969; Levorsen, 1967; Chapman, 1973).

The following section is a brief account of the composition, genesis and entrapment of hydrocarbons. This is followed by discussion of those aspects of sedimentology which are particularly relevant to the search for hydrocarbons.

A. Composition, Genesis and Entrapment of Hydrocarbons

Hydrocarbons are complex organic compounds which are found in nature within the pores and fractures of rocks. They consist primarily of hydrogen and carbon compounds with variable amounts of nitrogen, oxygen and sulphur, together with traces of elements such as vanadium and nickel. Hydrocarbons occur in solid, liquid and gaseous states (p. 69).

Solid hydrocarbons are variously known as asphalt, tar, pitch and gilsonite. Liquid hydrocarbons are termed crude oil or simply "crude". Gaseous hydrocarbons are loosely referred to as natural gas, ignoring inorganic natural gases such as those of volcanic origin.

It is a matter of observation that hydrocarbons occur in sedimentary basins; not in areas of vulcanism or of regional metamorphism. Most geologists conclude, therefore, that hydrocarbons are both generated and retained within sedimentary rocks rather than in those of igneous or metamorphic origin. (See Tissot and Welte, 1978, and Hunt, 1979, for expositions of the Western orthodox view, but see also Porfir'ev, 1974, for the Russian inorganic theory.)

The processes of hydrocarbon generation and migration are complex and controversial. Detailed analyses will be found in the references previously cited. Almost all sedimentary rocks contain some traces of hydrocarbons. The source of hydrocarbons is generally thought to be due to large accumulations of organic matter, vegetable or animal, in anaerobic subaqueous environments. During burial and compaction this source sediment becomes heated. Hydrocarbons are formed and migrate out of the source rock into permeable carrier beds. The hydrocarbons will then migrate upward, being lighter than the pore water. Ultimately the hydrocarbons will be dissipated at the earth's surface through natural seepages. In some fortunate instances, however, they are trapped by an impervious rock formation. They form a reservoir in the porous beds beneath, and await discovery by the oil industry. This brief summary of a complex and little understood sequence of events shows that a hydrocarbon accumulation requires a source rock, a reservoir rock, and a cap rock.

Source rocks are, according to the folklore of the industry, black shales. More sophisticated geochemical techniques can now detect source rocks and predict the type of hydrocarbon which they may generate. It is equally important for the source rock to have been heated sufficiently to generate and expel hydrocarbons, yet not to such a high temperature as to crack the hydrocarbon molecules into their constituent elements (Tissot and Welte, 1978; Hunt, 1979).

Any rock with permeability is a potential reservoir. Most of the world's reservoirs are in sandstones, dolomites and limestones. Fields do also occur in weathered granites and gneisses and in fractured shale, chert and metamorphic rocks.

The impermeable cap rocks which seal hydrocarbons in reservoirs are generally shales or evaporites. Less commonly "tight" (non-permeable) limestone and sandstone beds may be cap rocks.

Three main types of trapping mechanism are recognized (Table XXXV).

Table XXXV
A classification of hydrocarbon traps

I. *Structural*	folded faulted	IV. *Combination* traps involve both structural and stratigraphic elements
II. *Stratigraphic*	for classification see Table XXXVII	
III. *Hydrodynamic*		

Structural traps are those which are primarily controlled by the presence of folds, faults or salt diapirs (Fig. 185). Stratigraphic traps are due primarily to lateral permeability barriers caused by sedimentological factors.

Combination traps are those which are due to both structural configuration and lateral permeability changes.

Hydrodynamic traps are a fourth very rare group. These are traps in which the hydrocarbon accumulation is retained in place by a hydrodynamic gradient. Flow of water retains oil in a position from which it would leak in a hydrostatic environment. This may occur in monoclines. Genuine hydrodynamic traps are rare and seldom searched out. Tilted oil: water contacts in some oil fields demonstrate that hydrodynamic factors do play a role in hydrocarbon accumulation.

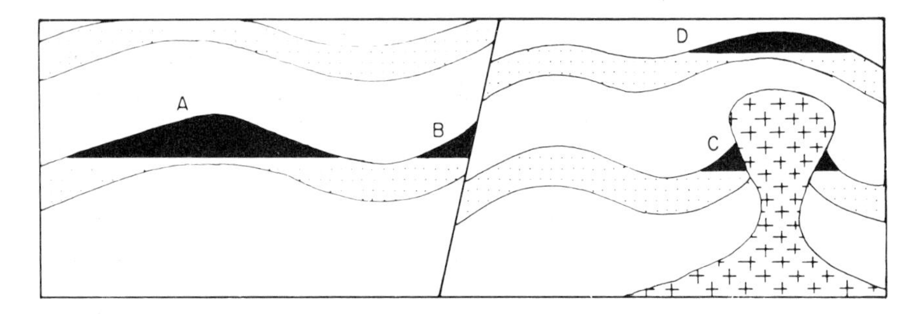

Fig. 185. Cross-section to illustrate structurally-trapped hydrocarbon accumulations (black). A: anticlinal trap, B: fault trap, C and D: traps associated with piercement salt domes.

Sedimentology assists in the search for oil in all types of trap, and in the subsequent effective exploitation of a reservoir. Sedimentology is, however, of prime importance in the location and development of stratigraphic traps.

The role of sedimentology will now be evaluated, firstly as to its broad applications, and then with regard to the search for stratigraphic traps when its application is essential.

B. The Changing Role of Sedimentology During the Exploitation of a Hydrocarbon-producing Basin

The exploitation of hydrocarbons from an area falls into a regular time sequence. Initially, broad regional stratigraphic studies are carried out to define the limits and architecture of sedimentary basins. With modern offshore exploration this is largely based on geophysical data. Preliminary magnetic and gravity surveys are followed by more detailed seismic shooting. This information is used to map marker horizons. Though the age and lithology of these strata are unknown at this stage, the basin can be broadly defined and prospective structures located within it. The first well locations will test structural traps such as anticlines, but even at this stage seismic data can detect such sedimentary features as delta fronts, reefs, growth faults and salt diapirs.

The first wells in a new basin provide a wealth of information. Regardless of whether these tests yield productive hydrocarbons, they give the age and lithology of the formations previously mapped seismically. Geochemical analysis tells whether source rocks are present and whether the basin has matured to the right temperature for oil or gas generation.

As more wells are drilled in a productive basin, so more sedimentological data are available for analysis. The initial main objective of this phase is to predict the lateral extent and thickness of porous formations adjacent to potential source rocks and within the optimum thermal envelope. Regional sedimentological studies may thus define productivity fairways such as a line of reefs, a delta front or a broad belt of shoal sands. The location of individual traps may be structurally controlled within such fairways.

Sedimentology becomes more important still when the structural traps have all been seismically located and drilled. The large body of data now available can be used to locate subtle stratigraphic oil fields. This is the major application of sedimentology in the oil industry and is discussed at length in the next section.

Concluding this review of the role of sedimentology in the history of the exploitation of a basin, it is worth noting the contribution it can make not just in finding fields, but in their development and production.

Few reservoir formations are petrophysicaly isotropic. Most oil fields show some internal variation not only in reservoir thickness, but also in its porosity and permeability (see p. 36). These differences may be both vertical or horizontal and in the case of permeability, there is often a preferred azimuth of optimum flow (see p. 37).

These variations are due either to primary depositional features or to secondary diagenetic changes. Primary factors are common in sandstone reservoirs. Gross variations in porosity within a reservoir formation may relate to the location of discrete clean sand bodies such as channels or bars, within an overall muddy sand. Variations in the direction of maximum flow potential (i.e. permeability) may relate to gross sand-body trend or to sand-grain orientation (Pryor, 1973).

In carbonate reservoirs on the other hand, depositional variations in porosity and permeability tend to be masked by subsequent diagenetic changes. Hence the interest shown by oil companies in carbonate diagenesis.

Understanding of the petrophysical variations within a reservoir assists development drilling of a field by predicting well locations which will produce the maximum oil and the minimum amount of water. Subsequently during the terminal phase of production, secondary recovery techniques can also utilize this knowledge. Selection of wells for water injection should take account of the direction of optimum permeability within the reservoir formation.

Concluding this review of the applications of sedimentology through the evolution of a productive oil basin, it is interesting to note how they change.

Initially during the first exploratory phase, close liaison is necessary with geophysics to elucidate gross structure and stratigraphy. Subsequent establishment of drillable prospects is almost purely geologic, and is based on stratigraphic studies and subsurface facies analysis. The final phase of development drilling and production necessitates close liaison between geologists and engineer. The relationship between petrography and petrophysics is most important at this time.

Sedimentology is most useful in the search for stratigraphic traps. This topic will now be described.

C. Stratigraphic Traps

> A stratigraphic trap is one in which the chief trap-making element is some variation in the stratigraphy or lithology, or both, of the reservoir rock, such as a facies change, variable local porosity and permeability, or an up-structure termination of the reservoir rock, irrespective of the cause.
>
> (Levorsen, 1967, p. 237.)

Table XXXVI
Giant oil and gas fields classified according to trapping mechanism

	Stratigraphic	Combination	Structural	Total
Oil	13	23	151	187
Gas	4	7	68	79
Total	17	30	219	266
Per cent	6	11	83	100

Giant oil fields are those with more than 5×10^8 recoverable barrels. Giant gas fields are those with more than $10{\cdot}6 \times 10^{18}$ m of recoverable gas. Data from Halbouty *et al.* (1970).

Table XXXVI shows that only about 6% of the world's known giant oil and gas fields are pure stratigraphic traps, while another 11% are trapped by a combination of stratigraphy and structure.

These figures probably reflect the fact that anybody can find structurally trapped fields by drilling all the geophysically defined highs. Traditionally stratigraphic fields are found either by accident or by careful analysis of sedimentological data from structural test wells.

Thus the low ratio of known stratigraphic to structurally trapped fields probably reflects lack of imaginative exploration rather than the true ratio. This is an exciting challenge for applied sedimentology. Extensive documentation on stratigraphic traps and methods of locating them is given in a volume edited by King (1972).

Traditionally stratigraphic traps were found either by chance or as a result of facies analysis of data gathered from wells drilled to test structural traps. The great improvements made in seismic geophysics in recent years have enabled stratigraphic traps to be directly located. The advent of seismic stratigraphy has done much to bring together geology and geophysics. Nowadays it is often possible to map channels, bars, reefs and submarine fans from seismic data, prior to drilling any wells (Payton, 1977).

The total number of potential stratigraphic trap situations is infinite. Nevertheless, several of the more common types may be defined and classified (Table XXXVII). Many stratigraphic oil fields are trapped by combination of these various types. The main varieties of stratigraphic trap will now be described.

Table XXXVII
A classification of stratigraphic traps

Traps unassociated with unconformities	Depositional traps	Channels Sand bars Reefs
	Diagenetic traps—	Porosity/permeability transition
Traps associated with unconformities	Trap above unconformity—	Palaeogeomorphic traps (strike valley and palaeochannel sands)
	Trap below unconformity—	Truncation traps

Based on Rittenhouse (1972).

Depositional traps occur in sedimentary sequences uninterrupted by major stratigraphic breaks where a reservoir formation with primary depositional porosity passes laterally into an impervious lithology. Depositional traps are most common in clastic sequence where lenses of porous sand are totally enclosed in impermeable silt or shale. Channel and bar sand bodies are the commonest examples of these. Facies change stratigraphic traps are less common in carbonate rocks due to their susceptibility to diagenetic changes. Nevertheless, reef and carbonate-shoal stratigraphic traps do occur, albeit with modified porosity.

The three main types of depositional trap; channels, bars and reefs, will now be described.

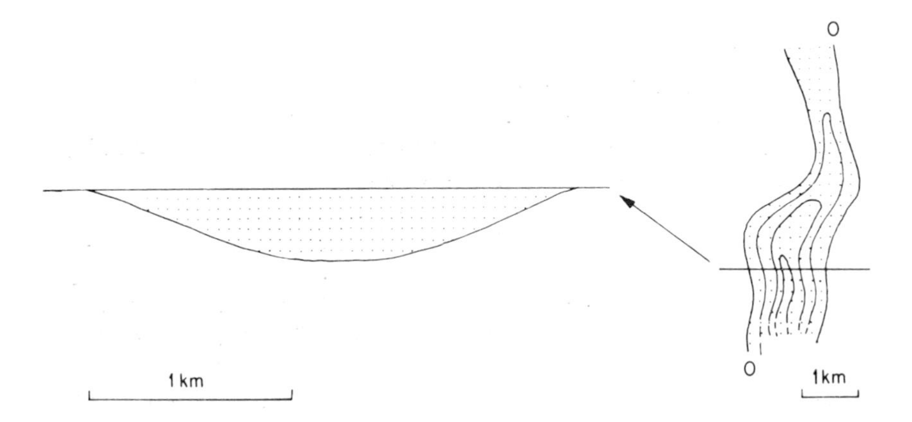

Fig. 186. Cross-section and isopach map of the Rosedale channel, Miocene, California (from Martin, 1963). Cross-section drawn true scale, isopach map with 122 m contours. This is interpreted as a submarine channel infilled by oil-bearing turbidite sands.

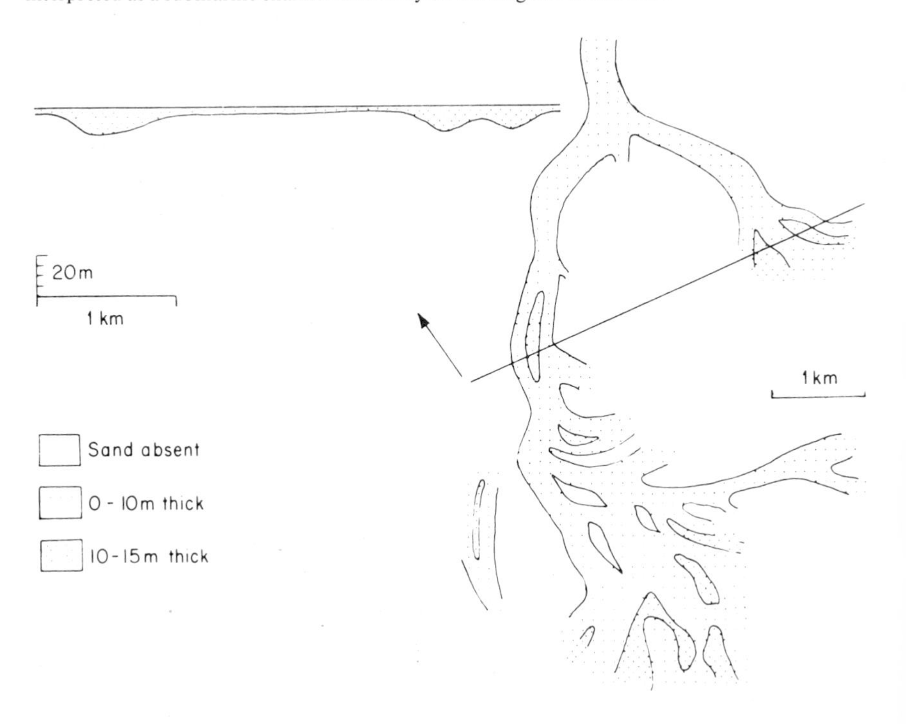

Fig. 187. Cross-section and isopach map of the oil-bearing sands of the Pennsylvanian Booch Formation, Oklahoma. (Busch, 1977.) These are interpreted as radiating deltaic distributary sands.

1. Channels

Chapter 8 demonstrated that channels of sand enclosed in shale occur in many environments. Examples range from submarine canyons, through tidal channels, deltaic distributaries to fluvial meander belts. Similarly, therefore, stratigraphic oil accumulations occur in channel traps in diverse situations. Examples of turbidite channel sand reservoirs have been extensively documented from the Tertiary basins of California, Fig. 186 (e.g. Martin, (1963). Oil accumulations in radiating delta distributary channel-sands occur in the Pennsylvanian of the Arkoma basin, Oklahoma (Fig. 187) (Busch, 1971); while Shannon and Dahl (1971) have described stratigraphic entrapment within irregular delta front sand bodies (Fig. 188). Stratigraphic traps in fluvial channels are more common than any other. The majority of these, however, are the fills of valleys cut into bedrock. They are thus more properly considered under the heading of traps associated with unconformities.

These examples show the variation in size and shape to be found in

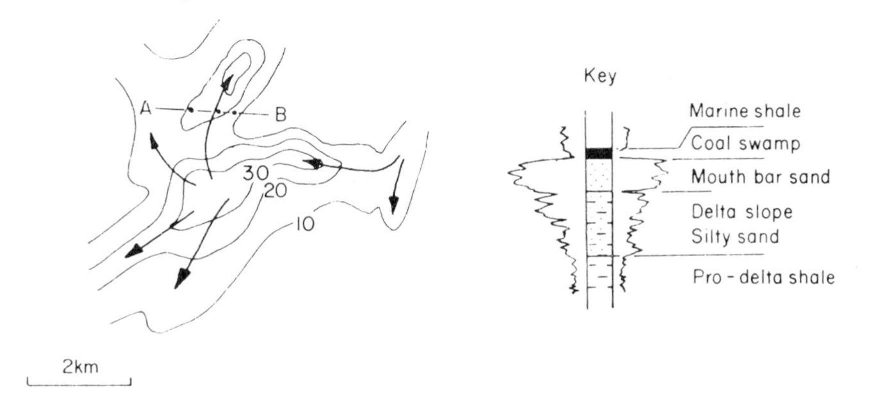

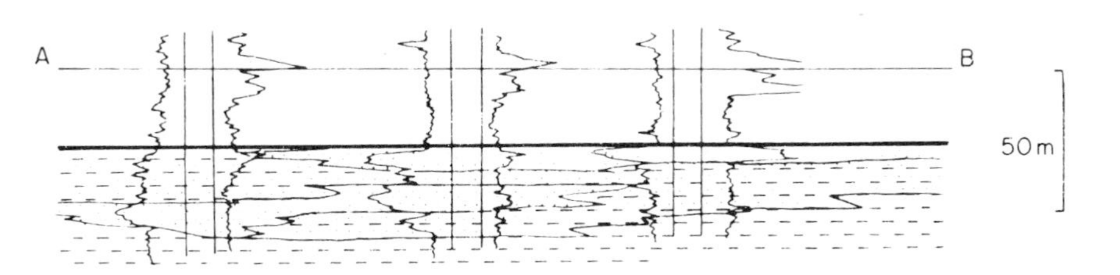

Fig. 188. Isopach map and log correlation (SP and resistivity) of Pennsylvanian sands of the West Tuscola oil field, Texas. (From Shannon and Dahl, 1971.) These sand bodies are interpreted as regressive delta-front sands; arrows indicate direction of inferred mouth bar migration.

channel sands of differing environments. Similarly there are wide ranges of porosity and permeability values and shale content. Few channels are infilled by sand alone. Temporary or total abandonment of channels may cause them to be partially or totally infilled by mud. This is as true of submarine canyons as of alluvial valleys.

2. *Bar sands*

Modern geomorphologists have paid great attention to the nomenclature and genesis of coastal sand bodies. Geologists tend to use the term bar sand for a elongate marine sand body enclosed in shale and striking subparallel to the palaeoshoreline (see p. 297).

The subtle differences which a geomorphologist would use to distinguish off-shore bars, barrier bars, spits and the like, are seldom clear from subsurface data. It is, however, extremely important to distinguish bar sands from channel sands when they contain oil fields. Basically a channel sand body will be expected to trend down the palaeodip, while a bar will be elongated perpendicular to the palaeodip. Criteria for distinguishing bar sands from subsurface data have been described at length by many geologists (e.g. Shelton, 1967).

Examples of bar sand stratigraphic traps are numerous. Particularly well-documented examples occur in the Texas Gulf coast Tertiary province, in Cretaceous and Eocene basins of the Rocky Mountains, and in Pennsylvanian strata of Oklahoma and Kansas.

In the Texas Gulf coast bar sand bodies were deposited in association with marine-dominated delta systems such as those of the Upper Wilcox, Vicksburg and Frio Formations. Isopachs of drilled-up fields reveal discrete sand bodies with shoe-string morphologies trending subparallel to the palaeostrike (e.g. Levorsen, 1967, p. 294). Pure stratigraphic traps occur by up-dip lensing out of the sands. Combination traps are also found where sand bars are closed up-dip by growth faults. Individual productive sands are seldom more than 1·6 km wide and 16 km long, but productive sand trends, such as the Frio, can be traced for hundreds of kilometres (Boyd and Dyer, 1966; Halbouty, 1969).

Some of the best examples of depositional stratigraphic traps occur in the late Cretaceous basins on the east side of the Rocky Mountains.

During the Late Cretaceous to Early Tertiary, a seaway extended from the Gulf of Mexico to the Arctic, separating eastern North America from the rising Rocky Mountains. Numerous stratigraphic oil accumulations occur in several sedimentary basins along this trend. These include many good examples of bar sand traps in the Upper Cretaceous, Muddy Sandstone on the north-east side of the Powder River basin (Fig. 189). In this fluvial channels drained the Black Hills arch and fed deltas which

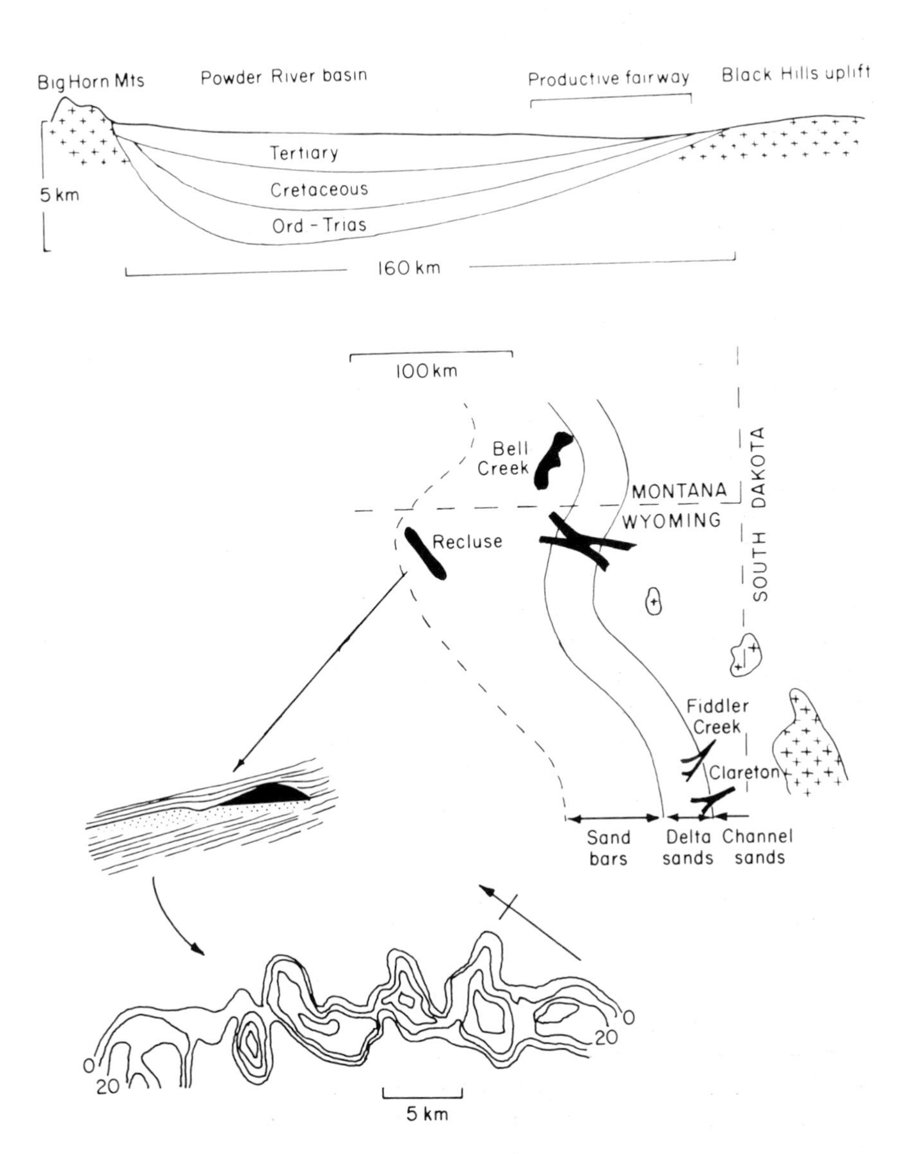

Fig. 189. Upper: cross-section of the Powder River basin, Rocky Mountains, USA. Lower: palaeogeographic map showing the productive fairways of stratigraphically trapped oil in marine bar, deltaic and fluvial channel sands. Inset: isopach map and cross-section of Recluse oil field showing stratigraphic trapping mechanism due to up-dip pinch out of this bar sand into impermeable shale. Compiled from data due to Mc Gregor and Biggs (1972) and Woncik (1972). Contour interval 3·23 m.

prograded into the basin. Barrier sands, reworked from the deltas, were deposited on their seaward edges and along-shore away from major lobes. Oil and gas are stratigraphically trapped both in the bar sand bodies and, further up-dip, in channels cut in older strata. The Bell Creek field is one of the most notable examples of these bar sand stratigraphic traps with ultimate recoverable reserves of some 114 million barrels of oil. The top of the Muddy Sand shows a uniform north-west dip with no structural closure. The sand body of the Bell Creek field has a maximum thickness of some 9 m, and measures about 19 by 6 km (McGregor and Biggs, 1972).

The Recluse field, some 48 km to the south-west, is of similar dimensions and genesis (Woncik, 1972).

In Saskatchewan the Muddy Sand equivalent is termed the Viking Sand. Close well control shows that this formation is composed of discrete

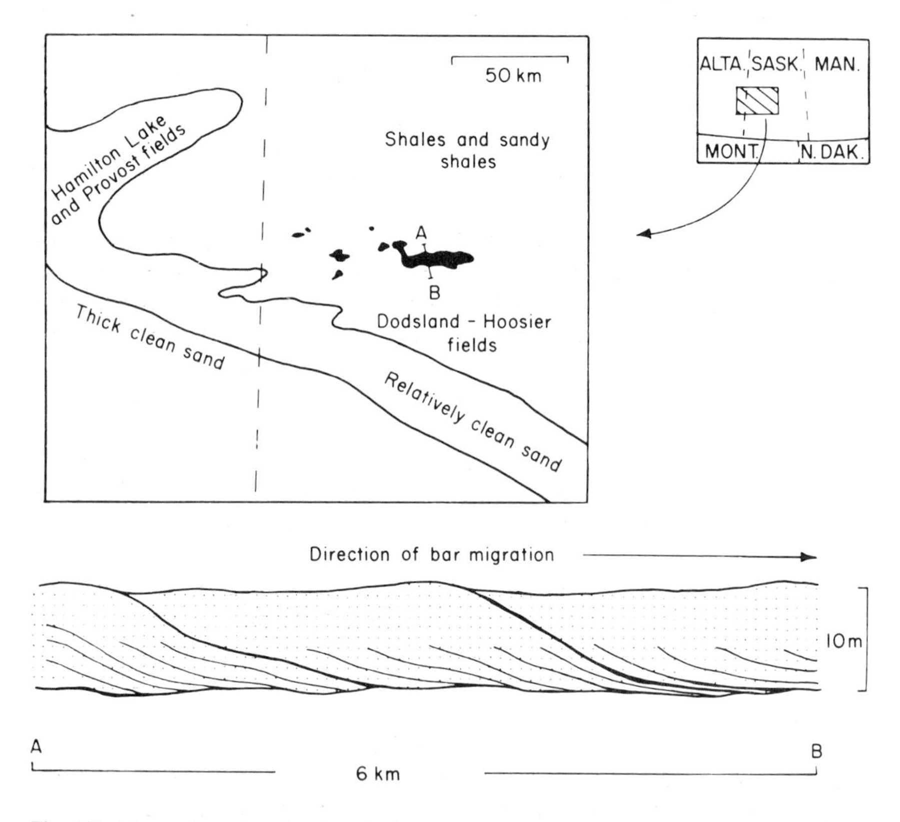

Fig. 190. Map and section showing the occurrence of oil and gas fields in the Viking Sands (Lower Cretaceous). These linear imbricate shoal sands increase in shale content downwards. Intimate log correlation in the Dodsland–Hoosier area, suggests a general southerly progradation. (After Evans, 1970.)

imbricately nested sand bodies which prograded southwards and can be mapped along strike for over a hundred kilometres (Fig. 190). Regional palaeogeographic studies suggest that these sands were deposited on offshore shoals trending oblique to the local shoreline (Evans, 1970).

An analogous example of closely associated bar and channel stratigraphic traps occurs in the Enid embayment of the Anadarko basin, Oklahoma (Withrow, 1968). Here oil is trapped in Pennsylvanian sandstone reservoirs. Some of these are bar sands elongated parallel to the shoreline, others are basinward-trending channels. This situation provides a classic problem for applied sedimentology. Environmental facies analysis is critical both in the prediction of new fields and in their subsequent development.

3. *Reefs*

Reefs are another group of important stratigraphic traps. As pointed out on p. 305 there are certain semantic problems in the nomenclature of lenticular organic carbonate formations. It is particularly difficult to differentiate the various types of reef, shoal and "build-up" in subsurface studies. This is partly because of the limited amount of data available from bore holes and partly because of the extensive diagenesis which often obliterates the biogenic features of these rocks.

For this reason too, reef stratigraphic traps span the two groups of facies and diagenetic traps in our classification. A reef, by most definitions, is a primary depositional feature and, if enclosed in shale or evaporites, is a facies trap. Nevertheless, because of extensive diagenesis, primary porosity is often obliterated and secondary porosity is generated which is different in scale, geometry and spatial distribution. In such cases, therefore, reefs may be classified as diagenetic stratigraphic traps.

Case histories of seismic reef-hunting have been given by Evans (1972), Bubb and Hatledid (1977) and Exploration Staff of Chevron Standard Ltd (1979), while the gravity meter has for long been used as a reef-finding tool (Ferris, 1972).

Facies analysis has an important role in several stages in the location and development of reef stratigraphic traps. Initially, regional palaeogeographic studies are essential in defining the existence of a reef play; such as, for example, defining a shelf/basin hinge which is a favourable site for reef growth.

Once a reef stratigraphic trap has been found to be productive, then it is important to locate development wells effectively. This needs a close integration of facies analysis, petrography and petrophysics. In the early stages of development drilling, facies must first be defined. Then it must be decided whether the distribution of porosity and permeability relate to facies, or whether there has been extensive diagenetic modification. In the latter case it is necessary to decipher the diagenetic history of the reef since the various reservoir zones which concern the production engineer may cross-cut facies.

Studies of this kind have played an important part in the development of reefal oil fields, notably in the Devonian of Alberta, Canada (e.g. Klovan,

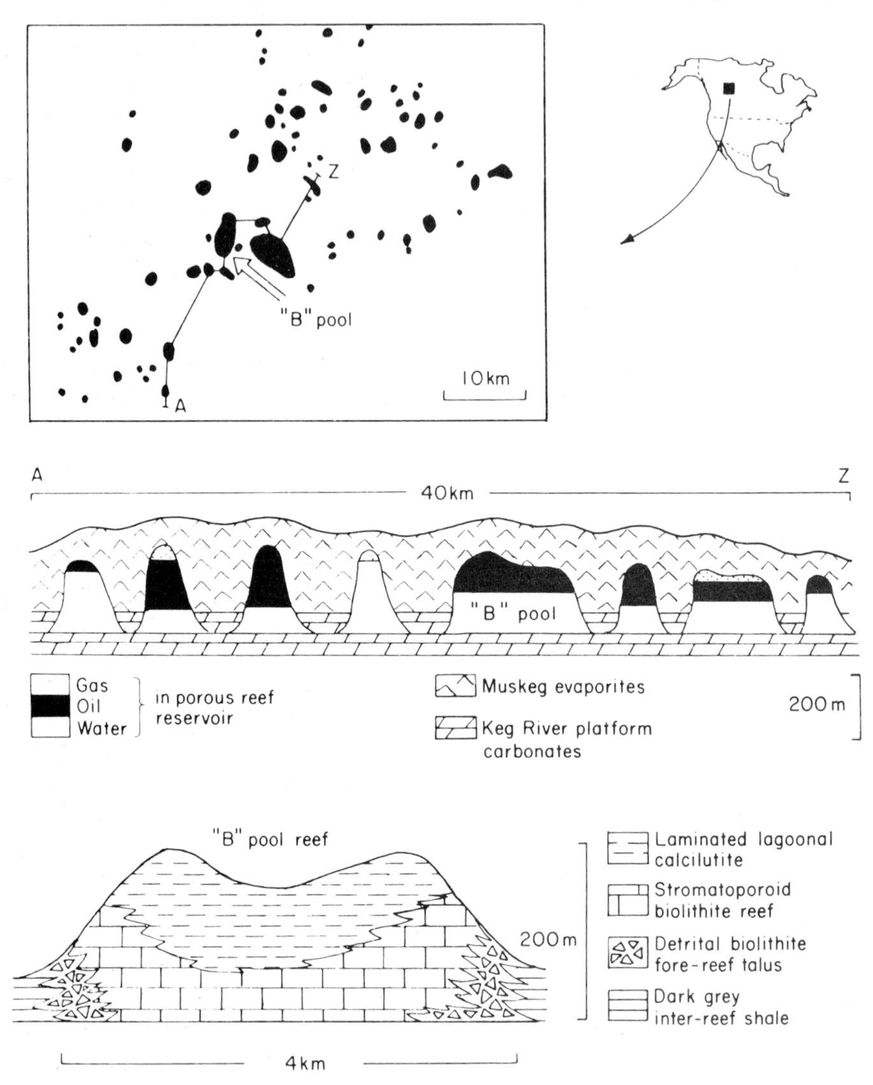

Fig. 191. Stratigraphically-trapped oil and gas fields in Mid-Devonian reefs of the Rainbow area, Alberta, Canada. (From Baars *et al.*, 1970.) These fields are trapped in an archipelago of pinnacle reefs with an evaporite cap rock.

1964; Jenik and Lerbekmo, 1968). Figure 191 shows a typical example of the geometry, facies and distribution of stratigraphic fields in the Devonian Rainbow Reef complex of Alberta.

4. Unconformity traps

It has long been known that hydrocarbons are often trapped adjacent to unconformities in both structural and stratigraphic settings. Regional stratigraphic and sedimentologic studies play an important role in defining the location of favourable unconformity zones in time and place (Halbouty, 1972).

The classification of stratigraphic traps into two groups in Table XXXVII delineates unconformity traps into two groups depending on whether the reservoir is above or below the unconformity surface. In the search for hydrocarbon traps in these situations it is obviously important to make a study of the actual geometry of the unconformity surface itself. This is termed palaeogeomorphology (Martin, 1966).

Considering first those accumulations which are trapped above an unconformity, three main types can be defined: wedge-outs, channels and strike valley fills.

Wedge-outs are the simplest form. They occur where a sand sheet pinches out against an essentially planar unconformity (Fig. 192).

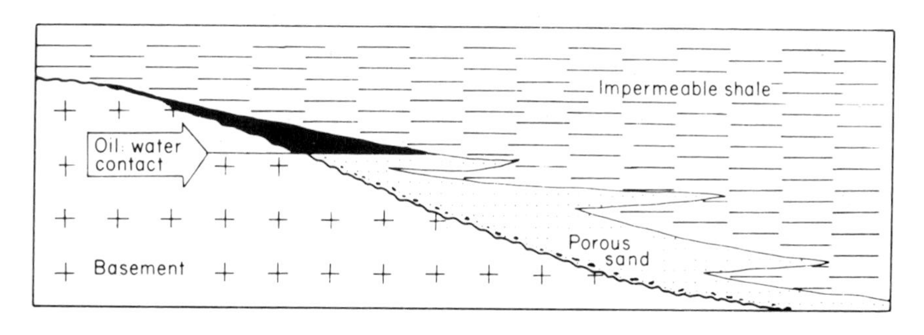

Fig. 192. Sketch section to illustrate a simple stratigraphic wedge-out trap.

More complex in geometry and, therefore harder to find, are reservoir sands which infill channels cut into unconformities. Channel stratigraphic traps, which occur solely due to facies changes away from unconformities, have been described already. There is little difference between the two types in scale, geometry or fill. There is, however, some difference in the approach to searching for these two channel types. Channels which are not cut into unconformities are governed only by syndepositional factors. Channels cut into unconformities, by contrast, will be strongly influenced, notably in orientation, by the various bedrocks through which they were cut.

Two major unconformity zones occur in North American stratigraphy which have been of great significance in the search for hydrocarbons. One occurs where the Pennsylvanian coal measures of Oklahoma and Kansas overlie Mississippian carbonates. The second, at the base of the Cretaceous, extends along the east side of the Rocky Mountains from New Mexico to Alberta. Amongst the many traps associated with these unconformities, good examples of sub-Pennsylvanian channels have been described by Kranzler (1966). This particular study demonstrated the importance of the juxtaposition of channel sands above the unconformity with a narrow band of limestone, the presumed source rock, cropping out beneath the unconformity.

Notably studies of sub-Cretaceous channel sand reservoirs have been made by Harms (1966) and Martin (1966).

More subtle still are the strike valley sand reservoirs. These are sand bodies which fill in strike valleys cut in gently dipping sediments of alternating hard and soft strata (Fig. 193). Strike valley sands, like channel

Fig. 193. Block diagram to illustrate the habitat of strike valley sands infilling a dissected monoclinal topography. (From Busch, 1960.)

sand traps, tend to be overlain by transgressive marine shales which often act as both hydrocarbon source and seal. They differ from channel sand traps, however, in that they are generally marine in origin and elongated along the local depositional strike. Strike valley sands described by Busch from the sub-Pennsylvania unconformity of Oklahoma are up to 64 km long, though only 0·8–1·6 km wide.

Cretaceous strike valley sands of the San Juan Basin are shorter and wider (McCubbin, 1969). In both these examples individual sand bodies are generally less than 30 m thick.

Let us now consider truncation traps where the entrapment of hydrocarbons occur sealed immediately beneath an unconformity surface.

Primary porosity is sometimes preserved beneath an unconformity but, as pointed on p. 104, extensive secondary porosity development (epidiagenesis) is the more general rule. In sandstones, weathering beneath an unconformity may remove non-siliceous cements, matrix and labile grains. In the same situation carbonates may develop extensive and often cavernous porosity due to solution and leaching. These processes of chemical weathering leading to porosity development also operate on igneous and metamorphic terrains. Similarly, fracturing due to stress release increases porosity and permeability of rocks beneath an unconformity. In searching for subunconformity traps, therefore, two factors are paramount: the topography of the unconformity surface and the disposition of the various subcropping strata. Thus in one situation porosity may be present in carbonate palaeotopographic highs. Elsewhere porosity may be found in porous soft sands which were easily eroded and floor palaeotopographic valleys flanked by resistant tight strata (Fig. 194).

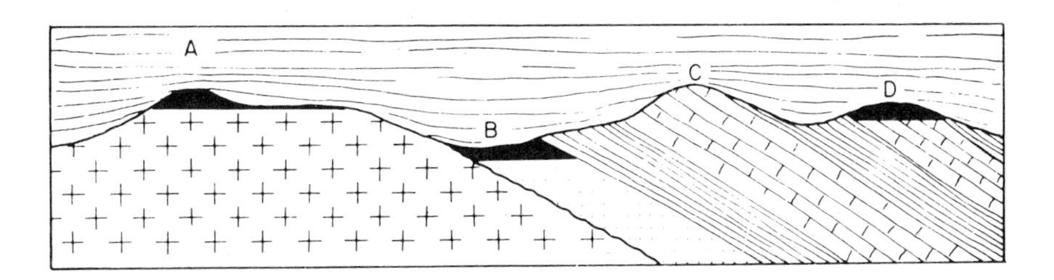

Fig. 194. Sketch section to indicate sub-conformity hydrocarbon traps. These can occur in weathered granitoid rocks (A), in porous pre-unconformity sands (B), and in leached carbonates (D). Solution porosity is not invariably present in carbonates, however, and location C, though structurally high, is tight and non-productive.

When a subunconformity trap has been located, therefore, it is important to use all available data to attempt to map the palaeotopography and to produce a palaeogeologic map of the pre-unconformity surface. Notable examples of truncation traps occur in Mississippian carbonates beneath the Pennsylvanian unconformity of the Illinois basin, in the Woodbine Sand beneath the Cretaceous Austin Chalk in East Texas, and in Eocene strata beneath the Miocene unconformity in the Bolivar coastal field of Venezuela (Levorsen, 1964).

D. Sedimentology and Hydrocarbon Exploration: Conclusion

The preceding account of the applications of sedimentology in the search for hydrocarbons needs to be put in perspective. A lot of oil was found before most oil men could even spell "sedimentology".

The study of sedimentary rocks is certainly important in locating oil and gas. As shown in this chapter, applications range from initial regional basin analysis down to petrophysics and production. Table XXXVI shows what a small percentage of known oil is actually found in pure stratigraphic traps. The total reserves in stratigraphic traps, found by accident or design, are infinitesimal compared with those in broad anticlinal structures found by geophysical techniques.

Most of the case histories which have been described are based on extremely close well-spacing of 0·8 km or less. It is always important to wonder at what point in the development drilling programme the true nature of the trap was realized, and to what extent this knowledge was used in subsequent development and wildcat drilling.

III. SEDIMENTARY ORES

A. Introduction

Sedimentology has never been used by the mining industry to the extent that it has been employed in the search for hydrocarbons. There are two good reasons for this. First, many metallic ores occur within, or juxtaposed to, igneous and metamorphic rocks. In such situations sedimentology can neither determine the genesis nor aid the exploitation of the ore body. A second reason for the inapplicability of sedimentology in metallic mining is because there are direct sensing methods of prospecting. There is at present no direct method of locating hydrocarbons. Indirect techniques such as sedimentology and seismic surveys can only define geologic structures which may be prospective. This guess work must be tested by the drill. Ores, however, can be located by direct geochemical and geophysical techniques. Geochemical techniques range from the old-style panning of alluvium to locate the mother-lode by gradually working up-stream, to modern stream sampling for trace elements. Direct geophysical methods of locating ore bodies include magnetometer and scintillometer surveys.

For these reasons sedimentology is not so useful in searching for sedimentary ores as it is for hydrocarbons. Its relevance in mining is twofold. First, it has a large contribution to make to the problems of ore genesis in sedimentary rocks. Secondly it is useful as a back-up tool in the search and development of ore bodies.

There are generally believed to be three main processes responsible for the genesis of sedimentary ores (Table XXXVIII). Detrital placer deposits are formed where current action winnows less-dense quartz grains away to leave a lag concentrate of denser grains. These heavy minerals may sometimes be of economic importance.

Table XXXVIII
Summary of the main sedimentary ores and their common modes or origin

Name	Process	Examples
I. Syngenetic	Originated by direct precipitation during sedimentation	Manganese nodules and crusts, some oolitic ironstones
II. Epigenetic	Originated by post-depositional diagenesis	Carnotite, copper–lead–zinc assemblages, some ironstones
III. Placer	Syndepositional detrital sands	Alluvial gold, cassiterite and zircon

Syngenetic sedimentary ores form by direct chemical precipitation within the depositional environment.

Epigenetic ores form by the diagenetic replacement of a sedimentary rock, generally limestone, by ore minerals.

The origin of sedimentary ores has always excited considerable debate. This generally hinges on two criteria. How may epigenetic and syngenetic ores be differentiated? And to what extent are the mineralizing fluids of epigenetic ores derived from normal sedimentary fluids, and to what extent are they hydrothermal in origin? In recent years it has been increasingly realized how ordinary sedimentary processes can generate sedimentary ores, both syngenetic and epigenetic. Conversely less credence is now given to a hydrothermal origin for many epigenetic ores (e.g. Sangster, 1971; Amstutz and Bernard, 1973).

The following account of detrital, syngenetic and epigenetic sedimentary ore attempts to show how sedimentology helps decipher problems of ore genesis and exploitation.

B. Placer Ores

Detrital sediments are generally composed of particles of varying grain density. For example, most sandstones contain a certain amount of clay and silt. The sand particles are largely grains of quartz, feldspar and mafic minerals with specific gravities of between two and three. At the same time most sands contain traces of minerals with specific gravities between four and five. The heavy minerals, as these are called, commonly consist of opaque iron ores, tourmaline, garnet, zircon and so on. More rarely they include ore minerals such as gold, cassiterite (SnO_2), or monazite. It is a matter of observation that the heavy mineral fraction of a sediment is much finer grained than the light fraction. There are several reasons for this.

First, many heavy minerals occur in much smaller crystals than do quartz and feldspar in the igneous and metamorphic rocks in which they form. Secondly, the sorting and composition of a sediment is controlled by both the size and density of the particles, this is spoken of as their hydraulic ratio. Thus, for example, a large quartz grain requires the same current velocity to move it as a small heavy mineral. It is for this reason that sandstones contain traces of heavy minerals which are finer than the median overall grain size of the less dense grains.

In certain flow conditions, however, the larger, less dense grains of a sediment are winnowed away to leave a residual deposit of finer grained heavy minerals.

The flow conditions which cause this separation are of great interest to the mining industry, both as a key to understanding the genesis of placer ores, and in the flotation method of mineral separation in which crushed rock is washed to concentrate the ore.

Placer deposits occur in nature both in alluvial channels, on beaches and on marine abrasion surfaces. Placer sands are frequently thinly laminated with occasional slight angular disconformities and shallow troughs. This is true of both fluvial and beach-sand placers. Penecontemporaneous deformation of laminae is sometimes found where the dense placer bands have sunk into mobilized quartzose sands of average density (Selley, 1964). In Quaternary alluvium placers occur in the present channel system, notably on the shallow riffles immediately down-stream of meander pools, and also in river terrace deposits and buried channel-fill alluvium below the floor of the present river bed.

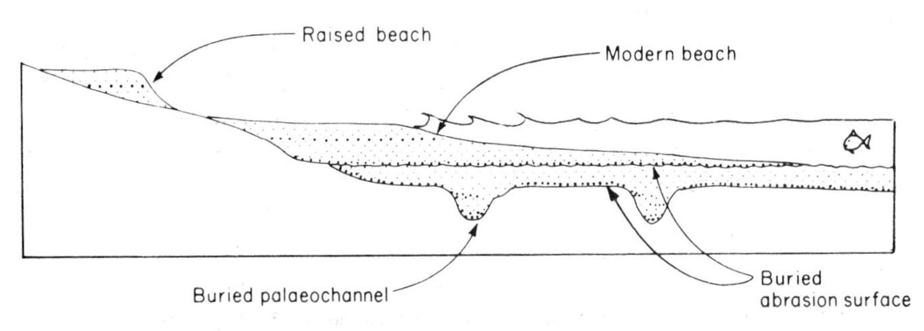

Fig. 195. Sketch section to illustrate the zones of heavy mineral concentration in coastal and and near-shore marine sediments. Placers shown by heavy stipple.

Notable examples of modern alluvial placer ores include the cassiterite deposits of Cornwall and Malaya (Newell, 1971). Gold is another characteristic alluvial placer ore mineral occurring in the Yukon and Australia. The Pre-Cambrian gold- and uranium-bearing fluvial conglomerates of the Witwatersrand appear to be palaeoplacers deposited on a braided alluvial fan though this has been disputed (Haughton, 1964; Pretorius, 1979).

Recent marine placers, like their fluvial counterparts occur at different topographic levels due to Pleistocene sea level changes (Fig. 195). Work on Quaternary beach placers of New South Wales and Queensland points to the importance of stable shorelines to give time of the concentration of heavy minerals to take place (Hails, 1972).

The optimum zone for marine heavy mineral separation to occur is within the tidal zone of the beach (Mero, 1965). Concentration can also take place on wave-cut terraces. The raised beaches which formed during high Pleistocene sea levels contain placer ores. The gold deposits of Nome, Alaska, are a case in point. Present-day beach placers include the zircon, ilmenite, rutile and monazite sands of Travancore, India. Submarine placers include the cassiterite deposits of the Sunda Shelf off Indonesia. Here the tin sands occur both exposed on modern submarine abrasion terraces, and associated with abrasion terraces and fluvial channels now buried beneath Recent marine sediment (Aleva, 1973). These deposits were formed during the glacial maxima when the sea level of the world dropped and much of the Sunda Shelf was dissected by fluvial channels.

Essentially, placer ores are controlled by the geochemistry of the sediment source, by the climate, and hence depth of weathering, by the geomorphology which controls the rate of erosion and topographic gradient, and by the hydrodynamics of the transportational and depositional processes.

The genesis of placer deposits is seldom in dispute because the textures and field relationships of the ores clearly indicate their detrital origin. Furthermore, the processes of placer formation can be observed at work on the earth today. The origins of syngenetic and epigenetic sedimentary ores are more equivocal.

C. Syngenetic Ores

Syngenetic ores are those which were precipitated during sedimentation. This phenomenon has been observed on the earth's surface at the present time. One of the best known examples of this process is the widespread formation of manganese nodules and crusts on large areas of the beds of oceans, seas and lakes (e.g. Glasby, 1972; Fryer and Hutchinson, 1976; Roy, 1981). Goethite ooliths are forming today in Lake Chad in Central Africa (Lemoalle and Dupont, 1971). More spectacular examples of modern sea-floor metallogenic processes occur associated with volcanic activity as at Santorini and in the "hot holes" of the Red Sea (Puchelt, 1971; Degens and Ross, 1969).

A large number of stratiform ore bodies have been attributed to a syndepositional or syngenetic origin, though an epigenetic (post-depositional replacement) origin has been argued for many of them. The criteria for differentiating these origins will be examined later.

Particularly cogent arguments have been advanced to support a syngenetic origin for ores of copper, manganese and iron in bedded sedimentary rocks (Bernard, 1974; Wolf, 1976).

Classic examples of syngenetic copper ores include the Permian Kupferschiefer of Germany and the Pre-Cambrian Copper belt of Katanga and Zambia. The Kupferschiefer is a thin black organic radioactive shale which is remarkable for its lateral uniformity across the North Sea basin. It overlies the eolian sands of the Rotliegende and is itself overlain by the Zechstein evaporites. Locally the Kupferschiefer is sufficiently rich in copper minerals such as chalcopyrite and malachite to become an ore. A syngenetic origin for the ore is widely accepted (e.g. Gregory, 1930; Brongersma-Sanders, 1967). In Zambia, relatively unmetamorphosed Pre-Cambrian sediments of the Roan Group overlie an igneous and metamorphic basement. Extensive copper mineralization in the Roan Group sediments includes pyrite, chalcopyrite, bornite and chalcocite. These occur within shallow marine shales and adjacent fluvial sands, but not in an eolian sand facies or basement rocks. The mineralization is closely related to a palaeocoastline. The ore bodies are mineralogically zoned with optimum mineralization in sheltered embayments (Fig. 196). Early workers favoured a hydrothermal epigenetic origin for the ore due to intrusion of the underlying granite. It was demonstrated, however, that the irregular

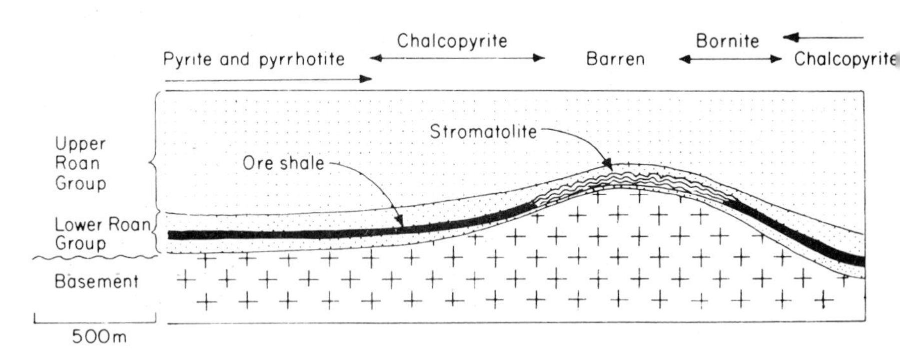

Fig. 196. Section of the Roan Group (Pre-Cambrian) in the Zambian copper belt. This shows a close correlation between the type of mineralization and palaeogeography. Barren algal stromatolites occur on topographic basement highs. Coeval shales in sheltered environments host zoned ore minerals. (After Garlick and Fleischer, 1972.)

granite/sediment surface is due to erosion, not intrusion. Criteria now quoted in favour of a syngenetic origin for the ore include its close relationship with palaeotopography and facies, and reworked ore grains in younger but penecontemporaneous sandstones (Garlick, 1969; Garlick and Fleischer, 1972).

Analogous deposits at Kamoto in Katanga have been interpreted as shallow marine in origin, but extensive diagenesis is invoked as the mineralizing agent (Bartholome *et al.*, 1973).

Though the source of the Copper belt ores is unknown, a syngenetic origin is now widely accepted, with the proviso that subsequent diagenesis has modified ore fabrics and mineralogy.

D. Epigenetic Sedimentary Ores

Epigenetic or exogenous ores in sedimentary rocks are those which formed later than the host sediment. Epigenetic ores can be generated by a variety of processes. These include the concentration of disseminated minerals into discrete ore bodies by weathering, diagenetic, thermal or metamorphic effects. Epigenesis also includes the introduction of metals into the host sediment by meteoric and hydrothermal solutions resulting in the replacement of the country rock by ore minerals.

Criteria for differentiating syngenetic and epigenetic ores have been mentioned in the preceding section. Epigenetic ores are characteristically restricted to modern topography, to unconformities or, if hydrothermal in origin, to centres of igneous activity. The ore may not contain a dwarfed and stunted fauna as anticipated in syngenetic deposits. Isotopic dating will show the ore to post-date the host sediment (Bain, 1968).

Once upon a time, epigenetic ores were very largely attributed to hydrothermal emanations from igneous sources, except for obvious examples of shallow supergene enriched ores. Within recent years it has been shown that many epigenetic ores lie far distant from any known igneous centre and show no evidence of abnormally high geothermal gradients. It has been argued that these ores formed from concentrated chloride-rich solutions derived from evaporites and from residual solutions derived from the final stages of compaction of clays (Davidson, 1965; Amstutz and Bubinicek, 1967). Two main groups of ores are commonly attributed to these low temperature (telethermal) processes. These are the lead–zinc sulphide ores and certain uraniferous carnotite deposits.

1. Lead–zinc strata-bound ores

The lead–zinc telethermal ores are found as the sulphides galena and sphalerite. They generally show obvious replacement textures such as veins and geodes associated with coarsely crystalline calcite, dolomite, fluorite and barite. Such sulphide ores are found in carbonate shelf facies, and they are especially characteristics of reefs. The geometry of individual ore bodies is generally irregular though they may lie along trends with some structural control.

Notable examples of lead–zinc mineralization occur in Lower Carboniferous (Mississippian) reefs of Oklahoma, Missouri and Kansas and have lead this type of ore to be referred to as of "Mississippi Valley" type (Brown, 1968).

Some of these strata-bound sulphides show obvious close relationships to overlying karstic weathered unconformities; notable examples occur in Pre-Cambrian limestones of Baffin Island (Geldsetzer, 1971) and in Sardinia (Padalino *et al.*, 1971). In other instances the source of the ore is less obvious, but its mode of emplacement is closely related to the facies variations of the host carbonate. This has been shown, for example, in the Devonian reefs of Belgium and in Lower Cretaceous reefs in Spain (Monseur and Pel, 1973). The ultimate location at which metalliferous fluids precipitate their ore is related to the petrophysical properties of the host rock, and hence its facies and previous diagenetic history.

2. *Uranium mineralization*

Uranium is an important mineral which appears in various ways. It occurs as a primary hydrothermal vein mineral, commonly as pitchblende; it occurs as a placer deposit, commonly as monazite; and as an epigenetic mineral, commonly as carnotite (potassium uranium vanadate).

Methods of uranium prospecting have been extensively documented (Bowie *et al.*, 1972; Armstrong, 1979), and many accounts have been given of the role of sedimentology in both explaining the genesis of epigenetic uranium and predicting its distribution (e.g. Gabelman, 1971).

Uranium mineral concentrations occur in certain black shales of both marine and non-marine origin. These are typically carbonaceous and often phosphatic. Examples include the Chattanooga shale of south-western US, the Green River Tertiary oil shales of Utah and of the Rum Jungle, Darwin, Australia (Dodson, 1972).

A second common mode of occurrence is in non-marine conglomerates, such as those of the Witwatersrand basin (Pre-Cambrian) of South Africa (Davidson, 1957).

Alternatively, fluvial sandstones are a typical host of epigenetic carnotite. Three such uranium provinces are known today in the US. These occur in the Colorado Plateau, the Wyoming basins and the Texas Gulf Coast plain (Jobin, 1962; Rackley, 1972; Fisher *et al.*, 1970, respectively). In these areas, uranium mineralization occurs in rocks ranging in age from Permian to Tertiary. It is generally absent from marine and eolian sands. Typically it occurs in poorly sorted arkosic fluvial sands which are rich in carbonaceous detritus. Plant debris and fossil tree trunks are not uncommon. Pyrite and carbonate cement are typical minor constituents. The host sandstones are generally cross-bedded with conglomeratic erosional bases and typical channel geometries. The usual ore body is in the form of a "role front" (Fig. 197). This is an irregular cup-shaped mass lying sideways with respect to the channel axis. The sandstone on the concave side of the roll is highly altered and bleached to a white colour. Pyrite, calcite and carbonaceous material are absent; matrix and feldspars are extensively kaolinized.

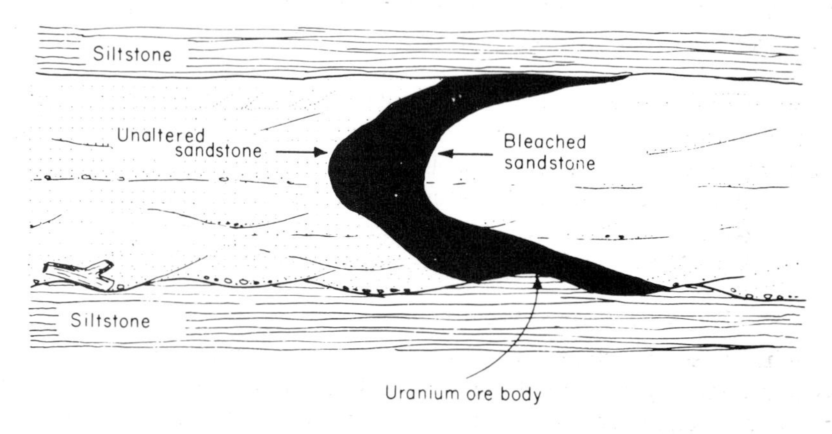

Fig. 197. Sketch to illustrate the occurrence of carnotite roll-front ore bodies in fluvial channel sandstones. The direction of front migration in this case was from right to left.

These observations have suggested to many geologists that mineralization occurred when a diagenetic front moving along a channel was stabilized by a change in the flow or chemistry of migrating pore fluids. The original source of the uranium has always been a source of debate but volcanic ashes and tuffs are often taken as the primary source, with subsequent migration and concentration by percolating acid ground water.

This close association of mineral migration and hydrology has lead to an understanding of the roll of permeability barriers in determining the locus of carnotite precipitation (Kovalev, 1972). Thus mineralization is absent in the laterally continuous sands of eolian and marine origin which possess near uniform permeability. It typically occurs adjacent to permeability barriers, such as unconformities, or in fluvial formations with sand:shale ratios of between 1:1 to 1:4.

In conclusion it should be noted that uranium deposits can be found by wandering vacantly around with a geiger counter, scintillation counter or gamma-ray spectrometer. Nevertheless, regional facies analysis can first select areas where permeability barriers may be expected. The geometry and trend of ore-bearing channels can be predicted by simple palaeocurrent analysis (Lowell, 1955).

E. Ores in Sediments: Conclusion

The preceding account shows that many ore bodies occur intimately associated with sedimentary rocks. They originated as detrital placers, syngenetically by direct precipitation and epigenetically by diagenetic replacement. Sedimentologic studies throw considerable light on these problems of ore genesis, notably by utilizing its geochemical and petrophysical aspects.

It can be argued that these problems of ore genesis are peripheral to the actual business of locating workable deposits. But if one believes, for example, that lead–zinc sulphide mineralization is an inherent feature of reef limestones, then the search can be extended beyond areas of known hydrothermal mineralization. Theories of metallogenesis thus play a part in deciding which areas may be prospective for a particular mineral.

Regardless of prevailing prejudices of metallogenesis, however, sedimentology can still be used as a searching tool. Facies analysis can map a complex of mineralized reefs, define a minette ore shoreline or locate permeability traps in carnotite-rich alluvium. Sedimentologic surveys can be carried out at the same time as direct geochemical and geophysical methods and should form an integral part of the total exploration effort.

IV. REFERENCES

Aleva, G. J. J. (1973). Aspects of the historical and physical geology of the Sunda shelf essential to the exploration of submarine tin placers. *Geologie Mijnb.* **52**, 79–91.

Amstutz, G. C. and Bernard, A. J. (1973). "Ores in Sediments" Springer-Verlag, Heidelberg. 350pp.

Amstutz, G. C. and Bubinicek, L. (1967). Diagenesis in sedimentary mineral deposits. *In* "Diagenesis in Sediments" (S. Larsen and G. V. Chilingar, Eds), 417–475. Elsevier, Amsterdam.

Armstrong, F. C. (1979). Estimation of uranium resources, Gas Hills District, Wyoming, U.S.A. *In* "Evaluation of Uranium Resources" 215–228. International Atomic Energy Commission, Vienna.

Bain, G. W. (1968). Syngenesis and epigenesis of ores in layered rocks. Rept. 23rd Int. Geol. Cong. Prague. Section 7, 119–136.

Baars, D. L., Copeland, A. B. and Ritchie, W. D. (1970). Geology of middle Devonian reefs, Rainbow area, Alberta, Canada. *In* "Geology of Giant Petroleum Fields" (M. T. Halbouty, Ed.), 19–49. Am. Ass. Petrol. Geol., Tulsa.

Bartholome, P., Evrard, P., Katekesha, F., Lopez-Ruiz, J. and Ngongo, M. (1973). Diagenetic ore-forming processes at Komoto, Katanga, Republic of Congo. *In* "Ores in Sediments" (G. C. Amstutz and A. J. Bernard, Eds), 21–41. Springer-Verlag, Heidelberg.
Bernard, A. J. (1974). Essai de revue des concentration métallifères dans le cycle sediméntaires. *Geol. Rdsch.* **63**, 41–51.
Bowie, S. H. U., Davis, M. and Ostle, D. (1972). "Uranium Prospecting Handbook" Inst. Min. Met., London. 346pp.
Boyd, D. R. and Dyer, B. F. (1966). Frio barrier bar system of South Texas. *Bull. Am. Ass. Petrol. Geol.* **50**, 170–178.
Brongersma-Sanders, M. (1967). Permian wind and the occurrence of fish and metals in the Kupferschiefer and Marl Slate. Proc. 15th Inter-Univ. Geol. Congr. Leicester, 61–71.
Brown, J. S. (Ed.) (1968). "Genesis of Stratiform Lead-Zinc-Barite-Fluorite Deposits" Econ. Geol. Monograph No. 3. Econ. Geol., Blacksburg, Virginia. 443pp.
Bubb, J. N. and Hatledid, W. G. (1977). Seismic Recognition of Carbonate Buildups. *In* "Seismic Stratigraphy: Applications to Hydrocarbon Exploration". (C. Payton, Ed.), 185–204. Am. Assoc. Petrol. Geol. Mem. No. 26. Tulsa.
Busch, D. A. (1960). Prospecting for stratigraphic traps. *In* "Geometry of Sandstone Bodies" (J. A. Peterson and J. C. Osmond, Eds), 220–232. Am. Ass. Petrol. Geol., Tulsa.
Busch, D. A. (1971). Genetic units in delta prospecting. *Bull. Am. Ass. Petrol. Geol.* **55**, 1137–1154.
Buschinski, B. I. (1964). On shallow-water origin of phosphorite sediments. *In* "Deltaic and Shallow Marine Deposits" (L. M. J. U. Van Straaten, Ed.), 62–70. Elsevier, Amsterdam.
Buzzalini, A. D., Adler, F. J. and Jodry, R. L. (Eds) (1969). Evaporites and petroleum. *Bull. Am. Ass. Petrol. Geol.* **53**, 775–1011.
Chapman, R. E. (1973). "Petroleum Geology" Elsevier, Amsterdam. 310pp.
Dapples, E. C. and Hopkins, M. E. (Eds) (1969). Environments of coal deposition. *Spec. Pap. geol. Soc. Am.* **114**.
Davidson, C. F. (1957). On the occurrence of uranium in ancient conglomerates. *Econ. Geol.* **52**, 668–693.
Davidson, C. F. (1965). A possible mode of strata-bound copper ores. *Econ. Geol.* **60**, 942–954.
Degens, E. T. and Ross, D. A. (Eds) (1969). "Hot Brines and Recent Heavy Metal Deposits in the Red Sea" Springer-Verlag, Berlin. 600pp.
Dodson, P. (1972). Some environments of formation of uranium deposits. *In* "Uranium Prospecting Handbook" (S. H. O. Bowie, M. Davis and D. Ostle, Eds), pp. 33–43. Inst. Min. Met. London.
Dott, R. H. and Reynolds, M. J. (1969). "Sourcebook for Petroleum Geology" Am. Ass. Petrol. Geol. Mem. No. 5. Tulsa. 471pp.
Evans, H. (1972). Zama—a geophysical case history. *In* "Stratigraphic Oil and Gas Fields" (R. E. King, Ed.), 440–452. Am. Ass. Petrol. Geol. Mem. No. 16.
Evans, W. E. (1970). Imbricate linear sandstone bodies of Viking formation in Dodsland-Hoosier area of southwestern Saskatchewan, Canada. *Bull. Am. Ass. Petrol. Geol.* **54**, 469–486.
Exploration Staff, Chevron Standard Ltd (1979). The Geology, Geophysics and Significance of the Nisku Reef discoveries, West Pembina area, Alberta, Canada. *Bull. Can. Pet. Geol.* **27**, 326–359.

Ferris, C. (1972). Use of gravity meters in search for traps. *In* "Stratigraphic Oil and Gas Fields" (R. E. King, Ed.), 252–270. Am. Ass. Petrol. Geol. Mem. No. 16.

Fisher, W. L., Proctor, C. V., Galloway, W. E. and Nagle, J. S. (1970). Depositional systems in the Jackson Group in Texas, their relationship to oil, gas and uranium. *In* "Exploration Concepts for the Seventies" 234–261. Trans. Gulf-Cst Ass. geol. Socs. No. 20.

Fryer, B. J. and Hutchinson, R. W. (1976). Generation of metal deposits on the sea floor. *Can. Jl Earth Sci.* **13**, 126–135.

Gabelman, J. W. (1971). Sedimentology and uranium prospecting. *Sedimentary Geol.* **6**, 145–186.

Garlick, W. G. (1969). Special features and sedimentary facies of stratiform sulphide deposits in arenites. *In* "Sedimentary Ores Ancient and Modern" (C. H. James, Ed.), 107–169. Spec. Pub. Geol. Dept Leicester University, No. 7.

Garlick, W. G. and Fleischer, V. D. (1972). Sedimentary environment of Zambian copper deposition. *Geologie Mijnb.* **51**, 277–298.

Geldsetzer, H. (1971). Syngenetic dolomitization and sulphide mineralization. *In* "Ores in Sediments" (G. C. Amstutz and A. J. Bernard, Eds), 115–127. Springer-Verlag, Heidelberg.

Glasby, G. P. (1972). The mineralogy of manganese nodules from a range of marine environments. *Mar. Geol.* **13**, 57–72.

Gregory, J. W. (1930). The copper-shale (Kupferschiefer) of Mansfeld. *Trans. Instn Min. Metall.* **40**, 3–30.

Hails, J. R. (1972). The problem of recovering heavy minerals from the sea floor —an appraisal of depositional processes. Rep. 24th Int. geol. Cong. Montréal. Section 8, 157–164.

Halbouty, M. T. (1969). Hidden trends and subtle traps in Gulf Coast. Bull. Am. Ass. Petrol. Geol. **53**, 3–29.

Halbouty, M. T. (1972). Rationale for deliberate pursuit of stratigraphic, unconformity and paleogeomorphic traps. *Bull. Am. Ass. Petrol. Geol.* **56**, 537–541.

Halbouty, M. T., Meyerhoff, A. A., King, R. E., Dott, R. H. S., Klemme, H. D. and Shabad, T. (1970). World's giant oil and gas fields, geologic factors affecting their formation, and basin classification. *In* "Geology of Giant Petroleum Fields" (M. T. Halbouty, Ed.), 502–556. Am. Ass. Petrol. Geol. Mem. No. 14.

Harms, J. C. (1966). Valley Fill, Western Nebraska. *Bull. Am. Ass. Petrol. Geol.* **50**, 2119–2149.

Haughton, S. H. (Ed.) (1964). The geology of some ore deposits in southern Africa. *Proc. Geol. Soc. S. Afr.*, 25–61.

Hedberg, H. D. (1964). Geologic aspects of origin of petroleum. *Bull. Am. Ass. Petrol. Geol.* **48**, 1755–1803.

Hunt, J. H. (1979). "Petroleum Geochemistry and Geology" Freeman, San Francisco. 617pp.

Jansa, L. (1972). Depositional history of the coal-bearing Upper Jurassic - Lower Cretaceous Kootenay Formation, southern Rocky Mountains, Canada. *Bull. geol. Soc. Am.* **83**, 3199–3222.

Jenik, A. J. and Lerbekmo, J. F. (1968). Facies and geometry of Swan Hills member of Beaverhill Lake Formation (Upper Devonian), Goose River field, Alberta, Canada. *Bull. Am. Ass. Petrol. Geol.* **52**, 21–56.

Jobin, D. A. (1962). Relation of transmissive character of the sedimentary rocks of the Colorado plateau to the distribution of Uranium deposits. *Bull. U.S. geol. Surv.* **1124**. 151pp.

King, R. E. (Ed.) (1972). "Stratigraphic Oil and Gas Fields—Classification, Exploration methods, and case histories" Am. Ass. Petrol. Geol. Mem. 16. 687pp.

Kirkland, D. W. and Evans, R. (1973). "Marine Evaporites" Benchmark papers in geology. Dowdon, Hutchinson and Ross, Stroudsburg, Pennsylvania. 426pp.

Klovan, J. E. (1964). Facies analysis of the Redwater reef complex, Alberta, Canada. *Bull. Can. Petrol. Geol.* **12**, 1-100.

Knill, J. (1970). Environmental geology. *Proc. Geol. Ass.* **81**, 529-537.

Kovalev, A. A. (1972). Polygenic character of uranium mineralization in coal-bearing deposits. *Int. Geol. Rev.* **14**, 345-353.

Kranzler, I. (1966). Origin of oil in Lower Member of Tyler Formation of central Montana. *Bull. Am. Ass. Petrol. Geol.* **50**, 2245-2259.

Lemoalle, J. and Dupont, B. (1971). Iron-bearing oolites and the present conditions of iron sedimentation in Lake Chad (Africa). *In* "Ores in Sediments" (G. C. Amstutz and A. J. Bernard, Eds), 167-178. Springer-Verlag, Heidelberg.

Levorsen, A. I. (1964). Big geology for big needs. *Bull. Am. Ass. Petrol. Geol.* **48**, 141-156.

Levorsen, A. I. (1967). "Geology of Petroleum" Freeman, London. 724pp.

Lowell, J. D. (1955). Applications of cross-stratification studies to problems of uranium exploration, Chuska Mountains, Arizona. *Econ. Geol.* **50**, 177-185.

MacKenzie, D. B. (1972). Primary stratigraphic traps in sandstone. *In* "Stratigraphic Oil and Gas Fields" (R. E. King, Ed.), 47-63. Am. Ass. Petrol. Geol. Mem. No. 16.

Martin, D. B. (1963). Rosedale channel: evidence for Late Miocene submarine erosion in Great Valley of California. *Bull. Am. Ass. Petrol. Geol.* **47**, 441-456.

Martin, R. (1966). Paleogeomorphology and its application to exploration for oil and gas (with examples from western Canada). *Bull. Am. Ass. Petrol. Geol.* **50**, 2277-2311.

McCubbin, D. G. (1969). Cretaceous Strike-Valley Sandstone reservoirs, northwestern New Mexico. *Bull. Am. Ass. Petrol. Geol.* **53**, 2114-2140.

McGregor, A. A. and Biggs, C. A. (1972). Bell Creek oil field, Montana. *In* "Stratigraphic Oil and Gas Fields" (R. E. King, Ed.), 367-375. Am. Ass. Petrol. Geol. Mem. No. 16.

McKelvey, V. E., Swanson, R. W. and Sheldon, R. P. (1953). The Permian phosphorite deposits of western United States. Int. Geol. Cong. Algiers, 1952. Comptes Rendus, Section 11, pt. 11, 45-64.

Mero, J. L. (1965). "The Mineral Resources of the Sea" Elsevier, Amsterdam. 312pp.

Monseur, G. (1974). Rhythme sedimentaire et mineralizations stratiformes dans l'environment recifal. *Geol. Rdsch.* **63**, 23-40.

Monseur, G. and Pel, J. (1973). Reef environment and stratiform ore deposits. *In* "Ores in Sediments" (G. C. Amstutz and A. J. Bernard, Eds), 195-207. Springer-Verlag, Heidelberg.

Newell, R. A. (1971). Characteristics of the stanniferous alluvium in the southern Kinta Valley, west Malaysia. *Bull. geol. Soc. Malaysia.* **4**, 15-37.

Padalino, G., and many others, (1971). Ore deposition in karst formations with examples from Sardinia. *In* "Ores in Sediments" (G. C. Amstutz and A. J. Bernard, Eds), 209-220. Springer-Verlag, Heidelberg.

Payton, C. F. (1977). Seismic Stratigraphy—applications to hydrocarbon exploration. *Am. Assoc. Petrol. Geol.* Blem. No. 26, Tulsa. 516pp.

Porfir'ev, V. B. (1974). Inorganic origin of petroleum. *Bull. Am. Ass. Petrol. Geol.* **58**, 3–33.

Pretorius, D. A. (1979). The Depositional Environment of the Witwatersrand Goldfields: A Chronological Review. *In* "Some Sedimentary Basins and Associated Ore Deposits of South Africa". Geol. Soc. S. Africa. Sp. Pub. No. 6, 33–56.

Pryor, W. A. (1973). Permeability-porosity patterns and variations in some Holocene sand bodies. *Bull. Am. Ass. Petrol. Geol.* **57**, 162–189.

Puchelt, H. (1971). Recent iron sediment formation at the Kameni Islands, Santorini (Greece). *In* "Ores in Sediments" (G. C. Amstutz and A. J. Bernard, Eds), 227–245. Springer-Verlag, Heidelberg.

Rackley, R. I. (1972). Environment of Wyoming Tertiary uranium deposits. *Bull. Am. Ass. Petrol. Geol.* **56**, 755–774.

Rainwater, E. H. (1972). The factors which control petroleum accumulation. *Trans. Gulf-Cst Ass. geol. Socs* **22**, 39–54.

Richter-Bernburg, G. (Ed.) (1972). "Geology of Saline Deposits" Proc. Hanover Symp., May, 1968. Unesco, Paris. 316pp.

Rittenhouse, G. (1972). Stratigraphic trap classification. *In* "Stratigraphic Oil and Gas Fields—Classification, Exploration Methods and Case Histories" (R. E. King, Ed.), 14–28. Am. Ass. Petrol. Geol. Mem. No. 16.

Roy, S. (1981). "Manganese Deposits." Academic Press, London and New York. 458pp.

Sangster, D. F. (1971). Geological significance of strata-bound sulphide deposits. *Proc. geol. Ass. Can.* **23**, 69–72.

Selley, R. C. (1964). The penecontemporaneous deformation of heavy mineral bands in the Torridonian sandstone of northwest Scotland. *In* "Deltaic and Shallow Marine Deposits" (L. M. J. U. Van Straaten, Ed.), 362–367. Elsevier, Amsterdam.

Shannon, J. P. and Dahl, A. R. (1971). Deltaic stratigraphic traps in west Tuscola field, Taylor county, Texas. *Bull. Am. Ass. Petrol. Geol.* **55**, 1194–1205.

Shelton, J. W. (1967). Stratigraphic models and general criteria for recognition of alluvial, barrier bar and turbidity current sand deposits. *Bull. Am. Ass. Petrol. Geol.* **51**, 2441–2460.

Tissot, B. P. and Welte, D. H. (1978). "Petroleum Formation and Occurrence" Springer-Verlag, Berlin. 538pp.

Williamson, I. A. (1967). "Coal Mining Geology" Oxford University Press, London. 304pp.

Withrow, P. C. (1968). Depositional environments of Pennsylvanian Red Fork Sandstone in N.E. Anadarko basin, Oklahoma. *Bull. Am. Ass. Petrol. Geol.* **52**, 1638–1654.

Wolf, K. A. (Ed.) (1976). "Handbook of Strata-Bound and Stratiform Ore deposits" Vols 1–7 Elsevier, Amsterdam.

Woncik, J. (1972). Recluse field, Campbell county, Wyoming. *In* "Stratigraphic Oil and Gas Fields" (R. E. King, Ed.), 376–382. Am. Ass. Petrol. Geol. Mem. No. 16.

Author Index

T

U

V

Subject Index